Cell Culture and Somatic Cell Genetics of Plants

VOLUME 4

Cell Culture in Phytochemistry

Editorial Advisory Board

Cell Culture and Somatic Cell Genetics of Plants

VOLUME 4

Cell Culture in Phytochemistry

Edited by

FRIEDRICH CONSTABEL
Plant Biotechnology Institute
National Research Council
Saskatoon, Saskatchewan, Canada

INDRA K. VASIL
Laboratory of Plant Cell
and Molecular Biology
University of Florida
Gainesville, Florida

ACADEMIC PRESS, INC.
Harcourt Brace Jovanovich, Publishers
San Diego New York Berkeley Boston
London Sydney Tokyo Toronto

ACADEMIC PRESS, INC.
1250 Sixth Avenue, San Diego, California 92101

United Kingdom Edition published by
ACADEMIC PRESS INC. (LONDON) LTD.
24–28 Oval Road, London NW1 7DX

Library of Congress Cataloging in Publication Data
(Revised for vol. 4)

Cell culture and somatic cell genetics of plants.

Vol. 4– edited by Friedrich Constabel and Indra K. Vasil.
Includes bibliographies and indexes.
Contents: v. 1. Laboratory procedures and their applications – v. 2. Cell growth, nutrition, cytodifferentiation, and cryopreservation – – v. 4. Cell culture in phytochemistry.
1. Plant cell culture–Collected works. 2. Plant cytogenetics–Collected works. I. Vasil, I. K. II. Constabel, F.
QK725.C37 1984 581'.07'24 83–21538
ISBN 0–12–715004–8 (v. 4 : alk. paper)

PRINTED IN THE UNITED STATES OF AMERICA

87 88 89 90 9 8 7 6 5 4 3 2 1

Contents

General Preface

Recent advances in the techniques and applications of plant cell culture and plant molecular biology have created unprecedented opportunities for the genetic manipulation of plants. The potential impact of these novel and powerful biotechnologies on the genetic improvement of crop plants has generated considerable interest, enthusiasm, and optimism in the scientific community and is in part responsible for the rapidly expanding biotechnology industry.

The anticipated role of biotechnology in agriculture is based not on the actual production of any genetically superior plants, but on elegant demonstrations in model experimental systems that new hybrids, mutants, and genetically engineered plants can be obtained by these methods, and the presumption that the same procedures can be adapted successfully for important crop plants. However, serious problems exist in the transfer of this technology to crop species.

Most of the current strategies for the application of biotechnology to crop improvement envisage the regeneration of whole plants from single, genetically altered cells. In many instances this requires that specific agriculturally important genes be identified and characterized, that they be cloned, that their regulatory and functional controls be understood, and that plants be regenerated from single cells in which such gene material has been introduced and integrated in a stable manner.

Knowledge of the structure, function, and regulation of plant genes is scarce, and basic research in this area is still limited. On the other hand, a considerable body of knowledge has accumulated in the last fifty years on the isolation and culture of plant cells and tissues. For example, it is possible to regenerate plants from tissue cultures of many plant species, including several important agricultural crops. These procedures are now widely used in large-scale rapid clonal propagation of plants. Plant cell culture techniques also allow the isolation of mutant cell lines and plants, the generation of somatic hybrids by protoplast fusion, and the regeneration of genetically engineered plants from single transformed cells.

Many national and international meetings have been the forums for discussion of the application of plant biotechnology to agriculture. Neither the basic techniques nor the biological principles of plant cell culture are generally included in these discussions or their published proceedings. Following the very enthusiastic reception accorded the two volumes entitled "Perspectives in Plant Cell and Tissue Culture" that were published as supplements to the *International Review of Cytology* in 1980, I was approached by Academic Press to consider the feasibility of publishing a treatise on plant cell culture. Because of the rapidly expanding interest in the subject both in academia and in industry, I was convinced that such a treatise was needed and would be useful. No comprehensive work of this nature is available or has been attempted previously.

The organization of the treatise is based on extensive discussions with colleagues, the advice of a distinguished editorial advisory board, and suggestions provided by anonymous reviewers to Academic Press. However, the responsibility for the final choice of subject matter included in the different volumes, and of inviting authors for various chapters, is mine. The basic premise on which this treatise is based is that knowledge of the principles of plant cell culture is critical to their potential use in biotechnology. Accordingly, descriptions and discussion of all aspects of modern plant cell culture techniques and research are included in the treatise. The first volume describes every major laboratory procedure used in plant cell culture and somatic cell genetics research, including many variations of a single procedure adapted for important crop plants. The second and third volumes are devoted to the nutrition and growth of plant cell cultures and to the important subject of generating and recovering variability from cell cultures. An entirely new approach is used in the treatment of this subject by including not only spontaneous variability arising during culture, but also variability created by protoplast fusion, genetic transformation, etc. Future volumes are envisioned to cover most other relevant and current areas of research in plant cell culture and its uses in biotechnology.

In addition to the very comprehensive treatment of the subject, the uniqueness of these volumes lies in the fact that all the chapters are prepared by distinguished scientists who have played a major role in the development and/or uses of specific laboratory procedures and in key fundamental as well as applied studies of plant cell and tissue culture. This allows a deep insight, as well as a broad perspective, based on personal experience. The volumes are designed as key reference works to provide extensive as well as intensive information on all aspects of plant cell and tissue culture not only to those newly entering the field but also to experienced researchers.

Indra K. Vasil

Preface to Volume 4

The three previous volumes of this treatise have provided comprehensive coverage of the wide variety of laboratory procedures used in plant cell culture, the fundamental aspects of cell growth and nutrition, and plant regeneration and variability. The accumulation of phytochemicals (secondary metabolites) in plant cell cultures has been studied for more than thirty years. However, in recent years there has been considerable interest and activity in the subject due to the expectation of biotechnological application and industrial production. Inasmuch as this expectation became a problem, attention turned toward analysis of the synthesis and accumulation of plant products. At present two important events are taking shape: the realization of industrial plant cell culture for the production of phytochemicals, and a molecular biological approach to understanding the regulation of product synthesis. For the expeditious advancement of these two concepts and components, it appeared desirable to compile and review phytochemistry as studied by employing plant cell cultures. A comprehensive treatment of the subject in the tradition of the earlier volumes of this treatise required two volumes: Cell Culture in Phytochemistry (Volume 4) and Phytochemicals in Cell Cultures (Volume 5). Plant physiologists and biochemists will forgive our taking the liberty—for the sake of brevity—of using the term phytochemistry in a broad sense to cover their respective disciplines.

The timeliness of the proposed volumes must have been recognized worldwide as the call for manuscripts was received with great enthusiasm. Reports at international conferences and workshops on phytochemistry and plant tissue culture had fallen far short of providing a comprehensive account of the remarkable progress made in the subject. Here we gratefully acknowledge the cooperation of all our colleagues who submitted up-to-date and thorough reviews of their fields of study. At one point we felt overwhelmed by the amount of material received, while at the same time we realized that a few groups of chemicals could not be included.

A science in flux is a fabric of differing thoughts, approaches, and interpretations, all in a state of evolution. A comprehensive treatise such as this should reflect this state, and thus we were anxious not to streamline the presentations. Some overlap in various chapters and some divergence of opinions should therefore be seen as helpful in a broad understanding of the subject. Students as well as colleagues in academia and industry will appreciate the overall effort and the diverse viewpoints presented.

We acknowledge the support of the Editorial Advisory Board in identifying this important area of plant cell culture research for these volumes. The assistance of our colleagues at the Plant Biotechnology Institute (PBI) in Saskatoon, Saskatchewan, Canada, particularly Drs. Balsevich, DeLuca, Eilert, Kurz, and Tyler, and the PBI secretarial staff, is gratefully acknowledged. Spouses of the editors deserve special thanks for enduring countless hours alone: thank you, Christa and Vimla!

Friedrich (Fred) Constabel
Indra K. Vasil

Contents of Previous Volumes

VOLUME 1

VOLUME 2

VOLUME 3

Part I Regeneration

PART **I**

Introduction

CHAPTER **1**

Cell Culture in Phytochemistry*

Friedrich Constabel

Plant Biotechnology Institute
National Research Council
Saskatoon, Saskatchewan, Canada S7N 0W9

I. HISTORY

Plant tissue culture was conceived and initiated as a method to demonstrate longevity of plant meristems (White, 1934, 1939; Gautheret, 1939). The genealogy of a *Nicotiana glauca* × *N. langsdorffii* and a carrot callus line indicated immortality and, indeed, callus material has subsequently been subcultured in various laboratories for over 20 years. In the wake of early experiments, however, not longevity but differentiation became the objective of tissue culture (Gautheret, 1959). Dedifferentiation which leads to callus formation with explanted tissue and redifferentiation of cells in newly formed callus were observed. Contributions to biology and biotechnology culminated in regeneration of plants from tissues (Reinert, 1958; Steward, 1958) and single cells (Vasil and Hildebrandt, 1965). The postulate derived from this experimentation, totipotency of nucleated parenchyma cells, primarily aimed at root, shoot, and embryo formation, does include the formation of cells specialized in accumulation of phytochemicals (secondary metabolites).

Early work on product accumulation in tissue and cell cultures was descriptive. The introduction of special media (Zenk *et al.*, 1977) and of conditions exerting stress (Wolters and Eilert, 1983; DiCosmo and Towers, 1984) facilitated the stimulation of metabolite accumulation and made the process controllable. Cloning permitted the isolation of high-yielding cell lines and marked the beginning of manipulation of specialized cells (Tabata *et al.*, 1978). Today the scene is set for identification of the critical controls and regulation of product formation. The demon-

*NRCC No. 26781.

stration of differential synthesis of enzymes by utilizing the *in situ* hybridization technique paves the way for its use as a powerful tool in relating cellular differentiation to gene expression during plant development (Martineau and Taylor, 1986).

II. CYTODIFFERENTIATION

A. Description

When culturing callus, what do we observe? Callus may spontaneously form tracheids, turn green, produce red pigments, or show cells with fluorescent vacuoles. In short, callus may exhibit cytodifferentiation. Over time, such changes may cease. In a number of cases, however, cultures retain the ability to undergo cytodifferentiation. *Catharanthus roseus* callus, for example, occasionally accumulates visible amounts of anthocyanins after more than 10 years of culture. On the other hand, and more often, phytochemicals specific for the source plant have never been detected in callus. Only regeneration of roots, shoots, and plants from calli restored the entire spectrum of compounds specific for the source plant. Analysis of this observation is presented throughout Volumes 4 and 5 of this treatise.

Callus functions as a meristem. As are the primary and secondary meristems in plants, callus too is void of phytochemicals, in particular during periods of high mitotic activity. Any deceleration of meristematic activity will progressively permit product accumulation (see also Sakuta and Komamine, Chapter 5, this volume). A synopsis of contributions in Volume 5 would show that compounds which appear first and with many kinds of callus are those that (1) are widespread in seed plants, i.e., phenolics and flavonoids including anthocyanins and lignin, (2) occur in cells which do not feature special structures, i.e., common parenchyma cells, and (3) generally are biologically less complex.

The three criteria which denote phytochemicals of common occurrence in callus may be illustrated by two examples. In parenchyma of red beet tubers we observe that cambium zones are clear, that youngest derivates are yellow due to betaxanthins, and that older derivatives are thoroughly red of betacyanins dominating the betaxanthines. In callus rapidly growing tissue is clear (white), less rapidly growing tissue is yellow or orange in color, and red pigment is found only in resting tissue (Constabel and Nassif-Makki, 1971). In various *Beta* spp. callus cultures betaxanthines, and not betacyanins, were found to be common; they occur in parenchyma cells, and would appear to be biologically less complex than betacyanins. A ranking according to biological complexity can also be deduced from the frequency with which *Catharanthus* alkaloids occurred in cell clones of one leaf. Of 93 clones the majority showed the biologically least complex corynanthe type alkaloids. The more complex alkaloid catharanthine never occurred alone as did the least complex alkaloids ajmalicine and strictosidine, and over several years of culture the capacity for catharanthine accumulation faded earlier than that for ajmalicine and strictosidine (Kurz *et al.*, 1985).

The search for species-specific compounds in callus has often been in vain. Reasons may be (1) lack of concomitant structural features, (2) lack of precursors, (3) absence of enzyme activity, or (4) degradation of products. In plants terpenoids of essential oils, resins, or latex and alkaloids of latex are generally accumulated in special compartments, i.e., in intercellular or subcuticular lumina, in ducts and vessels, in vacuoles, and in particular populations of vesicles (see also Guern *et al.*, Chapter 3, this volume). Callus has not been seen to generate such structures unless roots or shoots begin to form. Exceptions are laticiferous cells in callus of *Asclepias syriaca* (Biesboer, 1983), *Calotropis procera* (Dhir *et al.*, 1984), and *C. gigantea* (Datta and Sibaprasad, 1986) or idioblasts in *Ruta graveolens* callus (Eilert *et al.*, 1986a). Phytochemicals known to solely occur in such structures are, therefore, not expected to occur in callus (see Krikorian and Steward, 1969).

On close examination by using improved analytical techniques the statement made cannot be upheld as strongly. Some terpenoids and alkaloids restricted to special cells in plants have been demonstrated for callus at comparatively low concentrations (see also Banthorpe, Chapter 8, Volume 5) and can be extracted by 2-phase systems (see also Beiderbeck and Knoop, Chapter 13, this volume). The medium of callus, earlier described as a site of extracellular enzyme activity (Straus and LaRue, 1954; Constabel, 1960), has recently been seen as the site of extracellular (artificial) product accumulation as well (see also Guern *et al.*, Chapter 3, this volume). Berberine, a pigment of vacuoles and cell walls of xylem in various Magnoliales, has been found to crystallize in the medium or to be released into agar, permitting a bioassay for screening (Nakagawa *et al.*, 1984; Suzuki *et al.*, 1987). *Catharanthus* alkaloids released from cells on treatment with elicitors were extracted from the medium in substantial amounts (Eilert *et al.*, 1986b).

Despite thorough investigation a number of phytochemicals have not been reported for typical, i.e., unorganized callus, or are awaiting confirmation by reexamination; codeine (Constabel, 1985) or vinblastine (Miura and Okasaki, 1983) may be quoted as examples for many such compounds. In line with earlier discussions these chemicals would have to be regarded as biologically complex. Dimeric indole alkaloids, voafrin A and B, have been demonstrated for *Voacanga africana* callus (Stöckigt *et al.*, 1983), but why not vinblastine, a compound of similar complexity which is much sought after? Lack of activity of enzymes catalyzing the synthesis of products under investigation appears to be responsible. Interestingly, this assumption applies not only to mitotic cells but to resting and senescent cells as well. Molecular biological technology, it is hoped, will soon be used to describe the special status (maturity), which permits cells to accumulate phytochemicals known to occur in common parenchyma but absent even in nonmitotic callus.

B. Control

Plant cell culture is not so much the method of morphology as of morphogenesis, not so much the method of descriptive as of experimental approaches of cytodifferentiation and phytochemistry. The isolation of cells from cellular interaction as characterizes cells in plant tissues, strict control of environmental conditions, direct exposure to chemical and physical factors, and various cell manipulations contribute to the preferential use of cell cultures for studying the synthesis and accumulation of phytochemicals. Whether insights obtained by experimentation with callus and cell suspensions reflect product accumulation that accompanies plant development has to be assessed in each case. The fact that cells in callus cannot simulate position effects, and interaction of cells that leads to patterns in plant development, subepidermal pigmentation or idioblast spacing, for instance, is an inherent impediment. Results, however, are encouraging and will further drive ingenuity to accomplish control of product formation in cell cultures (see also Yeoman, Chapter 10, this volume).

Growth of callus and cytodifferentiation resulting in product accumulation is largely dependent on the composition and administration of culture media. Formulation of special production media for callus of alkaloid plants (Zenk *et al.*, 1977) and of naphthoquinone plants (Fujita and Tabata, 1986) have centered on the nitrate/reduced nitrogen ratio, on levels of phosphorus, carbohydrates (sucrose), and hormones. For results of experiments to optimize culture media, see Ibrahim (Chapter 4, this volume) and chapters on various chemical groups (Volume 5). Of all media components hormones have received greatest attention; changes in concentration may reduce or increase product accumulation drastically. Just as deletion of auxins may trigger the appearance of roots or shoots or embryos, so with phytochemicals. It is this phenomenon which supports the view of product accumulation as a morphogenetic effect. Furthermore, investigations of the effect of auxins on product accumulation have been performed mostly without proper examination of the cellular auxin concentration. Introduction of immunological kits will eliminate this flaw from now on. The need to account for internal auxin concentration before assessing the effect of administered auxin arose from the observation of callus growing in its absence having become habituated. At this point the effect of auxin on the accumulation of phytochemicals appeared in a new light. Eilert *et al.* (1987) reported that a habituated cell line from a 5-year-old stock culture of *Catharanthus roseus*, which contrary to stock material showed idioblasts (cells with fluorescent vacuoles) and alkaloid accumulation, lost these features on subculture in the presence of nutrient media with 2,4-dichlorophenoxyacetic acid (the effect was not due to changes in growth rate).

An unconventional control of the accumulation of phytochemicals in callus has been accomplished by way of physical and chemical stress; for results, see Eilert (Chapter 9, this volume). The simulation of pathological events with cells cultured *in vitro* by treatment with elicitors, fungal homogenates, for instance, is particularly noteworthy. The elucidation of cytodifferentiation, however, has not been greatly enhanced yet. Phytochemicals produced on elicitation had been detected before; critical chemicals like codeine or vinblastine or vindoline did not appear. Benefit, however, may still accrue from the significant stimulation of the activity of enzymes catalyzing the synthesis of phytochemicals. Elicitation of callus simulating fungal or bacterial attack on plants may indeed greatly extend the biochemistry of secondary metabolism to molecular biology as has been pioneered for flavonoid metabolism with parsley cultures (Hahlbrock *et al.*, 1982).

C. Manipulation

Cell manipulation is meant to lead to better understanding of cytodifferentiation, and to control and even regulate the process. The task may be to shift the metabolism of a callus from, for example, sanguinarine to codeine accumulation, from catharanthine to catharanthine plus vindoline and vinblastine accumulation, the formation of products not found so far.

The ability to isolate and culture protoplasts from leaf, hypocotyl, and other parenchyma and from callus and cell suspension cultures would permit monitoring of the development of structurally and chemically specialized cells (idioblasts), cloning of selected specialized cells, and recombination of genes encoding metabolite production and accumulation by cell fusion and transformation. Protoplasting as an approach of the physiology of idioblasts has been proposed (Constabel, 1983) but has not been pursued persistently. Still, the idea of isolating as protoplasts articulated and nonarticulated laticifers from seedlings for developmental studies has been pursued by S. K. Dhir *et al.* (personal communication).

Experimentation to regulate product accumulation at the cell level using chemostats awaits exploitation (see also Rokem, Chapter 14, this volume). As far as synchronization was to amplify metabolic events in cells, immunological and molecular biological technology seems to have afforded a more appropriate approach and permitted manipulation to regulate accumulation. Selection of chemovariant cells and cloning of callus has been practiced ever since Tabata *et al.* (1978) demonstrated its significance with *Lithospermum erythrorhizon* cultures (see also Dougall, Chapter 6, this volume). The importance of conventional mutation has evolved slowly and may gain increasing application in the future (see Widholm, Chapter 7, this volume). Genetic modification of cells appears to be the open-ended road to ultimate manipulation of cells and regulation of cytodifferentiation. Single cells as protoplasts or as members of callus and explants are conceptually the starting material. The objective will be to transform cells by mutating genes encoding the enzymes critical for specific synthetic pathways or mutating entire genomes for increased enzyme activity (see also Dudits, Chapter 8, this volume).

The assumption that the accumulation of many phytochemicals requires structural differentiation of cells and the observation that phytochemicals of cell cultures often reflect the metabolite composition of roots (see Wink, Chapter 2, this volume) has spawned renewed work on root instead of callus cultures. Interest has grown since it has been demonstrated that the Ri plasmid present in *Agrobacterium rhizogenes* causes the transformation of plant cells by introducing T-DNA of the Ri plasmid into genomic DNA of plant cells and that on a hormone-free medium the transformed cells give rise to massive root formation, so-called hairy roots (White and Nester, 1980). And already, a number of hairy root systems have successfully been used to enhance product formation *in vitro* (Kamada *et al.*, 1986).

TABLE I

Examples of Secondary Metabolite Production by Plant Tissue Culture

Plant	Metabolite	Culture method	Content (% dwt)	Content in plant (% dwt)	Ratio of content (cell culture/plant)	Reporter
Papaver somniferum	Sanguinarine	Liquid	2.9	—	—	Eilert *et al.*, 1985
Dioscorea deltoidea	Diosgenin	Liquid	2	2	1	Kaul and Staba, 1968
Coffea arabica	Caffeine	Agar	1.6	1.6	1	Frischknecht *et al.*, 1977
Coptis japonica	Berberine	Agar	7.4	7	1	Fukui *et al.*, 1982
Macleaya cordata	Protopin	Agar	0.4	0.32	1.25	Koblitz *et al.*, 1975
Coptis japonica	Berberine	Liquid	13	—	2	Sato and Yamada, 1984
Catharanthus roseus	Ajmalicine	Liquid	1	0.3	3	Zenk *et al.*, 1977
Coleus blumei	Rosmarinic acid	Liquid	15	3.6	5	Razzaque and Ellis, 1977
Panax ginseng	Ginsengoside	Agar	27	4.5	6	Furuya *et al.*, 1983
Lithospermum erythrorhizon	Shikonin	Agar	12	1.5	8	Tabata *et al.*, 1978
Morinda citrifolia	Anthraquinones	Liquid	18	2.5	8	Zenk *et al.*, 1975
Lithospermum erythrorhizon	Shikonin	Liquid	14	1.5	9.3	Fujita and Tabata, 1986
Cassia tora	Anthraquinone	Agar	6	0.6	10	Tabata *et al.*, 1975
Nicotiana tabacum	Ubiquinone-10	Liquid	0.18	0.003	60	Matsumoto *et al.*, 1981
Catharanthus roseus	Catharanthine	Liquid	0.24	0.002	77	Smith *et al.*, 1987

III. OUTLOOK

Within academia the accumulation of secondary metabolites remains as exciting a phenomenon as seen by Sperlich (1939). The excitement arises from a shift in methodology from description to manipulation and regulation. Elicitation certainly is one of today's techniques to stimulate synthesis and accumulation, thus encouraging enzymology. Immunochemical and protein purification techniques are ready to characterize enzymes for subsequent RNA and DNA synthesis. Directed expression of tissue- and cell-specific genes is becoming a not too distant goal. The size of the seed plant genome and the complexity of factors influencing gene expression, however, will be major barriers to rapid advancement. As a result it may take some time before the relationship between synthesis and accumulation of phytochemicals and cell specialization by structural features can be clarified convincingly, before the occurrence of compounds like codeine and morphine, cardenolides, or polyterpenes can be postulated or ruled out for plant cell cultures with reason.

Industrial application, meanwhile, will advance steadily. Table I showing *in vitro* culture systems with accumulation of phytochemicals exceeding their levels in source plants is an impressive document of technological progress. The feasibility of industrial application has already been demonstrated (Fujita and Tabata, 1986). The question whether such a process is profitable may become less relevant when society recognizes plant cell culture as a means to relieve plants, in particular endangered species, from being (over) harvested. Can we continue to collect *Arnica, Coptis, Drosera, Hydrastis, Orchis, Polygala,* and *Sanguinaria* plants in natural stands when plant cell culture is a feasible substitute and once *in vitro* grown cells are recognized by pharmacopoeia?

REFERENCES

Biesboer, D. D. (1983). The detection of cells with a laticifer-like metabolism in *Asclepias syriaca* L. *Plant Cell Rep.* **2,** 137–139.

Constabel, F. (1960). Zur Amylasesekretion pflanzlicher Gewebekulturen. *Naturwissenschaften* **47,** 17–18.

Constabel, F. (1983). Protoplast technology applied to metabolite production. *Int. Rev. Cytol., Suppl.* **16,** 209–217.

Constabel, F. (1985). Morphinan alkaloids from plant cell cultures. *In* "The Chemistry and

Biology of Isoquinoline Alkaloids" (J. D. Phillipson, M. F. Roberts, and M. H. Zenk, eds.), pp. 257–264. Springer-Verlag, Berlin and New York.
Constabel, F., and Nassif-Makki, H. (1971). Betalainbildung in Beta-Calluskulturen. *Ber. Dtsch. Bot. Ges.* **84,** 629–636.
Datta, S. K., and Sibaprasad, D. E. (1986). Laticifer differentiation of *Calotropis gigantea* R. Br. ex. Ait. in cultures. *Ann. Bot. (London)* [N.S.] **57,** 403–406.
Dhir, S. K., Shekhawat, N. S., Purohit, S. D., and Arya, H. C. (1984). Development of laticifer cells in callus cultures of *Calotropis procera* (Ait.) R. Br. *Plant Cell Rep.* **3,** 206–209.
DiCosmo, F., and Towers, G. H. N. (1984). Stress and secondary metabolism in cultured plant cells. *In* "Phytochemical Adaptations to Stress" (B. N. Timmermann, C. Steelink, and F. A. Loewus, eds.), pp. 97–175. Plenum, New York.
Eilert, U., Kurz, W. G. W., and Constabel, F. (1985). Stimulation of sanguinarine accumulation in *Papaver somniferum* cell cultures by fungal elicitors. *J. Plant Physiol.* **119,** 65–76.
Eilert, U., Wolters, B., and Constabel, F. (1986a). Ultrastructure of acridone alkaloid idioblasts in roots and cell cultures of *Ruta graveolens. Can. J. Bot.* **64,** 1089–1096.
Eilert, U., Constabel, F., and Kurz, W. G. W. (1986b). Elicitor stimulation of monoterpene indole alkaloid formation in suspension cultures of *Catharanthus roseus. J. Plant Physiol.* **126,** 11–22.
Eilert, U., DeLuca, V., and Kurz, W. G. W. (1987). Alkaloid formation by habituated and tumorous cell suspension culture of *Catharanthus roseus. Plant Cell Rep.* (in press).
Frischknecht, P. M., Baumann, T. W., and Tanner, H. (1977). Tissue culture of *Coffea arabica:* Growth and caffeine formation. *Planta Med.* **31,** 344–350.
Fujita, Y., and Tabata, M. (1986). Secondary metabolites from plant cells: Pharmaceutical application and progress in commercial production. *In* "Plant Tissue and Cell Culture" (D. A. Somer, B. G. Gegenbach, D. D. Biesboer, W. P. Hackett, and C. E. Green, eds.), pp. 169–185. Univ. of Minneapolis Press, Minneapolis, Minnesota.
Fukui, H., Nakagawa, K., Tsuda, S., and Tabata, M. (1982). Production of isoquinoline alkaloids by cell suspension cultures of *Coptis japonica. In* "Plant Tissue and Cell Culture 1982" (A. Fujiwara, ed.), pp. 313–314. Maruzen Co., Tokyo.
Furuya, F., Yoshikawa, T., Orihara, Y., and Oda, H. (1983). Saponin production in cell suspension cultures of *Panax ginseng. Planta Med.* **48,** 83–87.
Gautheret, R. J. (1939). Sur la possibilité de réaliser la culture indéfinie des tissus de tubercules de carotte. *C. R. Hebd. Seances Acad. Sci.* **208,** 118–121.
Gautheret, R. J. (1959). "La culture des tissus végétaux. Techniques et réalisation." Masson, Paris.
Hahlbrock, K., Kreuzaler, F.. Ragg, H., Fautz, E., and Kuhn, D. N. (1982). Regulation of flavonoid and phytoalexin accumulation through mRNA and enzyme induction in cultured plant cells. *Colloq. Ges. Biol. Chem., 33rd, 1982,* pp. 34–43.
Kamada, H., Okamura, N., Satake, M., Harada, H., and Shimomura, K. (1986). Alkaloid production by hairy root cultures in *Atropa belladonna. Plant Cell Rep.* **5,** 239–242.
Kaul, B., and Staba, J. E. (1968). Dioscorea tissue cultures: I. Biosynthesis and isolation of diosgenin from *Dioscorea deltoidea* callus and suspension cells. *Lloydia* **31,** 171–176.
Koblitz, H., Schumann, U., Böhm, H., and Franke, J. (1975). Gewebekulturen aus Alkaloidpflanzen: IV. *Macleaya microcarpa* (Maxim.) Fedde. *Experientia* **31,** 768–769.
Krikorian, A. I., and Steward, F. C. (1969). Biochemical differentiation: The biosynthetic potentialities of growing and quiescent tissue. *In* "Plant Physiology" (F. C. Steward, ed.), Vol. 5B, pp. 227–326. Academic Press, New York.
Kurz, W. G. W., Chatson, K. B., and Constabel, F. (1985). Biosynthesis and accumulation

of indole alkaloids in cell suspensions cultures of *Catharanthus roseus* cultivars. *In* "Primary and Secondary Metabolism of Plant Cell Cultures" (K.-H. Neumann, W. Barz, and E. Reinhard, eds.), pp. 143–153. Springer-Verlag, Berlin and New York.

Martineau, B., and Taylor, W. C. (1986). Cell-specific photosynthetic gene expression in maize determined using cell separation techniques and hybridization in vitro. *Plant Physiol.* **82,** 613–618.

Matsumoto, T., Kanno, H., Ikeda, T., Obi, Y., Kisaki, T., and Noguchi, M. (1981). Selection of cultured tobacco cell strains producing high levels of ubiquinone 10 by a cell cloning technique. *Agric. Biol. Chem.* **45,** 1627–1633.

Miura, Y., and Okasaki, M. (1983). Production process for vinblastine. Japanese Pat. (Kokai) 83/201982.

Nakagawa, K., Konagai, A., Fukui, H., and Tabata, M. (1984). Release and crystallization of berberine in the liquid medium of *Thalictrum minus* cell suspension cultures. *Plant Cell Rep.* **3,** 254–257.

Razzaque, A., and Ellis, B. E. (1977). Rosmarinic acid production in *Coleus blumei. Planta* **137,** 287–292.

Reinert, J. (1958). Untersuchungen über die Morphogenese an Gewebekulturen. *Ber. Dtsch. Bot. Ges.* **71,** 15.

Sato, F., and Yamada, Y. (1984). High berberine producing cultures of *Coptis japonica* cells. *Phytochemistry* **23,** 281–285.

Smith, J. I., Smart, N. J., Misawa, M., Kurz, W. G. W., Tavelli, S. G., and DiCosmo, F. (1987). Stimulation of indole alkaloid production in *Catharanthus roseus* (L.) G. Don by vanadyl sulfate. *Plant Cell Rep.* **6,** 142–145.

Sperlich, A. (1939). Exkretionsgewebe. *In* "Handbuch der Pflanzenanatomie" (K. Linsbauer, ed.), Vol. 6, Part B, pp. 1–184. Bornträeger, Berlin.

Steward, F. C. (1958). Growth and development of cultivated cells. III. Interpretations of the growth from free cell to carrot plant. *Am. J. Bot.* **45,** 709–713.

Stöckigt, J., Pawelka, K. H., Tanahashi, F., Danielli, B., and Hull, W. E. (1983). Voafrine A and voafrine B, new dimeric indole alkaloids from cell suspension cultures of *Voacanga africana* Stapf. *Helv. Chim. Acta* **66,** 2525–2533.

Straus, J., and LaRue, C. D. (1954). Maize endosperm tissue grown *in vitro:* I. Culture requirements. *Am. J. Bot.* **41,** 687–694.

Suzuki, T., Yoshioka, T., Hara, Y., Tabata, M., and Fujita. Y. (1987). A new bioassay system for screening high berberine-producing cell colonies of *Thalictrum minus. Plant Cell Rep.* **6,** 194–196.

Tabata, M., Hiraoka, N., Ikenone, M., Sano, Y., and Konoshima, M. (1975). The production of anthraquinones in callus cultures of *Cassia tora. Lloydia* **38,** 131–134.

Tabata, M., Ogino, F., Yoshioka, K., Yoshikawa, N., and Hiraoka, N. (1978). Selection of cell lines with higher yield of secondary products. *In* "Frontiers of Plant Tissue Culture 1978" (T. A. Thorpe, ed.), pp. 213–222. University of Calgary, Calgary, Alberta, Canada.

Vasil, V., and Hildebrandt, A. C. (1965). Differentiation of tobacco plants from single isolated cells in microculture. *Science* **146,** 76–77.

White, F. F., and Nester. E. W. (1980). Hairy root: Plasmid encodes virulence traits in *Agrobacterium rhizogenes. J. Bacteriol.* **141,** 1134–1141.

White, P. R. (1934). Potentially unlimited growth of excised tomato root tips in a liquid medium. *Plant Physiol.* **9,** 585–600.

White, P. R. (1939). Potentially unlimited growth of excised plant callus in an artificial nutrient. *Am. J. Bot.* **26,** 59–64.

Wolters, B., and Eilert, U. (1983). Elicitoren-Auslöser der Akkumulation von Pflanzenstoffen. Ihre Anwendung zur Produktions Steigerung in Zellkulturen. *Dtsch. Apoth.-Ztg.* **123,** 659–667.

Zenk, M. H. (1978). The impact of plant cell culture on industry. *In* "Frontiers of Plant Tissue Culture 1978" (T. A. Thorpe, ed.), pp. 1–13. University of Calgary, Calgary, Alberta, Canada.

Zenk, M. H., El-Shagi, H., and Schulte, U. (1975). Anthraquinone production by cell suspension cultures of *Morinda citrifolia. Planta Med., Suppl.* **75,** 79–81.

Zenk, M. H., El-Shagi, H., Arens, H., Stöckigt, J., Weiler. E. W., and Deus, B. (1977). Formation of the indole alkaloids serpentine and ajmalicine in cell suspension cultures of *Catharanthus roseus. In* "Plant Tissue Culture and its Biotechnological Application" (W. Barz, E. Reinhard, and M. H. Zenk, eds.), pp. 27–43. Springer-Verlag, Berlin and New York.

PART **II**

Accumulation of Phytochemicals

CHAPTER **2**

Physiology of the Accumulation of Secondary Metabolites with Special Reference to Alkaloids

Michael Wink

Genzentrum der Universität München
Institut für Pharmazeutische Biologie
D-8000 München 2, Federal Republic of Germany

I. INTRODUCTION

A special feature of higher plants is their capacity to produce a large number of organic chemicals of high structural diversity, the so-called secondary metabolites. Since many pharmaceuticals and other industrial products are based on plant products, much effort has been invested in the biotechnological production of secondary metabolites by plant cell cultures. Over 30 cell culture systems, which are better producers than the respective plants, are now available (Tabata, 1977; Zenk, 1978, 1982; Constabel *et al.*, 1982; Curtin, 1983; Berlin, 1984; Balandrin *et al.*, 1985; Staba, 1985; Wink, 1987a). However, many of the economically important products—such as morphine, codeine, hyoscyamine, scopolamine, vinblastine, vincristine, emetine, reserpine, tubocurarine, sennosides, steroids, digitalis glycosides, pyrethroids, spearmint oil, and other fragrances and flavors—are either not formed in sufficiently large quantities or not at all by plant cell cultures.

We have to be aware that the production of secondary products by plants is a highly complex and well-coordinated process in which the elements biosynthesis, storage, and degradation, and their temporal and spatial expression, are of prime importance. We have to understand the basic principles of the physiology and biochemistry of product formation in order to manipulate plant secondary metabolism according to our needs.

In this chapter I have tried to summarize the physiological background of secondary product formation in plants and its implications in the production of natural compounds by plant cell cultures. I have focused my review on alkaloids, since much information is available on this class of natural products. Furthermore, alkaloids—of which more than 7000 structures have been reported (Raffauf, 1970)—are economically important and are therefore of interest to biotechnology. Whenever appropriate, reference to other groups of natural products is made.

II. BIOSYNTHESIS

Alkaloids form a structurally heterogeneous class of secondary compounds that are derived from basically four amino acids, namely, ornithine, lysine, phenylalanine (tyrosine), and tryptophan (Mothes and Schütte, 1969; Mothes *et al.*, 1985; Dalton, 1979). There are many alkaloids in which part of the skeleton stems from other pathways, e.g., the terpenoid pathway. The regulation of alkaloid formation is certainly more complicated in these instances as compared to those pathways for which only one precursor is needed.

A. Enzymology of Alkaloid Pathways

The pathways of alkaloid biosynthesis have been studied by a number of investigators during the last three to four decades, chiefly by experiments with radioactive tracers (Spenser, 1968; Mothes and Schütte, 1969; Saxton, 1971–1975; Grundon, 1976–1983; Dalton, 1979; Robinson, 1981; Mothes *et al.*, 1985). However, tracer experiments have a number of inherent problems. The ultimate proof that a biogenetic scheme is correct can only be obtained from studies with purified enzymes which catalyze the respective steps of alkaloid biosynthesis. Plant cell cultures have proven to be very well suited for the isolation of enzymes, and the progress in the field of alkaloid enzymology was largely due to the use of plant cell cultures (Zenk, 1980, 1985). Well studied in this respect are the pathways of indole and benzylisoquinoline alkaloids (Stöckigt, 1981; Zenk, 1980, 1985; Zenk *et al.*, 1985) and to some degree those of quinolizidine alkaloids (Wink and Hartmann, 1985), rutacridone alkaloids (Baumert *et al.*, 1986), and *Conium* alkaloids (Roberts, 1981; see also Waller and Dermer, 1981; Mothes *et al.*, 1985; Anderson *et al.*. 1985; Grundon, 1976–1983).

Once the enzymes of a pathway are known it is possible to study their regulation. Enzymes of alkaloid biosynthesis are well expressed in cell cultures with a high product yield, and are therefore a good source for the isolation of these enzymes (Zenk 1980, 1985). If a culture fails to produce a certain chemical it may be that (a) all enzymes of the pathway are repressed, (b) some but not all enzymes are expressed, or (c) the enzymes are expressed, but there is no product formation due to other factors, which can be poor precursor supply, lack of storage capacity, product degradation, etc.

Unfortunately, few authors have looked into these possibilities in detail, probably because only a few research teams have studied alkaloid enzymology so far. Without further proof, repression of all enzymes of the pathway (a, above) is often assumed to be the most likely case. Berlin and co-workers postulated that the expression of the first enzyme of a pathway should be the most important factor and thus the "bottleneck" for alkaloid production (Berlin, 1984). However, this idea is not backed by conclusive evidence.

There is some evidence that expression of some but not all enzymes (b, above) is also possible in plant cell cultures: Specific biotransformation steps have been reported even from those cell cultures which do not produce secondary metabolites endogenously (Reinhard and Alfermann, 1980).

Some data indicate that expression of enzymes without product formation (c, above) also occurs: The limitation of product formation by endogenous precursor supply could be demonstrated in a number of feeding experiments in which the product yield could be enhanced substantially by the addition of exogenous precursors to the culture medium (Zenk *et al.*, 1977; Wink *et al.*, 1980; Hay *et al.*, 1986).

Many plants contain an L-alanine : aldehyde aminotransferase, an enzyme responsible for amine formation, irrespective of whether the plants actually produce amines or not (C. Wink and Hartmann, 1981). In this case the supply with adequate precursors seems to be the controlling step.

It has also been reported the lupine cell cultures fail to produce alkaloids in quantities comparable to those of the intact plants (Wink *et al.*, 1983; Wink, 1987d), although the enzymes of alkaloid biosynthesis were expressed in the cell cultures (Wink and Hartmann, 1982a). Alkaloids formed by lupine cells were not stored but were rapidly degraded (Wink and Witte, 1985; Wink, 1985a).

These data indicate that the expression of the biosynthetic capacity is the ultimate requirement for product formation, but that obviously other processes are also involved which are equally important. The following sections will focus on these processes.

B. Sites of Alkaloid Biosynthesis

Although all cells of a plant, for example, of *Papaver somniferum*, harbor the genes for morphine biosynthesis, only a limited number of cells express these enzymes (Kutchan *et al.*, 1983). The temporal and spatial expression of genes is a typical feature of eukaryotes. This holds naturally true for plants and plant secondary metabolism. For example, alkaloids and other products usually are not formed in all cells of a plant, rather their synthesis is restricted to a specific organ, such as the formation of colored flavonoids and anthocyanins in flowers (see Section II,B,1). Additionally, synthesis is often bound to a specific stage of plant development (James, 1950; Mothes, 1955; Wiermann, 1981). Even within a single cell a high degree of organization (compartmentation) can be observed (see Section II,B,2).

1. Tissue- and Organ-Specific Localization

If alkaloids can be detected in an organ or in a tissue, it does not necessarily mean that they have been formed there. The possibility of long-distance transport should also be taken into consideration (see Section III,D). The site of biosynthesis can be determined by localizing the enzymes of alkaloid biosynthesis in a given plant organ or by showing that a tissue is competent to convert a labeled precursor into the product. In some instances grafting experiments have been helpful (James, 1950; Mothes, 1955; Mothes and Schütte, 1969). Examples for the site of alkaloid biosynthesis are given in Table I.

2. Intracellular Compartmentation

Information on the enzymology of secondary metabolites is scanty. It is therefore not surprising that our knowledge of the intracellular compartmentation of secondary metabolism is even more limited.

Lupine alkaloids are formed in the green aerial parts, especially the

TABLE I

Examples of Organ-Specific Biosynthesis and Accumulation of Alkaloids and Other Secondary Metabolites

Compound	Species	Main sites of biosynthesis	Accumulation	Ref.[a]
Alkaloids				
Tropane	*Atropa, Datura*	Roots (shoots?)	Whole plant	1
Cocaine	*Erythroxylon*	Green parts	Green parts	1
Coniine	*Conium*	Green parts	Whole plant	1
Quinolizidine	*Lupinus*	Green parts	Whole plant	2
Pyrrolizidine	*Senecio*	Roots, green parts	Whole plant	1, 3
Emetine	*Cephaelis*	Roots	Roots	1
Morphine	*Papaver*	Leaves, latex	Latex	1
Berberine	*Berberis*	Rhizomes	Rhizomes, stems	1
Chelidonine	*Chelidonium*	Aerial parts	Latex	1
Sanguinarine	*Sanguinaria*	Roots	Roots	4
Quinine	*Cinchona*	Aerial parts	Aerial parts	1
Caffeine	*Coffea*	Green parts	Fruits, leaves	1
Solanidines	*Solanum*	Shoots	Whole plant	1
Betalaines	*Beta*	Roots (stems)	Roots (stems)	1
Terpenes				
Monoterpenes	Labiateae, etc.	Aerial parts	Aerial parts	5
Phenolics				
Flavonoids	Many species	Aerial parts	Aerial parts	5
Anthocyanins	Many species	Aerial parts	Aerial parts	5
Cyanogenic glycosides	*Sorghum*, Rosaceae	Aerial parts	Aerial parts	5
Glucosinolates	*Brassica*	Whole plant	Whole plant	5

[a] Key to references: 1, Mothes (1955), Mothes and Schütte (1969), Mothes *et al.* (1985); 2, Wink and Hartmann (1981a, 1985); 3, R. J. Molineux and T. Hartmann (personal communication); 4, Neumann and Müller (1972); 5, Molisch (1923).

leaves of lupines and other legumes (Wink and Hartmann, 1981a). Roots of intact plants and *in vitro*-cultured roots of *Lupinus polyphyllus* do not incorporate labeled cadaverine into the lupanine skeleton, which is readily observed if lupine leaves are used (Mothes, 1955; Wink, 1987b). The enzymes of alkaloid biosynthesis are localized in the chloroplast stroma (Wink and Hartmann, 1982a), where also the biosynthesis of the alkaloid precursor lysine takes place.

The only other example for the synthesis of alkaloids in plastids seems to be the piperdine alkaloid coniine. In *Conium maculatum* leaves (Roberts, 1981) the enzyme of coniceine formation was found in both chloroplasts and mitochondria. The latter finding is in agreement with results of C. Wink and Hartmann (1981) who localized the L-alanine : aldehyde aminotransferase in mitochondria of *Arum maculatum*. This enzyme catalyzes the formation of aliphatic amines but can also convert α-ketooctanal to coniceine (Roberts, 1981; C. Wink and Hartmann, 1981).

It should be recalled that the chloroplast is not only the compartment of photosynthesis but also that of lipid, amino acids (especially the essential amino acids), and terpenoid biosynthesis (Goodwin and Mercer, 1983; Schultz *et al.*, 1985). It is likely, however, that alkaloid formation is usually restricted to the cytoplasm and only in special instances to the plastids. It has been shown recently that berberine biosynthesis takes place in special vesicles inside the cytoplasm (Zenk *et al.*, 1985; Amann *et al.*, 1986).

3. Relevance of the Site of Synthesis for Product Formation in Cell Cultures

For lupine alkaloids it was possible to show that the knowledge of alkaloid compartmentation was crucial for the production of alkaloids by lupine cell cultures. Cell cultures of *L. polyphyllus* and *Cytisus scoparius* do not synthesize alkaloids when they are kept in the dark as heterotrophic cultures. When they are grown in the light alkaloid formation was correlated with chlorophyll content. Green photomixotrophic cultures, which contained functional chloroplasts (B. Hansen and M. Wink. unpublished), were able to synthesize alkaloids, such as lupanine or sparteine (Wink and Hartmann, 1980; Wink *et al.*, 1981). Furthermore, alkaloid formation followed a diurnal cycle when the cultures were maintained under day–night conditions, similar to the situation in the intact plants (Wink and Hartmann, 1982b; Wink and Witte, 1984). The light-dependent control of alkaloid formation could be explained by (a) a better lysine supply in the light, (b) that the enzymes of alkaloid biosynthesis have a pH optimum around pH 8, which is created in the chloroplast stroma only in the light, and (c) that the enzymes are subject to activation by reduced thioredoxin, which is regenerated only in the light (Wink and Hartmann, 1981b, 1985). Therefore, light and photomixotrophy are essential requirements for alkaloid formation in lupine cell cultures.

In the case of *Peganum harmala*, which forms alkaloids in the roots, the opposite phenomenon was observed (Barz *et al.*, 1980; Barz and Hüsemann, 1982). Whereas photoautotrophic and photomixotrophic suspension cultures failed to produce harman alkaloids, their synthesis could be recorded, however, in heterotrophic cultures kept in the dark.

Another interesting example has been reported for *Morinda lucida* cell suspension cultures (Igbavboa *et al.*, 1985). The synthesis of lipoquinones, which are formed in the chloroplast (Schultz *et al.*, 1985), can be detected only in green photoautotrophic cultures and not in heterotrophic cultures. On the other hand, the synthesis of anthraquinones, which are normally formed in nonphotosynthetic tissues such as the roots, is suppressed in green cultures but active in heterotrophic cultures cultivated in the dark.

Genetic analysis of the few plant genes studied so far clearly indicates the presence of tissue-specific promoters (Lamppa *et al.*, 1985; Sengupta-Gopalan *et al.*, 1985; Eckes *et al.*, 1985; Kaulen *et al.*, 1986; Rosahl *et al.*, 1986). This would mean that the genes of a specific organ are expressed only when their promoters are activated by tissue-specific factors, which are themselves coded by a control or "master" gene. We should, therefore, expect that a biosynthesis which takes place in the leaf is active only in those cultured cells which resemble leaf mesophyll cells, because only there may the respective genes be activated by a mesophyll-specific factor. On the other hand, a root synthesis is unlikely to occur under phototrophic culture conditions but more likely under heterotrophic growth which resembles root cells to some degree. This concept easily explains the examples mentioned in this chapter. One can also predict whether it will be appropriate to culture the cells in the dark or as green cultures in the light if one knows the site of biosynthesis in the plant.

These data, although rather limited, clearly indicate the need to first understand the basic biochemistry and physiology of the biosynthesis of secondary products before it is possible to trigger its metabolism in a controlled fashion. In the following sections evidence will be presented to show that, beside our knowledge of the biosynthesis, we have to study the mechanisms of product accumulation and storage, because they are of prime importance.

III. ACCUMULATION OF SECONDARY METABOLITES

Characteristically plant secondary metabolites are accumulated and stored in relatively large quantities, which can be explained by their role as chemical signals or defense compounds (see Section III,D). Since accumulation cannot be a random process we have to look for the underlying biochemical mechanisms and physiological parameters. Before considering the mechanisms of alkaloid accumulation, data are summarized on the intra- and intercellular sites of product storage (Tables I, II, and III).

A. Sites of Alkaloid Accumulation

1. Tissue and Organ Specificity

Secondary metabolites are usually not distributed uniformly within the whole plant (Wiermann 1981; Strack *et al.*, 1985). Some are restricted to specific organs, e.g., the roots or seeds, others to specific tissues such as the epidermis. While we have rather extensive knowledge of the biosynthetic pathways of secondary compounds, information on the precise sites of accumulation is more scanty. In Table I information is gathered on the sites of alkaloid storage in plants.

Within these organs alkaloids are often stored in specific storage cells or in defined cell layers (Table II). This feature was intensely studied at the turn of this century, however, only with the aid of histochemical methods (for detailed information and references, see James, 1950, Molisch, 1923, Thunmann and Rosenthaler, 1931), which do not provide unequivocal evidence. Therefore these findings should be rechecked by modern analytical means, such as spectrophotometry, electron microscopy, capillary gas–liquid chromatography (Wink, 1986), HPLC, laser desorption mass spectrometry (Wink *et al.*, 1984), or radioimmunoassay (Weiler and Zenk 1976; Weiler, 1980).

The storage of a compound in a specific cell or cell layer does not necessarily imply that the compound has also been synthesized by these cells (see Section III,C). For example, lupine alkaloids are accumulated in epidermal cells (Wink, 1986; Wink *et al.*, 1984), but they are formed in the chloroplasts of leaf cells (Wink and Mende, 1987). After synthesis the alkaloids are transported to the epidermis via the phloem (Wink and Witte, 1984). Accumulation also depends very much on the season and on the developmental stage of a plant (James, 1950; Mothes, 1955).

TABLE II

Examples of Tissue- and Cell-Specific Accumulation of Alkaloids and Other Secondary Metabolites

Tissue	Organ	Compound	Species	Ref.[a]
Epidermis	Stems, leaves	Tropane alkaloids	*Atropa, Datura*	1
		Cocaine	*Erythroxylon*	1
		Colchicine	*Colchicum*	1
		Aconitine	*Aconitum*	1
		Delphinine	*Delphinium*	1
		Steroid alkaloids	*Solanum*	1
		Nicotine	*Nicotiana*	1
		Morphinane	*Papaver*	1
		Veratrine	*Veratrum*	1
		Nupharidine	*Nuphar*	1
		Buxine	*Buxus*	1
		Coniine	*Conium*	1
		Lupanine	*Lupinus*	2, 3
		Sparteine	*Cytisus*	3
		Anthocyanins	Many species	1
		Flavonoids	Many species	1, 4
		Cyanogenic glucosides	*Sorghum*	5
		Glucosinolates	*Brassica*	1
Glandular hairs	Shoots	Essential oils and other terpenes	Many species	7
		Flavonoids	Many species	7
Cuticle	Shoots	Terpenes, fats, waxes, flavonoids	Many species	7
Laticifers	Shoots	Morphinane alkaloids	*Papaver*	12
		Isoquinoline alkaloids	*Chelidonium*	6
	Fruits	Vindolinine	*Catharanthus*	11
Ducts	Shoots	Phenolics and terpenes	*Pinus*	7
		Benzofurans	*Enzelia*	8
		Flavonoids	*Adenostoma*	14
		Essential oils and gums	Various species	7
Idioblasts				
Alkaloid cells	Roots	Corydaline	*Corydalis*	1
	Rhizomes	Sanguinarine	*Sanguinaria*	9
	Roots	Rutacridones	*Ruta*	13
	Leaves	Indole alkaloids	*Catharanthus*	10
	Leaves	Protopine	*Macleaya*	16
Phenolic cells	Stems	Tannins	*Juniperus*	15

[a] Key to references: 1, Molisch (1923), Thunmann and Rosenthaler (1931); results obtained from simple histochemical studies (see text); 2, Wink (1986); 3, Wink *et al.* (1984); 4, Wiermann (1981), Knogge and Weissenböck (1986); 5, Kojima *et al.* (1979); 6, Matile (1976); 7, Wiermann (1981); 8, Proksch *et al.* (1985); 9, Neumann and Müller (1972); 10, Neumann *et al.* (1983); 11, Eilert *et al.* (1985a); 12, Homeyer and Roberts (1984); 13, Verzar-Petri *et al.* (1976); 14, Proksch *et al.* (1982); 15, Constabel (1969); 16, Neumann and Müller (1967).

TABLE III

Examples of Vacuolar Storage of Secondary Metabolites

Compound	Concentration (m*M*)	Species	Ref.[a]
Alkaloids			
Nicotine		*Nicotiana*	1
Serpentine		*Catharanthus*	2
Ajmalicine		*Catharanthus*	2
(*S*)-Reticuline		*Fumaria*	9
Lupanine	>40	*Lupinus*	3
Sparteine	>200	*Cytisus*	3
Morphine		*Papaver*	2
Atropine		*Atropa*	3
Dihydrocoptisine	254	*Chelidonium*	4
Chelyerythrine	61	*Chelidonium*	4
Sanguinarine	126	*Chelidonium*	4
Berberine	35	*Chelidonium*	4
Betalaines		*Beta*	7
Coumaryl glucosides		Barley	5
Cyanogenic glucosides		*Sorghum*	6
Glucosinolates		Brassicaceae	7
Anthocyanins	100	*Petunia*	8
Shikimic acid		*Fagopyrum*	10
Flavonoids		Many species	7
Capsaicine		*Capsicum*	7

[a] Key to references: 1, Saunders (1979); 2, Deus-Neumann and Zenk (1984); 3, Mende and Wink (1987), Wink *et al.* (1984); 4, Matile (1976); 5, Werner and Matile (1985); 6, Saunders and Conn (1978); 7, Matile (1984); 8, Aerts and Schramm (1985); 9, Deus-Neumann and Zenk (1986); Holländer-Czytko and Amrhein (1983).

2. Intracellular Sites of Product Accumulation

The sites of synthesis and accumulation are usually separated in cells by compartmentation. As a rule lipophilic compounds are accumulated in membranes, vesicles, dead cells, or extracellular sites such as the cell wall. Hydrophilic compounds are stored in an aqueous environment, i.e., the vacuole. The tendency of plants to form glycosides of many secondary metabolites, which renders a molecule more hydrophilic, can be seen in this context. Due to improved methods of protoplast and vacuole isolation, the number of secondary products which have been localized in vacuoles has increased considerably (Table IV) (Matile, 1978, 1984; Marty *et al.*, 1980).

B. Mechanisms of Alkaloid Accumulation

The central vacuole of differentiated cells can be considered as the main storage compartment for secondary metabolites in plants (Table III; Matile, 1978, 1984). The concentrations of secondary metabolites in vacuoles can be higher than 500 mmol/liter. For lupine alkaloids we have recorded a value greater than 200 mmol/liter for sparteine in *Cytisus scoparius* epidermal cells (Wink *et al.*, 1984), which is certainly concentrated in the vacuole (Mende and Wink, 1987). It is obvious that alkaloids and other natural products must be concentrated in the vacuole against a concentration gradient. When we look for the underlying mechanism we have to explain how alkaloids are transferred from the cytoplasm across the tonoplast into the vacuole and why alkaloid storage can be observed in specialized cells ("alkaloid cells") or tissues, e.g., the epidermis (Table II).

The vacuoles also store sucrose and other carbohydrates, malate, amino acids, and inorganic ions (Matile, 1978; Marty *et al.*, 1980). In recent years strong evidence has been obtained that the tonoplast contains a number of carrier proteins, which are specific for their substrate (Willenbrink and Doll, 1979; Thom *et al.*, 1982; Matile, 1984; Lüttge and Smith, 1985). The tonoplast has active proton-translocating ATPases and pyrophosphatases (Sze, 1984), and the cell sap and ultimately the vacuolar sap has an acidic pH which can vary between pH 2 and 6.5. The import of compounds into the vacuole is achieved by an H^+–substrate antiport mechanism (Sze, 1984; Thom and Komor, 1984).

TABLE IV

Examples of Phloem or Xylem Transport of Secondary Metabolites

Compound	Species	Vessel	Ref.[a]
Tropane alkaloids	*Atropa*	Xylem	1
Nicotine	*Nicotiana*	Xylem	1
Lupanine	*Lupinus*	Phloem	2
Sparteine	*Cytisus*	Phloem	3
Cytisine	*Petteria*	Phloem	4
Pyrrolizidine alkaloids	*Senecio*	Phloem	5
Swainsonine	*Astragalus*	Phloem	6

[a] Key to references: 1, Mothes and Schütte (1969), James (1950); 2, Wink and Witte (1984); 3, Wink *et al.* (1982); 4, Wink and Witte (1985); 5, R. J. Molyneux (personal communication); 6, Dreyer *et al.* (1985).

We have to be aware that the biochemical mechanisms which work in the primary metabolism of a cell are the same for secondary metabolism. For example, enzymes of alkaloid biosynthesis show the same kinetic properties as enzymes of primary pathways, i.e., they are substrate specific, have a high affinity toward their substrate, and are sensitive to ionic strength, temperature, hydrogen ion concentration, and to activators or inhibitors. Since carrier transport exists for primary metabolites, it is reasonable to investigate whether alkaloid transport too is mediated by specific carrier proteins.

Deus-Neumann and Zenk (1984, 1986) have studied this question recently. They showed that vacuoles isolated from cell suspension cultures of different alkaloidal plants show a remarkable specificity for alkaloid import into the vacuole. In general only those alkaloids are translocated which are synthesized in the respective cell culture. Structurally different alkaloids were discriminated. In case of *Fumaria capreolata* vacuoles it was shown that the (*S*) enantiomers of reticuline and scoulerine are transported whereas the (R) forms, which do not occur in the plant, were not taken up by the vacuoles. Alkaloid uptake displayed saturation kinetics, dependence on hydrogen ion concentration, and sensitivity to inhibitors. It is thus plausible to postulate the presence of carrier proteins, which are specific for their substrates. The transport is activated by ATP, which would indicate that an H^+–alkaloid antiport is the mechanism of alkaloid transport against a concentration gradient. However, Guern and co-workers (Renaudin *et al.*, 1985, 1986) consider it more likely that lipophilic alkaloids of a low pK_a pass the tonoplast in the uncharged form by simple diffusion and that the alkaloids are concentrated in the vacuole by an ion-trap mechanism.

Working with alkaloids, which under physiological conditions are present as charged species by 99%, we could recently confirm the proposal made by Zenk and co-workers and could rule out that diffusion or the ion-trap mechanism is responsible for alkaloid accumulation in lupine vacuoles (Mende and Wink, 1987). Vacuoles of suspension cultured cells of *L. polyphyllus* are highly specific and selective for alkaloid uptake in that they accumulate sparteine and lupanine (both alkaloids occur in that species) but discriminate other alkaloidal types. Interestingly, cells and protoplasts were much less specific and took up a wide variety of compounds (Mende and Wink, 1987; Wink and Mende 1987). The vacuolar transport system has a high affinity for lupanine (K_m 94 μM), is pH and temperature dependent and can be activated by ATP and KCl. In the presence of ATP we could measure an activation energy of 53 kJ/mol. Since simple diffusions are characterized by an activation energy of about 17 kJ/mol and enzyme-catalyzed processes by values between 29 and 150 kJ/mol, we exclude simple diffusion as the relevant mechanism. These data show that under physiological conditions alkaloid transport into vacuoles is mediated by specific carrier proteins. Few data indicate the presence of carrier proteins for the transport of other secondary metabolites: coumaryl glucosides (Werner and Matile, 1985), flavonoids (Matern *et al.*, 1986), and cardenolides.

Since carrier molecules are proteins they are coded for by genes, which would mean that we need the concomitant expression of the genes for biosynthesis and accumulation before we can expect any significant production. If both groups of genes are regulated by the same master gene, concomitant gene expression should be possible under cell culture conditions. More difficult will be the situation in those instances where we have long-distance transport (see Section III,C) between the site of synthesis and the site of storage, since concomitant gene expression in two organs will be required. Hyoscyamine, lupanine, and morphine, which have not been produced in cell cultures in large quantities, fall in the last category.

C. Long-Distance Transport of Secondary Metabolites

If we consider the sites of biosynthesis and storage the following situations may occur:

1. The compound is stored in the same cell as it is produced.
2. The compound is stored in the neighboring cells which would mean an intercellular transport.
3. The compound is stored in an organ different from that of its synthesis which would require long-distance transport.

Mechanisms discussed above seem to be appropriate for cases (1) and (2). In the case of long-distance transport (3) we have to distinguish between phloem and xylem transport and transport in laticifers or resin channels. For secondary metabolites respective data (Tables III and IV) are rather limited, which is due to the difficulty in obtaining phloem or xylem sap, but is certainly not as rare as one would assume from Table IV. The mechanisms involved in transport and in crossing membrane barriers have not been studied so far.

It is also unclear how this phenomenon influences product formation in cell cultures. Interestingly, there is hardly any cell culture system which produces a secondary metabolite that is subject to long-distance transport in the differentiated plant (Table V).

TABLE V

Production of Secondary Metabolites by Plant Cell Cultures in Relation to Their Site of Biosynthesis and Accumulation in the Differentiated Plant

Compound	Species	Main site of biosynthesis	Accumulation	Ref.[a]
Shikonine	*Lithospermum*	Roots	Roots	1
Alkannine	*Echium*	Roots	Roots	2
Berberine	*Coptis*	Rhizomes	Rhizomes	3
Jatrorhizine	*Berberis*	Roots	Roots	4
Sanguinarine	*Papaver*	Roots	Roots	5
Dihydrochelirubine	*Eschscholtzia*	Whole plant(?)	Whole plant(?)	6
Nicotine	*Nicotiana*	Roots	Whole plant	7
Ajmalicine	*Catharanthus*	Roots	Roots	8
Harmane alkaloids	*Peganum*	Roots	Roots	9
Anthraquinones	*Cassia*	Roots	Roots	10
	Morinda	Roots	Roots	11
	Rubia	Roots	Roots	12
(*R*)-Ginsenoside	*Panax*	Roots	Roots	13
Glycyrrhizine	*Glycyrrhiza*	Roots	Roots	13
Diosgenin	*Dioscorea*	Rhizomes	Rhizomes	13
Rosmarinic acid	*Coleus*	Shoots	Shoots	13
Cryptotanshinone	*Salvia*	Roots	Roots	14

[a] Key to references: 1, Fukui *et al.* (1983b); 2, Fukui *et al.* (1983a); 3, Sato and Yamada (1984); 4, Hinz and Zenk (1981); 5, Eilert *et al.* (1985b); 6, Berlin *et al.* (1983); 7, Röper *et al.* (1985); 8, Zenk *et al.* (1977); 9, Barz and Hüsemann (1982); 10, Tabata (1977); 11, Leistner (1975); Zenk *et al.* (1975); 12, Schulte *et al.* (1984); 13, Zenk (1978); 14, Nakanishi *et al.* (1983).

In a number of cell cultures it was observed that alkaloids were released from the cells into the culture medium (Böhm, 1978; Böhm and Franke, 1982; Wink and Hartmann, 1982c; Renaudin *et al.*, 1985; see also Chapter 3, this volume). Whether this observation reflects transport which would take place under physiological conditions remains unclear at the present time. If the cell culture does not contain cells which specifically take up and store the metabolites it is often observed that the compounds disappear from the medium. This is probably due to degradation by exoenzymes (hydrolases, oxidases) which are released from the cells into the culture medium (Wink, 1984, 1985c). The medium might thus act as an extracellular lytic compartment.

D. Degradation of Secondary Metabolites

Secondary metabolites are usually not end products of metabolism but show a high degree of turnover (Barz, 1977; Barz and Köster, 1981;

Wiermann, 1981). Nitrogen is a limiting factor for plants. It is not surprising that plants use their nitrogen "economically" and do not form nitrogen-containing waste products, as do heterotrophic organisms, e.g., animals. This is also true for nitrogen-containing secondary metabolites, such as alkaloids (Neumann and Tschöpe, 1966; Robinson, 1974; Waller and Nowacki, 1978; Eltayeb and Roddick, 1985).

In many plant species alkaloids are accumulated in seeds in relatively high concentrations, which can be interpreted as a means of chemical defense (see Section III,E). During germination lupine alkaloids, for example, are metabolized and their nitrogen is obviously reused for the seedlings' metabolism (Wink and Witte, 1985). A number of plants contain alkaloids in all their organs. It has been observed that during senescence at the end of the growing season the alkaloids disappear, i.e., they are metabolized (James, 1950; Waller and Nowacki, 1978; Wink and Hartmann, 1981a).

Evidence for alkaloid turnover can also be obtained from studies on the daily fluctuations of alkaloid concentrations in a plant. In *Conium, Papaver, Atropa* (Waller and Nowacki, 1978), *Datura* (Flück, 1963), *Nicotiana* (Bünning, 1963), *Cytisus, Baptisia,* and *Lupinus* (Wink and Hartmann, 1982b; Wink and Witte, 1984), diurnal fluctuations of alkaloids have been observed, indicating a high degree of alkaloid turnover. For *Nicotiana,* it has been calculated that about 15% of all CO_2 assimilated in a day is channeled through nicotine (Robinson, 1974).

Plants and especially plant cell suspension cultures are able to degrade exogenously added alkaloids (Mothes, 1955; Robinson, 1974; Wink, 1985a; Wink and Witte, 1985). In case of cell suspension cultures of *Lupinus polyphyllus* we could recently show that the cells can even survive on sparteine as the sole nitrogen source for more than 6 months (Wink and Witte, 1985). Many other compounds can also be degraded by plant cell cultures (Barz, 1977; Barz and Köster, 1981) which can be due to breakdown by intra- or extra-cellular enzymes. The mechanisms and pathways remain to be elucidated in most instances (Barz, 1985).

In plants the production of secondary metabolites seems to be an equilibrium between synthesis, storage, and degradation, and it depends on the developmental stage as to which component dominates. The spatial and temporal interactions of these components are largely unknown. It is obvious that these processes are very important for the production of secondary metabolites by plant cell cultures and that it is necessary to control the turnover of the products formed by the cells. This could be achieved (a) by establishing adequate storage capacities, which can be storage cells or an absorbing lipophilic matrix (Beiderbeck, 1982), or (b) by selecting cell lines with low degrading activities, or (c) by inhibition of degrading enzymes.

E. Accumulation in Relation to Environmental Factors

1. Biological Function of Secondary Metabolites

There is hardly any plant which does not produce a secondary metabolite. An obvious question is why do plants invest so much effort in the production of secondary compounds? About a century ago the formation of these compounds was interpreted in a Darwinian sense: Stahl (1888) formulated the hypothesis that secondary metabolites either serve to attract pollinating and seed-dispersing animals or more importantly to repel phytophagous species. Then, plant physiologists tried to find a physiological role for these compounds. Since they could not discover an obvious function and since they disliked the Stahl hypothesis, the prevailing idea for the next 50 years was that secondary metabolites have no general function. They were considered either as waste products or useless (Mothes, 1955, 1976; Paech, 1950).

Since 1960 this question has been reconsidered. An increasing body of evidence suggests that Stahl (1888) was essentially correct in his ecological ideas. According to the work of Fraenkel (1959), Ehrlich and Raven (1964), Whitacker and Feeny (1971), Swain (1977), Levin (1976), Rosenthal and Janzen (1979), Zenk (1968), Levinson (1976), Harborne (1977, 1982), and Wink (1985b, 1987c), the following picture emerges. Plant secondary metabolites may play a role in the physiology of some plants, e.g., flavonoids and other phenols may protect against UV light (Harborne, 1982); nonprotein amino acids, lectins, and alkaloids may serve as nitrogen storage compounds (Rosenthal, 1982; Wink, 1985a,b) or as nitrogen transport substances (Wink and Witte, 1984; Wink, 1985b). More importantly, however, secondary metabolites play a role in an ecological context: They serve as defense compounds against microorganisms (Levin, 1976; Harborne, 1982; Deverall, 1977) (viruses, bacteria, or fungi), phytophagous animals, and other plants (allelopathy). Since plants cannot run away in case of danger and since they have no immune system, they evolved other means of defense, which can be mechanical (thorns, spikes, trichomes) or chemical (secondary metabolites, hydrolytic or oxidative enzymes). This defense system is not absolute but was overcome by a number of specialists (similar to our immune system being overcome by a few adapted viruses or parasites). In only a limited number of cases the biochemical mechanisms which are used by the specialists have been elucidated. To attract pollinating insects and other seed-dispersing animals plants also developed chemicals such as carbohydrates, organic acids, and secondary metabolites (colors and fragrances).

2. Induction of Product Formation

At first view the above-mentioned evolutionary and ecological background seems to be unrelated to the production of biochemicals in cell cultures. However, there are two important conclusions to be drawn:

1. The structures of secondary metabolites are not just random structures but have been shaped and optimized during evolution, because of their biological functions. It is therefore not surprising that many secondary metabolites are used by man as pharmaceuticals, spices, fragrances, pesticides, poisons, halucinogens, stimulants, colors, or even as lead structures for the chemical synthesis of even more active molecules. This background is thus the basis for the exploitation of plant secondary metabolites.
2. Because of their function as defense substances, secondary product formation can respond to environmental stress, such as predation by insects or infection by microorganisms. It has been reported that wounding of plant tissue or grazing could induce the increased synthesis of a number of compounds, including phenolics (Baldwin and Schulz, 1983), anthocyanins, terpenoid resins, alkaloids (Tanaka *et al.*, 1983; Wink, 1983), and protease inhibitors (Walker-Simmons and Ryan, 1984). On infection by microorganisms the formation of a large number of metabolites, the so-called phytoalexins, is promoted (Deverall, 1977; Grisebach and Ebel, 1978; Stoessl, 1980; Harborne, 1982).

It has been shown during the last decade that we can also trigger secondary product formation in cell culture by environmental stress. Treatment of cell cultures with fungal, bacterial, or plant cell walls resulted in the increased formation of flavonoids, stilbenes, terpenoids, anthraquinones, rutacridone alkaloids, sanguinarine, and gossypol (Chappell and Hahlbrock, 1984; DiCosmo and Misawa, 1985; Eilert *et al.*, 1985b). In some instances chemically more defined "elicitors," including heavy metal salts, bioregulators, and DNA-active substances such as alkaloids, resulted in similar inductions of product formation (Hadwiger and Schwochau, 1971; Wink and Witte, 1983; Stossel, 1984; Frischknecht and Baumann, 1985; Wink, 1985a). In most cases the increase of products was only transient. It is evident that we need more information on the basic principles of the defense response, such as structures and perception of defense signals, transduction, and translation of these signals into a response.

IV. CONCLUSIONS

The production of secondary metabolites in plants is a complex process highly coordinated in space and time. Its main components are *biosynthesis* and *accumulation* which are usually modified by tissue- and cell-specific compartmentation. Depending on the developmental stage, transport and degradation can be additional factors. Since most if not all of these processes are controlled by genes, understanding of the respective gene expression is crucial (see Section II,B,3). In view of the complexity of secondary metabolism, it is not surprising that most cell culture systems have failed to produce a given compound in large quantities. Thus, it is exciting that more than 30 cell cultures have been established which produce a higher amount of a secondary metabolite than the respective differentiated plant. Is there any factor which is common to all of them? According to Table V, nearly all successful compounds are synthesized by root tissue and the site of accumulation is either within the producing cell or the adjacent cell. On the other hand, compounds made in the roots (e.g., tropane alkaloids), which are transported via the xylem to other plant parts, usually fail to accumulate in plant cell suspension cultures. Compounds formed by leaf tissue are only produced by photomixo- or photoautotrophic cultures. Also here only the compounds which are accumulated within the producing cell were successful. This means that secondary metabolites whose biology shows a higher degree of complexity are usually not yet produced by plant cell cultures. It is interesting that when cultured plant cells differentiate into tissues and organs their capacity to produce alkaloids, such as morphine, hyoscyamine, or cardenolides, sets in immediately, i.e., on differentiation the coordinate gene expression takes place again.

Why are the compounds made in root tissue (Table V) so successful in plant cell culture? As discussed earlier (see Section II,B,3), the genes for a root biosynthesis are turned on only if the master gene for root cells is activated. Thus the suspension culture cells would no longer be undifferentiated but be biochemically in the state of root cells. I have speculated recently that we select for cells whose state of differentiation resembles that of root cells through our culture conditions (Wink, 1985c). For example, suspension culture cells are supplied 3–6% sucrose, are usually kept in the dark or dim light as heterotrophic cultures, and the oxygen supply in the cultures quickly becomes limited, especially when the cultures enter the stationary phase of growth. At this stage lupine cells break down sucrose to ethanol (Wink, 1985c), which has also been observed in other cell cultures (Thomas and Murashige, 1979). Formation of ethanol is typical for root cells under oxygen limitation (Davies, 1980). Thus without knowing, we select for root cells, because cells with the biochemical differentiation of root cells will survive and multiply better under the culture conditions applied. Further experimentation must show if this assumption is correct.

ACKNOWLEDGMENT

Support by the Deutsche Forschungsgemeinschaft and a Heisenberg fellowship are gratefully acknowledged.

REFERENCES

Aerts, J. M. F. G., and Schramm, A. W. (1985). Isolation of vacuoles from the upper epidermis of *Petunia hybrida* petals. I. A comparison of the isolation procedures. *Z. Naturforsch., C: Biosci.* **40C,** 189–195.

Amann, M., Wanner, G., and Zenk, M. H. (1986). Intracellular compartmentation of two enzymes of berberine biosynthesis in plant cell cultures. *Planta* **167,** 310–320.

Anderson, L. A., Phillipson, J. D.. and Roberts, M. F. (1985). Biosynthesis of secondary products by cell cultures of higher plants. *Adv. Biochem. Eng.* **21,** 1–36.

Balandrin, M. F., Klocke, J. A., Wurtele, E. S., and Bollinger, W. H. (1985). Natural plant chemicals: Sources of industrial and medicinal materials. *Science* **228,** 1154–1160.

Baldwin, I. T., and Schultz, J. C. (1983). Rapid changes in tree leaf chemistry induced by damage: Evidence for communication between trees. *Science* **221,** 277–278.

Barz, W. (1977). Catabolism of endogenous and exogenous compounds by plant cell cultures. *In* "Plant Tissue Culture and Its Biotechnological Application" (W. Barz, E. Reinhard, and M. H. Zenk, eds.), pp. 153–171. Springer-Verlag, Berlin and New York.

Barz, W. (1985). Metabolism and degradation of nicotinic acid in plant cell cultures. *In* "Primary and Secondary Metabolism of Plant Cell Cultures" (K.-H. Neumann, W. Barz, and E. Reinhard, eds.). pp. 186–195. Springer-Verlag, Berlin and New York.

Barz, W., and Hüsemann, W. (1982). Aspects of photoautotrophic cell suspension cultures. *In* "Plant Tissue Culture 1982" (A. Fujiwara, ed.), pp. 245–247. Maruzen, Tokyo.

Barz, W., and Köster, J. (1981). Turnover and degradation of secondary (natural) products. *In* "The Biochemistry of Plants" (E. E. Conn, ed.), Vol. 7, pp. 35–83. Academic Press, New York.

Barz, W., Herzbeck, H., Hüsemann, W., Schneiders, G., and Mangold, H. K. (1980). Alkaloids and lipids of heterotrophic, photomixotrophic and photoautotrophic cell suspension cultures of *Peganum harmala*. *Planta Med.* **40,** 137–148.

Baumert, A., Schneider, G., and Gröger, D. (1986). Biosynthesis of acridone alkaloids. A cell-free system from *Ruta graveolens* cell suspension cultures. *Z. Naturforsch., C: Biosci.* **41C,** 187–192.

Beiderbeck, R. (1982). Zweiphasenkultur- ein Weg zur Isolierung lipophiler Substanzen aus pflanzlichen Suspensionskulturen. *Z. Pflanzenphysiol.* **108,** 27–30.

Berlin, J. (1984). Plant cell cultures—A future source of natural products? *Endeavour* [N.S.] **8,** 5–8.

Berlin, J., Forche, E., Wray, V., Hammer, J., and Hösel, W. (1983). Formation of benzophenanthridine alkaloids by suspension cultures of *Eschscholtzia californica. Z. Naturforsch., C: Biosci.* **38C,** 346–352.

Böhm, H. (1978). Regulation of alkaloid production in plant cell cultures. *In* "Frontiers of Plant Tissue Culture 1978" (T. A. Thorpe, ed.), pp. 201–211. University of Calgary, Calgary, Canada.

Böhm, H., and Franke, J. (1982). Accumulation and excretion of alkaloids by *Macleaya microcarpa* cell cultures. I. Experiments on solid medium. *Biochem. Physiol. Pflanz.* **177,** 345–356.

Bünning, E. (1963). "Die physiologische Uhr. Zeitmessung in Organismen mit ungefähr tagesperiodischen Schwankungen." Springer-Verlag, Berlin and New York.

Chappell, J., and Hahlbrock, K. (1984). Transcription of plant defense genes in response to UV light or fungal elicitors. *Nature (London)* **311,** 76–78.

Constabel, F. (1969). Über die Entwicklung von Gerbstoffzellen in Calluskulturen von *Juniperus communis* L. *Planta Med.* **80,** 103–115.

Constabel, F., Kurz, W. G. W., and Kutney, J. P. (1982). Variation in cell cultures of periwinkle, *Catharanthus roseus. In* "Plant Tissue Culture 1982" (A. Fujiwara, ed.), pp. 301–304. Maruzen, Tokyo.

Curtin, B. (1983). *Bio/Technology* **1,** 649–657.

Dalton, D. R. (1979). "The Alkaloids. The Fundamental Chemistry." Dekker, New York.

Davies, D. D. (1980). Anaerobic metabolism and the production of organic acids. *In* "The Biochemistry of Plants" (D. D. Davies, ed.), Vol. 2, p. 581. Academic Press, New York.

Deus-Neumann, B., and Zenk, M. H. (1984). A highly selective alkaloid uptake system in vacuoles of higher plants. *Planta* **162,** 250–260.

Deus-Neumann, B., and Zenk, M. H. (1986). Accumulation of alkaloids in plant vacuoles does not involve an ion-trap mechanism. *Planta* **167,** 44–53.

Deverall, B. J. (1977). "Defense Mechanisms of Plants." Cambridge Univ. Press, London and New York.

DiCosmo, F., and Misawa, M. (1985). Eliciting secondary metabolism in plant cell cultures. *Trends Biotechnol.* **3,** 318–322.

Dreyer, D., Jones, K. C., and Molyneux, R. J. (1985). Feeding deterrency of some pyrrolizidine, indolizidine, and quinolizidine alkaloids towards pea aphid (*Acyrthosiphon pisum*) and evidence for phloem transport of indolizidine alkaloid swainsonine. *J. Chem. Ecol.* **11,** 1045–1051.

Eckes, P., Schell, J., and Willmitzer, L. (1985). Organ-specific expression of three leaf/stem specific cDNA from potato is regulated by light and correlated with chloroplast development. *Mol. Gen. Genet.* **199,** 216–224.

Ehrlich, P. R., and Raven, P. H. (1964). Butterflies and plants: A study of coevolution. *Evolution (Lawrence, Kans.)* **18,** 586–608.

Eilert, U., Nesbitt, R. L., and Constabel, F. (1985a). Laticifers and latex in fruits of periwinkle. *Catharanthus roseus. Can. J. Bot.* **63,** 1540–1546.

Eilert, U., Kurz, W. G. W., and Constabel, F. (1985b). Stimulation of sanguinarine accumulation in *Papaver somniferum* cell cultures by fungal elicitors. *J. Plant Physiol.* **119,** 65–76.

Eltayeb, E., and Roddick, J. D. (1985). Biosynthesis and degradation of tomatine in developing tomato fruits. *Phytochemistry* **24,** 253–257.

Flück, H. (1963). Intrinsic and extrinsic factors affecting the production of secondary products. *In* "Chemical Plant Taxonomy" (T. L. Swain, ed.), pp. 167–186. Academic Press, New York.

Fraenkel, G. (1959). The raison d'etre of secondary substances. *Science* **129,** 1466–1470.

Frischknecht, P. M., and Baumann, T. (1985). Stress induced formation of purine alkaloids in plant tissue culture of *Coffea arabica. Phytochemistry* **24,** 2255–2257.

Fukui, H., Tsukuda, M., Mizukami, H., and Tabata, M. (1983a). Formation of stereoisomeric mixtures of naphthoquinone derivatives in *Echium lycopsis* callus cultures. *Phytochemistry* **22,** 453–456.

Fukui, H., Yoshikawa, N., and Tabata, M. (1983b). Induction of shikonin formation by agar in *Lithospermum erythrorhizon* cell suspension cultures. *Phytochemistry* **22,** 2451–2453.

Goodwin, T., and Mercer, E. I. (1983). "Introduction to Plant Biochemistry." Pergamon, Oxford.

Grisebach, H., and Ebel, J. (1978). Phytoalexine, chemische Abwehrstoffe höherer Pflanzen? *Angew. Chem.* **90,** 668–681.

Grundon, M. F. (1976–1983). "The Alkaloids. A Specialists Report," Vols. 6–13. Chemical Soc., London.

Hadwiger, L. A., and Schwochau, M. E. (1971). Specificity of DNA intercalating compounds in the control of PAL and pisatin levels. *Plant Physiol.* **47,** 346–351.

Harborne, J. B. (1977). Chemosystematics and coevolution. *Pure Appl. Chem.* **49,** 1403–1421.

Harborne, J. B. (1982). "Introduction to Ecological Biochemistry," 2nd ed. Academic Press, New York.

Hay, C. A., Anderson, L. A., Roberts, M. F., and Phillipson, J. D. (1986). In vitro cultures of *Cinchona* species. Precursor feeding of *C. ledgeriana* root organ suspension cultures with L-tryptophan. *Plant Cell Rep.* **5,** 1–4.

Hinz, H., and Zenk, M. H. (1981). Production of protoberberine alkaloids by cell suspension cultures of *Berberis* species. *Naturwissenschaften* **68,** 620–621.

Holländer-Czytko, H., and Amrhein, N. (1983). Subcellular compartmentation of shikimic acid and phenylalanine in buckwheat cell suspension cultures grown in the presence of shikimate pathway inhibitors. *Plant Sci. Lett.* **29,** 89–96.

Homeyer, B. C., and Roberts, M. F. (1984). Alkaloid sequestration by *Papaver somniferum* latex. *Z. Naturforsch., C: Biosci.* **39C,** 876–881.

Igbavboa, U., Siewecke, H.-J., Leistner, E., Röwer, I., Hüsemann, W., and Barz, W. (1985). Alternative formation of anthraquinones and lipoquinones in heterotrophic and photoautotrophic cell suspension cultures of *Morinda lucida. Planta* **166,** 537–544.

James, J. O. (1950). Alkaloids in the plants. *In* "The Alkaloids" (R. H. F. Manske and H. L. Holmes, eds.), Vol. 1, pp. 15–90. Academic Press, New York.

Kaulen, H., Schell, J., and Kreuzaler, F. (1986). Light-induced expression of the chimeric chalcon synthase NPT II gene in tobacco cells. *EMBO J.* **5,** 1–8.

Knogge, W., and Weissenböck, G. (1986). Tissue-distribution of secondary phenolic biosynthesis in developing primary leaves of *Avena sativa* L. *Planta* **167,** 196–205.

Kojima, M., Poulton, J. E., Thayer, S., and Conn, E. E. (1979). Tissue distribution of dhurrin and of enzymes involved in its metabolism in leaves of *Sorghum bicolor. Plant Physiol.* **63,** 1022–1128.

Kutchan, T. M., Ayabe, S., Krueger, R. J., Coscia, E. M., and Coscia, C. J. (1983). Cytodifferentiation and alkaloid accumulation in cultured cells of *Papaver bracteatum*. *Plant Cell Rep.* **2,** 281–284.

Lamppa, G., Nagy, F., and Chua, N.-H. (1985). Light-regulated and organ-specific expression of a wheat *Cab* gene in transgenic tobacco. *Nature (London)* **316,** 750–752.

Leistner, E. (1975). Isolierung, Identifizierung und Biosynthese von Anthrachinonen in Zellsuspensionskulturen von *Morinda citrifolia*. *Planta Med., Suppl.*, pp. 214–224.

Levin, D. A. (1976). The chemical defenses of plants to pathogens and herbivores. *Annu. Rev. Ecol. Syst.* **7,** 121–159.

Levinson, H. Z. (1976). The chemical defensive role of alkaloids in insects and plants. *Experientia* **32,** 406–411.

Lüttge, U., and Smith, J. A. C. (1985). Transport of malic acid in cells of CAM plants. *In* "Biochemistry and Function of Vacuolar Adenosine-Triphosphatase in Fungi and Plants" (B. P. Marin, ed.), pp. 227–238. Springer-Verlag, Berlin and New York.

Marty, F., Branton, D., and Leigh, R. A. (1980). Plant vacuoles. *In* "The Biochemistry of Plants" (P. K. Stumpf and E. E. Conn, eds.), Vol. 1, pp. 625–658. Academic Press, New York.

Matern, U., Reichenbach, C., and Heller, W. (1986). Efficient uptake of flavonoids into parsley (*Petroselinum hortense*) vacuoles requires acylated glycosides. *Planta* **167,** 183–189.

Matile, P. (1976). Localization of alkaloids and mechanism of their accumulation in vacuoles of *Chelidonium majus* laticifers. *Nova Acta Leopold. Suppl.*, 7, pp. 139–155.

Matile, P. (1978). Biochemistry and function of vacuoles in plants. *Annu. Rev. Plant Physiol.* **29,** 193–213.

Matile, P. (1984). Das toxische Kompartiment der Pflanzenzelle. *Naturwissenschaften* **71,** 18–24.

Mende, P., and Wink, M. (1987). Uptake of the quinolizidine alkaloid lupanine by protoplasts and vacuoles of *Lupinus polyphyllus* cell suspension cultures. Diffusion or carrier-mediated transport? *J. Plant Physiol.* **129,** 229–242.

Molisch, H. (1923). "Mikrochemie der Pflanze." Fischer, Jena.

Mothes, K. (1955). Physiology of alkaloids. *Annu. Rev. Plant Physiol.* **6,** 393–432.

Mothes, K. (1976). Secondary plant substances as materials for chemical high quality breeding in higher plants. *Recent Adv. Phytochem.* **10,** 385.

Mothes, K., and Schütte, H. R. (1969). "Biosynthese der Alkaloide." VEB Dtsch. Verlag Wiss., Berlin.

Mothes, K., Schütte, H. R., and Luckner, M. (1985). "Biochemistry of Alkaloids." Verlag Chemie, Weinheim.

Nakanishi, T., Miyasaka, H., Nasu, M., Hashimoto, H., and Yoneda, K. (1983). Production of cryptotanshinone and ferruginol in cultured cells of *Salvia miltiorrhiza*. *Phytochemistry* **22,** 721–722.

Neumann, D., and Müller, E. (1967). Intrazellulärer Nachweis von Alkaloiden in Pflanzenzellen im licht-und elektronenmikroskopischen Massstab. *Flora (Jena), Abt. A* **158,** 479–491.

Neumann, D., and Müller, E. (1972). Beiträge zur Physiologie der Alkaloide. III. *Chelidonium majus* L. und *Sanguinaria canadensis* L.: Ultrastruktur der Alkaloidbehälter, Alkaloidaufnahme und -verteilung. *Biochem. Physiol. Pflanz.* **163,** 375–391.

Neumann, D., and Tschöpe, K. H. (1966). Abbau von Solanaceen-Alkaloiden in höheren Pflanzen, *Flora (Jena), Abt. A* **156,** 521–542.

Neumann, D., Krauss, G., Hieke, M., and Gröger, D. (1983). Indole alkaloid formation and storage in cell suspension cultures of *Catharanthus roseus*. *Planta Med.* **48,** 20–23.

Paech, K. (1950). "Biochemie und Physiologie der sekundären Pflanzenstoffe." Springer-Verlag, Berlin and New York.

Proksch, M., Proksch, P., Seissenböck, G., and Rodriguez, E. (1982). Flavonoids from the leaf resin of *Adenostoma sparsifolium*. *Phytochemistry* **21,** 1835–1836.

Proksch, P., Proksch, M., Weck, W., and Rodriguez, E. (1985). Localization of chromenes and benzofurans in the genus *Encelia* (Asteraceae). *Z. Naturforsch., C: Biosci.* **40C,** 301–304.

Raffauf, R. F. (1970). "A Handbook of Alkaloids and Alkaloid-Containing Plants." Wiley (Interscience), New York.

Reinhard, E., and Alfermann, W. (1980). Biotransformation by plant cell cultures. *Adv. Biochem. Eng.* **16,** 49–83.

Renaudin, J. P., Brown, S. C., and Guern, J. (1985). Compartmentation of alkaloids in a cell suspension of *Catharanthus roseus:* A reappraisal of the role of pH gradients. *In* "Primary and Secondary Metabolism of Plant Cell Cultures" (K.-H. Neumann, W. Barz, and E. Reinhard, eds.), pp. 124–132. Springer-Verlag, Berlin and New York.

Renaudin, J. P., Brown, S. C., Barbier-Brygoo, H., and Guern, J. (1986). Quantitative characterization of protoplasts and vacuoles from suspension-cultured cells of *Catharanthus roseus*. *Physiol. Plant.* (in press).

Roberts, M. F. (1981). Enzymic synthesis of coniceine in *Conium maculatum* chloroplasts and mitochondria. *Plant Cell Rep.* **1,** 10–13.

Robinson, T. (1974). Metabolism and function of alkaloids in plants. *Science* **184,** 430–435.

Robinson, T. (1981). "The Biochemistry of Alkaloids." Springer-Verlag, Berlin and New York.

Röper, W., Schulz, M., Chaouiche, E., and Meloh, K. A. (1985). Nicotine production by tissue cultures to tobacco as influenced by various culture parameters. *J. Plant Physiol.* **118,** 463–470.

Rosahl, S., Eckes, P., Schell, J., and Willmitzer, L. (1986). Organ-specific gene expression in potato. Isolation and characterization of tuber-specific cDNA sequences. *Mol. Gen. Genet.* **202,** 368–373.

Rosenthal, G. A. (1982). "Plant Nonprotein Amino and Imino Acids." Academic Press, New York.

Rosenthal, G. A., and Janzen, D. H., eds. (1979). "Herbivores: Their Interaction with Secondary Plant Metabolites." Academic Press, New York.

Sato, F., and Yamada, Y. (1984). High berberine-producing cultures of *Coptis japonica* cells. *Phytochemistry* **23,** 281–285.

Saunders, J. A. (1979). Investigations of vacuoles isolated from tobacco. *Plant Physiol.* **64,** 74–78.

Saunders, J. A., and Conn, E. E. (1978). Presence of the cyanogenic glycoside dhurrin in isolated vacuoles from *Sorghum*. *Plant Physiol.* **61,** 154–157.

Saxton, J. E. (1971–1975). "The Alkaloids: A Specialists Report," Vols. 1–5. Chemical Soc., London.

Schulte, U., El-Shagi, H., and Zenk, M. H. (1984). Optimization of 19 Rubiaceae species in cell culture for the production of anthraquinones. *Plant Cell Rep.* **3,** 51–54.

Schultz, G., Soll, J., Fiedler, E., and Schulze-Siebert, D. (1985). Synthesis of prenylquinones in chloroplasts. *Physiol. Plant.* **64,** 123–129.

Sengupta-Gopalan, C., Reichert, N. A., Barber, R. F., Hall, T. C., and Kemp, J. D. (1985). Developmentally regulated expression of the bean β-phaseolin gene in tobacco seeds. *Proc. Natl. Acad. Sci. U.S.A.* **82,** 3320–3324.

Spenser, I. D. (1968). The biosynthesis of alkaloids and of other nitrogenous secondary metabolites. *In* "Comprehensive Biochemistry" (M. Florkin and E. H. Stotz, eds.), Vol. 20, pp. 231–413, Elsevier, Amsterdam.

Staba, E. J. (1985). Milestones in plant tissue culture systems for the production of secondary products. *J. Nat. Prod.* **48,** 203–209.

Stahl, E. (1888). Pflanzen und Schnecken. *Jena. Z. Med. Naturwiss.* **22,** 557.

Stöckigt, J. (1981). The biosynthesis of heteroyohimbine-type alkaloids. *In* "Indole and Biogenetically Related Alkaloids" (J. D. Phillipson and M. H. Zenk, eds.), p. 113. Academic Press, New York.

Stoessl, A. (1980). Phytoalexins—A biogenetic perspective. *Phytopathol. Z.* **99,** 251–272.

Stoessel, P. (1984). Regulation by sulfhydryl groups of glyceollin accumulation in soybean hypocotyls. *Planta* **160,** 314–319.

Strack, D., Pieroth, M., Scharf, H., and Sharma, V. (1985). Tissue distribution of phenylpropanoid metabolism in cotyledons of *Raphanus sativus* L. *Planta* **164,** 507–511.

Swain, T. L. (1977). Secondary compounds as protective agents. *Annu. Rev. Plant Physiol.* **28,** 479–501.

Sze, H. (1984). H^+-translocating ATPases of the plasma membrane and tonoplast of plant cells. *Physiol. Plant.* **61,** 683–691.

Tabata, M. (1977). Recent adavances in the production of medicinal substances by plant cell cultures. *In* "Plant Tissue Culture and Its Biotechnological Application" (W. Barz, E. Reinhard, and M. H. Zenk, eds.), p. 326. Springer-Verlag, Berlin and New York.

Tanaka, Y., Data, E., Hirose, S., Taniguchi, T., and Uritani, I. (1983). Biochemical changes in secondary metabolites in wounded and deteriorated cassava roots. *Agric. Biol. Chem.* **47,** 693–700.

Thom, M., and Komor, E. (1984). H^+–sugar antiport as mechanism of sugar uptake by sugarcane vacuoles. *FEBS Lett.* **173,** 1–4.

Thom, M., Komor, E., and Maretzki, A. (1982). Vacuoles of sugar cane suspension cultures. II. Characterization of sugar uptake. *Plant Physiol.* **69,** 1320–1325.

Thomas, D., and Murashige, T. (1979). Volatile emissions of plant tissue culture. I. Identifications of the major components. *In Vitro* **15,** 654–658.

Thunmann, O., and Rosenthaler, L. (1931). "Pflanzenmikrochemie," 2nd ed. Borntraeger, Berlin.

Verzar-Petri, G., Csedö, K., Möllmann, H., Szendri, K., and Reisch, J. (1976). Fluoreszenzmikroskopische Untersuchungen über die Lokalisation der Acridonalkaloide in Geweben von *Ruta graveolens. Planta Med.* **29,** 370–375.

Walker-Simmons, M., and Ryan, C. A. (1984). Proteinase inhibitor synthesis in tomato leaves. *Plant Physiol.* **76,** 787–790.

Waller, G. R., and Dermer, O. C. (1981). Enzymology of alkaloid metabolism in plants and microorganisms. *In* "The Biochemistry of Plants" (E. E. Conn, ed.), Vol. 7, pp. 317–402. Academic Press, New York.

Waller, G. R., and Nowacki, E. (1978). "Alkaloid Biology and Metabolism in Plants." Plenum, New York.

Weiler, E. W. (1980). Radioimmunassays for the differential and direct analysis of free and conjugated abscisic acid in plant extracts. *Planta* **148,** 262–272.

Weiler, E. W., and Zenk, M. H. (1976). Radioimmunassay for the determination of digoxin and related compounds in *Digitalis lanata. Phytochemistry* **15,** 1537–1545.

Werner, C., and Matile, P. (1985). Accumulation of coumaroylglucosides in vacuoles of barley mesophyll protoplasts. *J. Plant Physiol.* **118,** 237–249.

Whitacker, R. H., and Feeny, R. P. (1971). Allelochemics: Chemical interaction between species. *Science* **171,** 757–770.

Wiermann, R. (1981). Secondary products and cell and tissue differentiation. *In* "The Biochemistry of Plants" (E. E. Conn, ed.), Vol. 7, pp. 85–115. Academic Press, New York.

Willenbrink, J., and Doll, S. (1979). Characteristics of the sucrose uptake system of vacuoles isolated from red beet tissue. Kinetics and specificity of the sucrose uptake system. *Planta* **147,** 159–162.

Wink, C., and Hartmann, T. (1981). Properties and subcellular localization of L-alanine : aldehyde aminotransferase: Concept of an ubiquitous plant enzyme involved in secondary metabolism. *Z. Naturforsch., C: Biosci.* **36C,** 625–632.

Wink, M. (1983). Wounding-induced increase of quinolizidine alkaloid accumulation in lupine leaves. *Z. Naturforsch., C: Biosci.* **38C,** 905–909.

Wink, M. (1984). Evidence for an extracellular lytic compartment of plant cell suspension cultures. The cell culture medium. *Naturwissenschaften* **71,** 635–636.

Wink, M. (1985a). Metabolism of quinolizidine alkaloids in plants and cell suspension cultures: Induction and degradation. *In* "Primary and Secondary Metabolism of Plant Cell Cultures" (K.-H. Neumann, W. Barz, and E. Reinhard, eds.), pp. 107–116. Springer-Verlag, Berlin and New York.

Wink, M. (1985b). Chemische Verteidigung der Lupinen: Zur biologischen Bedeutung der Chinolizidinalkaloide. *Plant Syst. Evol.* **150,** 65–81.

Wink, M. (1985c). Composition of the spent culture medium. I. Time-course of ethanol formation and the excretion of hydrolytic enzymes into the culture medium of suspension-cultured cells of *Lupinus polyphyllus. J. Plant Physiol.* **121,** 287–293.

Wink, M. (1986). Storage of quinolizidine alkaloids in epidermal tissues. *Z. Naturforsch., C: Biosci.* **41C,** 375–380.

Wink, M. (1987a). Production of plant secondary metabolites by plant cell cultures in relation to the site and mechanism of their accumulation. (in preparation).

Wink, M. (1987b). Sites of lupine alkaloid biosynthesis. *Z. Naturforsch., C: Biosci.* **42e** (in press).

Wink, M. (1987c). Chemical ecology of quinolizidine alkaloids. *ACS Symp. Ser.* **330,** 524–533.

Wink, M. (1987d). Why do cell suspension cultures of lupines fail to produce alkaloids in large quantities? *Plant Cell, Tissue Organ Cult.* **8,** 103.

Wink, M., and Hartmann, T. (1980). Production of quinolizidine alkaloids by photomixotrophic cell suspension cultures: Biochemical and biogenetic aspects. *Planta Med.* **40,** 149–155.

Wink, M., and Hartmann, T. (1981a). Sites of enzymatic synthesis of quinolizidine alkaloids and their accumulation in *Lupinus polyphyllus. Z. Pflanzenphysiol.* **102,** 337–344.

Wink, M., and Hartmann, T. (1981b). Activation of chloroplast-localized enzymes of quinolizidine alkaloid biosynthesis by reduced thioredoxin. *Plant Cell Rep.* **1,** 6–9.

Wink, M., and Hartmann, T. (1982a). Localization of the enzymes of quinolizidine alkaloid biosynthesis in leaf chloroplasts of *Lupinus polyphyllus. Plant Physiol.* **70,** 74–77.

Wink, M., and Hartmann, T. (1982b). Diurnal fluctuations of quinolizidine alkaloid accumulation in legumes and cell suspension cultures. *Z. Naturforsch., C: Biosci.* **37C,** 369–375.

Wink, M., and Hartmann, T. (1982c). Physiological and biochemical aspects of quinolizidine alkaloid formation in cell suspension cultures. *In* "Plant Tissue Culture 1982" (A. Fujiwara, ed.), pp. 333–334. Maruzen, Tokyo.

Wink, M., and Hartmann, T. (1985). Enzymology of quinolizidine alkaloid biosynthesis. *In* "Natural Products Chemistry 1984" (R. I. Zalewski and J. J. Skolik, eds.), pp. 511–520. Elsevier, Amsterdam.

Wink, M., and Mende, P. (1987). Uptake of lupine alkaloids by epidermal cells and suspension-cultured cells of *Lupinus polyphyllus. Planta Med.* (in press).

Wink, M., and Witte, L. (1983). Evidence for a widespread occurrence of the genes of quinolizidine alkaloids biosynthesis. Induction of alkaloid accumulation in cell suspension cultures of alkaloid-"free" species. *FEBS Lett.* **159,** 196–200.

Wink, M., and Witte, L. (1984). Turnover and transport of quinolizidine alkaloids: Diurnal variation of lupanine in the phloem sap, leaves and fruits of *Lupinus albus* L. *Planta* **161,** 519–524.

Wink, M., and Witte, L. (1985). Quinolizidine alkaloids as nitrogen source for lupine seedlings and cell suspension cultures. *Z. Naturforsch., C: Biosci.* **40C,** 767–775.

Wink, M., Hartmann, T., and Witte, L. (1980). Biotransformation of cadaverine and of potential intermediates of lupanine biosynthesis by plant suspension cultures. *Planta Med.* **40,** 31–39.

Wink, M., Witte, L., and Hartmann, T. (1981). Quinolizidine alkaloids of cell suspension cultures and plants of *Sarothamnus scoparius* and of its root parasite *Orobanche rapum-genistae. Planta Med.* **43,** 342–352.

Wink, M., Hartmann, T., Witte, L., and Rheinheimer, J. (1982). Interrelationship between quinolizidine alkaloid producing legumes and infesting insects: Exploitation of the alkaloid-containing phloem sap of *Cytisus scoparius* by the Broom aphid *Aphis cytisorum. Z. Naturforsch. C. (Biosci)* **37,** 1081–1086.

Wink, M., Witte, L., Hartmann, T., Theuring, C., and Volz, V. (1983). Accumulation of quinolizidine alkaloids in plants and cell suspension cultures: Genera *Lupinus, Cytisus, Baptisia, Genista, Laburnum,* and *Sophora. Planta Med.* **48,** 253–257.

Wink, M., Heinen, H. J., Vogt, H., and Schiebel, H. M. (1984). Cellular localization of quinolizidine alkaloids by laser desorption mass spectrometry. *Plant Cell Rep.* **3,** 230–233.

Zenk, M. H. (1968). Biochemie und Physiologie sekundärer Pflanzenstoffe. *Ber. Dtsch. Bot. Ges.* **80,** 573–591.

Zenk, M. H. (1978). The impact of plant cell culture on industry. *In* "Frontiers of Plant Tissue Culture 1978" (T. A. Thorpe, ed.), pp. 1–13. University of Calgary, Calgary, Canada.

Zenk, M. H. (1980). Enzymatic synthesis of ajmalicine and related indole alkaloids. *J. Nat. Prod.* **43,** 438–451.

Zenk, M. H. (1982). Pflanzliche Zellkulturen in der Arzneimittelforschung. *Naturwissenschaften* **69,** 534–536.

Zenk, M. H. (1985). Enzymology of benzylisoquinoline alkaloid formation. *In* "The Chemistry and Biology of Isoquinoline Alkaloids" (J. D. Phillipson, M. F. Roberts, and M. H. Zenk, eds.), pp. 240–256. Springer-Verlag, Berlin and New York.

Zenk, M. H., El-Shagi, H., and Schulte, U. (1975). Anthraquinone production by cell suspension cultures of *Morinda citrifolia. Planta Med., Suppl.,* pp. 79–101.

Zenk, M. H., El-Shagi, H., Arens, H., Stöckigt, J., Weiler, E. W., and Deus. B. (1977). Formation of the indole alkaloids serpentine and ajmalicine in cell suspension cultures of *Catharanthus roseus. In* "Plant Tissue Culture and its Biotechnological Application" (W. Barz, E. Reinhard, and M. H. Zenk, eds.), pp. 27–44. Springer-Verlag, Berlin and New York.

Zenk, M. H., Rüffer, M., Amann, M., and Deus-Neumann, B. (1985). Benzylisoquinoline biosynthesis by cultivated plant cells and isolated enzymes. *J. Nat. Prod.* **48.** 725–738.

CHAPTER **3**

The Compartmentation of Secondary Metabolites in Plant Cell Cultures

J. Guern
J. P. Renaudin[1]
S. C. Brown

Laboratoire de Physiologie Cellulaire Végétale
CNRS-INRA
F 91190 Gif-sur-Yvette, France

I. INTRODUCTION

Over the past 15 years, plant cell cultures have enabled a better understanding of the complex organization of secondary metabolism at the cell and tissue levels (Dougall, 1981). This progress has notably concerned knowledge of enzyme activities and the regulation of biosynthetic pathways (Zenk, 1980, 1985), two of several parameters determining the net content of a tissue at a given time. It has been repeatedly emphasized that the various components of the secondary metabolism such as biosynthetic pathways, transport mechanisms, storage processes at specific sites, conjugation, turnover, and catabolic reactions are organized and coordinated at the cell, tissue, and plant levels, and are also organized and coordinated with time (Luckner, 1980; Luckner *et al.*, 1980; Barz and Köster, 1981; Wiermann, 1981; Conn, 1984).

Some reviews have recently dealt with these interactive facets of the metabolism of given phytochemicals (Boudet *et al.*, 1984; Conn, 1984). However, the numerous recent publications on the topic of compartmentation of products at the (sub)cellular level now merit reviewing. This topic has been incorporated previously either in general articles about the organization of secondary metabolism (Luckner, 1980; Luckner *et al.*, 1980; Wiermann, 1981) or in reviews about plant organelles such as vacuoles, the main site of accumulation of phytochemicals with-

[1]Present address: INRA, Station de Physiopathologie Végétale, BV No. 1540, F 21034 Dijon, France.

in cells (Matile, 1978, 1982, 1984: Alibert and Boudet, 1982; Wagner, 1982; Boller and Wiemken, 1986).

Compartmentation means limitation within space and/or time. It implies that devices exist to limit or preclude random diffusion of solutes, substrates, and enzymes inside the cell. Such devices are first of all membranes which limit cell territories or compartments; translocators associated with energizing systems create driving forces, through electric and ionic gradients, which control the exchange of solutes between compartments and the development of concentration gradients. Microenvironments, and notably metabolic channeling through the formation of complexes between enzymes, substrates, and intermediates, are also significant elements of the spatial organization of secondary metabolism (Stafford, 1981).

Understanding compartmentation of phytochemicals in plant cell cultures implies: (i) an analytical description of the subcellular localization of secondary metabolites and of their biosynthetic pathways, (ii) a study of the mechanisms by which cell compartmentation is attained with a special emphasis on the role of membranes, (iii) an appreciation of the biological significance of compartmentation and of the peculiarities of plant cell cultures, and (iv) a concern for the biotechnological aspects of cell compartmentation. The first two aspects will be discussed in the first two sections of this chapter. The simple scheme depicted in Fig. 1 will be used as a guide to describe the analytical basis of the distinction between metabolic, storage, and extracellular pools of secondary metabolites and to emphasize the role of the membrane systems separating the metabolic and storage compartments (i.e., usually the vacuolar membrane), and the metabolic and extracellular compartments (i.e., the plasmalemma).

The biological significance of the compartmentation of secondary metabolites will be reviewed briefly. It insures high local concentrations of precursors and intermediates at the sites of biosynthesis. Conversely, feedback inhibition of biosynthesis is prevented by the spatial separation of synthesis and storage spaces. In addition, the removal of potentially harmful metabolites from the cytoplasm has a detoxifying effect (Matile, 1984). Moreover, the sites of accumulation can act as a reserve, with controlled release of nutrients, regulatory molecules, or defensive substrates.

Plant cell cultures may serve as an alternative industrial source of phytochemicals (see recent reviews of Heinstein, 1985; Staba, 1985). The study of compartmentation mechanisms together with metabolic studies, improvement of culture media, and selection of cell lines, is particularly relevant here in order to increase the production of phytochemicals, as illustrated by three examples. First, one should expect that

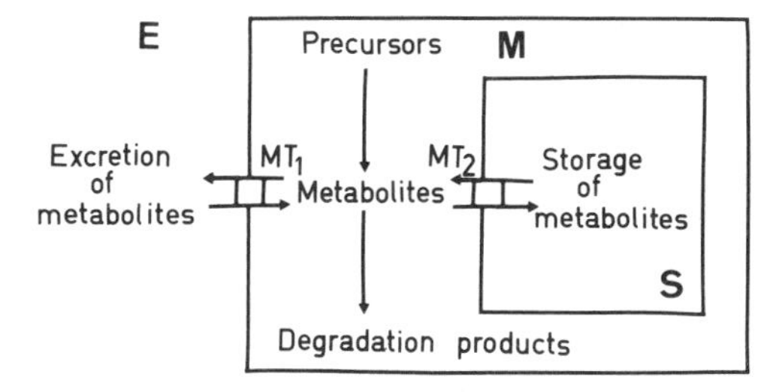

Fig. 1. Main compartments involved in the dynamics of secondary metabolism in plant cells. The metabolic compartment (M) is physically separated from the storage compartment (S) by a membrane system housing translocators (MT_2) which control the exchanges between the two compartments. Symmetrically, exchanges of the metabolic compartment with the extracellular compartment (E) are controlled by translocators at the plasmalemma (MT_1).

to attain a high production it is necessary to lower feedback inhibition of the biosynthesis. This can be reached by continuous removal of regulatory metabolites. Second, the production of secondary metabolites by a cell culture is not solely determined by its biosynthetic capacity. For example, the absence of phytochemicals in cell cultures, or presence in low amounts, may be due to the absence of storage sites where accumulation without degradation occurs, and not to low or inexistent biosynthetic capacity (Wink, 1985; Banthorpe *et al.*, 1986). Third, the rationale of biotechnological processes such as immobilized cells, continuous culture, or even bioreactors is to recover the products in the extracellular medium, the proportion of which is far greater in cell suspensions than in plants. It is therefore crucial to learn and to control the mechanisms which relate to the extracellular compartment and which determine its properties.

II. THE ANALYTICAL APPROACH TO CELL COMPARTMENTATION

A. Methodology Concerning Metabolic and Storage Compartments of Phytochemicals

The common methods to assess the subcellular localization of phytochemicals have been reviewed by Luckner *et al.* (1980). They involve cytochemistry, tracer experiments, and, mainly, cell fractionation. Their sensitivity is moderate so that they mainly indicate major sites of accumulation and contribute to the characterization of the storage compartment. Minor pools, such as those associated with the metabolic compartment or with transport, are more difficult to study. These methods strictly apply only to products whose localization does not change during the experiment, which precludes small amphiphilic molecules such as some alkaloids or phenolic aglycones that can diffuse passively through membranes.

A decisive step in the study of cell compartmentation has been the improved technology for the preparation and purification of vacuoles (reviewed by Alibert and Boudet, 1982; Wagner, 1983). These techniques have allowed description of large vacuolar pools for a number of solutes including secondary metabolites (Matile, 1982, 1984; Leigh, 1983; Boller and Wiemken, 1986).

Our knowledge of the metabolic compartment is derived mainly from the subcellular localization of the enzymes involved in the biosynthetic pathways. This aspect of cell compartmentation will not be detailed here, and the reader is referred to the relevant chapters in this volume. Examples of this strategy to identify the subcellular localization of the metabolic compartment will be found in Alibert *et al.* (1977) for phenolic metabolism, Hrazdina *et al.* (1980) for flavonoids, Stafford (1981) for a review of the phenylpropane pathway, Conn (1984) for cyanogenic glucosides and coumarin glucosides, and Zenk (1985) for benzylisoquinoline alkaloids.

This recent progress in our knowledge of the subcellular localization of some enzymes of secondary metabolism still needs to be strongly developed to achieve a more adequate picture of the spatial organization and properties of the metabolic compartment. On the other hand, a more strictly quantitative analysis of the compartmentation of secondary metabolites is necessary. Data are needed on the size of the pools in different compartments in order to estimate the existence and intensity of transmembrane concentration gradients. Unfortunately, most often the analyses concern the size of the pool in a compartment relative to the whole cell content, without integrating the relative volumes of the compartments and the distribution of the metabolites in terms of concentration. Showing that 90% of a metabolite is present in the vacuole is not strong evidence for an actual vacuolar accumulation when this compartment represents 90% of the cell volume.

B. Subcellular Localization of Secondary Metabolites

1. Phenolic Compounds

Some cinnamic acid derivatives and flavonoids have been reported to be located in plastids (Weissemböck and Schneider, 1974; Saunders and

McClure, 1976; Luckner, 1980; Wiermann, 1981). The flavonoids in plastids are lipophilic aglycones (Charrière-Ladreix, 1977) but mainly hydrophilic glucosides. Several cases have been reported in which only one of several glucosides of flavonoids is located in chloroplasts: the apigenin-6-C-glucoside, isovitexin, in *Avena sativa* (Weissemböck and Schneider, 1974) and the 7-O-glucosyl isovitexin, saponarin, in *Hordeum vulgare* (Saunders and McClure, 1976). In the latter case, the UDPglucose isovitexin 7-O-glucosyltransferase was found in the cytosol, not in chloroplasts (Blume *et al.*, 1979). In any case, the relative amount of a given glucosylated flavonoid in chloroplasts represents 0 to 20–30% of the total amount in whole tissue, and this can change according to environmental conditions (Saunders and McClure, 1976). These and other results (Hrazdina *et al.*, 1980) suggest that there is specific transport, after glycosylation, of some flavonoids into the chloroplasts. This could be the consequence of the channeling of the biosynthetic pathway by multienzyme complexes associated with chloroplasts (Stafford, 1981).

Phenolic compounds have been reported in several instances to be present almost exclusively in the vacuoles of many tissues, and have then been used as convenient vacuolar markers (Grob and Matile, 1979). An apparent exception is the stilbene prelunularic acid of which only 50% was associated with the vacuoles of suspension culture cells of *Marchantia polymorpha* (Abe Imoto and Ohta, 1985). But these results do not discriminate between the degree of artifactual leakage from the vacuole of this small and lipophilic molecule and its actual localization elsewhere, e.g., in alkaline compartments such as cytosol or mitochondria.

Specific examples of the exclusive location of phenolics in vacuoles include shikimic acid in cell cultures of *Fagopyrum esculentum* when the shikimate pathway is inhibited (Hollander-Czytko and Amrhein, 1983); rosmarinic acid in suspension culture cells of *Anchusa officinalis* and *Coleus blumei* (Chaprin and Ellis, 1984); coumaric acid glucosides in leaves of *Melilotus alba* (Oba *et al.*, 1981); hydroxycinnamic acid esters of tartaric acid in *Vitis* fruits (Moskowitz and Hrazdina, 1981); hydroxycinnamic acid esters of malic acid and of glucose or sucrose in *Raphanus sativus* cotyledons and leaves (Sharma and Strack, 1985; Strack and Sharma, 1985); apigenin 7-O-(6-O-malonylglucoside) in cultured cells of *Petroselinum hortense* (Matern *et al.*, 1983); anthocyanins in *Tulipa* and *Hippeastrum* petals (Wagner, 1979) and in cultured cells of *Daucus carota* (Hopp *et al.*, 1985); anthocyanins and flavonol glycosides in *Vitis* fruits (Moskowitz and Hrazdina, 1981); and capsaicin, the pungent flavor of *Capsicum*, in protoplasts of *C. annuum* fruit (Fujiwake *et al.*, 1980). Anthocyanoplasts are intensely pigmented membrane-bound organelles, probably involved in anthocyanin biosynthesis, located within large vacuoles (Pecket and Small, 1980). They have been recorded in more than 70 species. Plant tannins have also been found repeatedly in vacuoles. Several reports dealing with tissue cultures of coniferous species have demonstrated the biosynthesis of tannins in small vacuoles derived from endoplasmic reticulum and further coalescing with the large vacuole (Chafe and Durzan, 1973; Baur and Walkinshaw, 1974; Parham and Kaustinen, 1977) Some phenolics are also located in cell walls, mainly as lignin but also in simpler molecules such as flavonoids (Charrière-Ladreix, 1977) and esterified ferulic acid (Fry, 1979).

The naphthoquinone pigment shikonin of *Lithospermum erythrorhizon* has the same compartmentation in the periderm of the root and in cultured cells (Tsukada and Tabata, 1984). In the two cases it accumulates at the outer surface of the cell wall as granules (1-μm diameter) including shikonin, proteins, and lipids.

2. Isoprenoids

Lower terpenoids and essential oils are assumed to be synthesized in plants in specialized secretory structures such as glandular cells which often excrete them into subcuticular spaces (Schnepf, 1976; Wiermann, 1981). Contradictory reports about the production of such compounds by cultures *in vitro* are commonly attributed to the need for appropriate cytodifferentiation predisposing to biosynthesis. However, a high terpenoid biosynthesis potential has been demonstrated in tissue culture, but in addition oxidation and other degradative processes were occurring (Banthorpe *et al.*, 1986). Witte *et al.* (1983) interestingly have shown that suspension culture cells of *Thuja occidentalis* produce several mono- and diterpenoids. Whereas the diterpenoids were accumulated in cells, the monoterpenoids of the menthane type were recovered only in the culture medium.

Glucosylated isoprenoids are thought to accumulate in vacuoles. This was demonstrated for cardenolides in leaves of *Convallaria majalis* (Löffelhardt *et al.*, 1979); for purpureaglycoside A in *Digitalis lanata* cells (Pfeiffer *et al.*, 1982); and for gentiopicroside, a glucosylated monoterpenoid iridoid, in root of *Gentiana lutea* (Keller, 1986). Carotenoids are found exclusively in the grana of chloroplasts in leaves, and mainly in chromoplasts (e.g., in plastoglobuli) differentiating from chloroplasts in other organs (Wiermann, 1981).

3. Alkaloids

In whole plants, some qualitative evidence has been presented for the strong binding of alkaloids to lignified cell walls, e.g., nicotine in

Nicotiana leaves (Müller *et al.*, 1971) and berberine in *Berberis darwinii* (Mothes, 1955). However, many data substantiate the principal localization of alkaloids in vacuoles.

Numerous examples have been reported in the family Papaveraceae, in which laticifers or specialized cells are the main sites of accumulation of alkaloids (Neumann and Müller, 1972; Neumann. 1976). In the latex of *Chelidonium majus*, sanguinarine, chelerythrine, berberine, and coptisine are markedly enriched in the vacuoles (Matile, 1976). The laticifers of *Papaver somniferum* have numerous types of vacuolar vesicles, some of which containing nearly all the morphinan alkaloids (Fairbairn *et al.*, 1974; Roberts *et al.*, 1983; Homeyer and Roberts, 1984). Similar results have been found in *Papaver bracteatum* suspension cultures presenting some cytodifferentiation correlated with thebaine production (Kutchan *et al.*, 1983, 1985, 1986). Thebaine was uniquely recovered in a special class of vacuoles present only in laticifer cells (Kutchan *et al.*, 1985). Dopamine and sanguinarine occurred also in vacuoles differing from one another by their size and density and from thebaine-containing vacuoles (Kutchan *et al.*, 1985, 1986).

In other plant families, the vacuolar localization of alkaloids has been observed for nicotine in *Nicotiana* leaves (Saunders, 1979), (*S*)-reticuline in suspension culture cells of *Fumaria capreolata* (Deus-Neumann and Zenk, 1986), and serpentine in suspension culture cells of *Catharanthus roseus* (Deus-Neumann and Zenk, 1984a; Renaudin *et al.*, 1986). In the last material, intracellular ajmalicine was only 41% in isolated vacuoles, but this result was attributed to the leakage of the alkaloid from the vacuoles during their isolation (Renaudin *et al.*, 1986).

4. Miscellaneous

The cyanogenic glucoside dhurrin is totally vacuolar in leaves of *Sorghum bicolor* (Saunders and Conn, 1978). Glucosinolates are also completely vacuolar in horseradish root cells (Grob and Matile, 1979), as for betanin in red beet root (Leigh *et al.*, 1979).

III. COMPARTMENTATION AS A RESULT OF MEMBRANE TRANSPORT AND ACCUMULATION PROCESSES AT SPECIFIC SITES

Translocation processes at the limiting membranes are the major key points of cell compartmentation. The activity of permeation systems determines how tight the membrane barrier is and consequently how

TABLE I

Mechanisms of Transport at the Tonoplast and Mechanisms of Intravacuolar Accumulation

Metabolite	Plant species	Postulated transport system	Postulated accumulation system	References
Essential metabolites				
Sucrose	*Beta vulgaris*	Concentrative, energy-dependent carrier	None	Doll *et al.* (1979); Willenbrink and Doll (1979)
		Group translocation of UDPglucose	None	Thom *et al.* (1986)
	Hordeum vulgare	Nonconcentrative, energy-independent carrier	None	Kaiser and Heber (1984)
	Saccharum sp.	Group translocation of UDPglucose with a multienzyme complex	None	Thom and Maretzki (1985)
O-Methylglucose	*Saccharum* sp.	H^+-antiport carrier	None	Thom and Komor (1984)
O-Methylglucose, glucose	*Pisum sativum*	H^+-antiport carrier	—	Guy *et al.* (1979)
Malate	*Bryophyllum daigremontiana*	Nonconcentrative, energy-independent carrier	None	Buser-Suter *et al.* (1982)
	Hordeum vulgare	Energy-dependent carrier coupled to pH or ddp	None	Martinoia *et al.* (1985)
Citrate	*Hevea brasiliensis*	H^+-antiport carrier	Trapping as Mg^{2+} complex	Marin and Chréstin (1985)
Arginine	*Saccharomyces cerevisiae*	Permease (?), facilitated diffusion carrier, or cation antiport	Binding to polyphosphate and unknown anions	Boller *et al.* (1975); Boller (1985); Dürr *et al.* (1979); Matile (1978)
		H^+-antiport carrier[a]	—	Ohsumi and Anraku (1981)
Phenylpropanoids				
Apigenin malonylglucoside	*Petroselinum hortense*	Specific permease	pH-induced conformation changes with restricted affinity at acidic pH	Matern *et al.* (1986)

Esculine	*Hordeum vulgare*	Specific H^+-antiport carrier	None	Werner and Matile (1985)
O-coumaric acid glucoside	*Melilotus alba*	pH-dependent uptake[b] of the trans isomer	Conformation change with trapping of the cis isomer[a]	Rataboul *et al.* (1985)
Alkaloids				
Sanguinarine, chelerythrine	*Chelidonium majus*	Noncatalyzed transport	Specific binding to vacuolar sites such as phenolics	Matile (1976)
Tryptamine	*Catharanthus roseus*[c]	Diffusion of the neutral base	Restricted diffusion of the charged form	Courtois *et al.* (1980)
Nicotine	*Acer pseudoplatanus*[c], *Nicotiana tabacum*[c]	Diffusion of the neutral base	Restricted diffusion of the charged form and binding to vacuolar sites	Kurkdjian (1982)
Ajmalicine, tabernanthine	*Catharanthus roseus*[c]	Diffusion of the neutral base	Restricted diffusion of the charged form and binding to vacuolar sites	Renaudin (1981); Renaudin and Guern (1982); Renaudin *et al.* (1985)
Ajmalicine	*Catharanthus roseus*	Diffusion of the neutral base	Restricted diffusion of the charged form and binding to vacuolar sites	Renaudin and Guern (1987)
Morphine, codeine, thebaine	*Papaver somniferum* latex	Highly specific channel protein	Unknown trapping mechanism	Homeyer and Roberts (1984)
Vindoline	*Catharanthus roseus*	Highly specific energy-dependent carrier	None	Deus-Neumann and Zenk (1984a)
(*S*)-Reticuline, (*S*)-scoulerine	*Fumaria capreolata*	Highly specific H^+-antiport carrier	None	Deus-Neumann and Zenk (1986)

[a] Experiments on tonoplast vesicles.
[b] Experiments on protoplasts.
[c] Experiments on cells.

distinct the compartments separated by the membrane are. Accumulation, i.e., the building of concentration gradients, is often a complex interaction between the membrane transport system itself and various "trapping" mechanisms.

A. Transport and Concentration Processes as a Basis of Compartmentation in Vacuoles

Recently significant progress has been made concerning the molecular mechanisms of transport of solutes across the tonoplast (Alibert and Boudet, 1982; Leigh, 1983; Boller, 1985; Boller and Wiemken, 1986). Table I gives an outline of the transport systems tentatively identified at the tonoplast for essential metabolites and secondary compounds with, where possible, information on the systems responsible for the intravacuolar accumulation. The main strategies used to characterize the transport of solutes through the tonoplast are listed in Table II and will serve for a critical analysis of the results in Table I.

Saturation kinetics and competition for uptake between substrate and analogs are customary criteria to argue in favor of carrier-mediated transport. The ATP dependency of uptake has been shown for a variety of solutes whose transport appears energy dependent. However, quite often, the residual uptake measured in the absence of ATP has not been clearly explained. Furthermore, caution must be exercised in interpreting results obtained with ATP and Mg^{2+} in the presence of rather high

TABLE II

Main Strategies Used to Characterize the Mechanisms Involved in the Translocation of Solutes through the Tonoplast

Study of the kinetics of uptake as a function of substrate concentration
Description of the time course of uptake (measurement of the initial rate, presence and origin of a plateau phase)
Study of possible competitions between substrate and analogs
Control of the existence of a *net* uptake building a concentration gradient
Determination of the dependence of the uptake toward ATP–Mg^{2+}
Determination of the sensitivity of the uptake to inhibitors of ATPases or ionophores dissipating ΔpH or ΔEm
Measurement of the influence of ΔpH and ΔEm on the intensity of the uptake; measurement of the influence of uptake on ΔpH and ΔEm values
Study of the characteristics of the efflux of substrate from preloaded vacuoles
Control of possible intravacuolar binding or metabolization
Control of the specificity of solute uptake in homologous and nonhomologous systems

levels of EDTA. ATPase inhibitors and ionophores such as CCCP, FCCP, valinomycin, or nigericin have been widely used to demonstrate that the tonoplast ATPase could be involved directly or indirectly (through the electrochemical gradient of protons) in the uptake process. However, interpretation of the results is not always straightforward, as illustrated by a few examples. A proper use of ionophores necessitates knowledge of the ionic gradients between vacuoles and their suspension medium as exemplified by Martinoia *et al.* (1985). Interpretation of the FCCP sensitivity but NH_4Cl insensitivity of morphinan uptake (Homeyer and Roberts, 1984) is not evident. The sensitivity to DCCD of the uptake of vindoline cannot be construed as proof of the involvement of an ATPase in a situation where the uptake is not stimulated by ATP (Deus-Neumann and Zenk, 1984a). The cross-linking properties of DCCD suggest a need to check the specificity of its action.

Coupling solute uptake to a proton transfer through a H^+-antiport carrier has been quite often postulated. However, while in most cases this assumption appears reasonable, one must admit that direct evidence to support it are lacking. Testing the sensitivity of the uptake to external pH is simple but cannot replace measurement of the trans-tonoplast ΔpH and its modifications induced by solute uptake, as exemplified by Thom and Komor (1984).

The concentrative character of the uptake has rarely been checked when using radiolabeled substrates, and this makes hazardous the interpretation in terms of driving forces. The efflux of substrate from preloaded vacuoles has unfortunately been rarely characterized. This explains for one part why we know so little about what limits the accumulation of metabolites in the vacuole. Equilibrium between pump and leak systems or exchange mechanisms with different affinities for the substrate at the external and internal surfaces are each models which can be hypothesized from published data. This illustrates our meagre understanding of the processes by which metabolites can be discharged from the accumulation compartment and further metabolized.

As to the mechanisms by which metabolites are *concentrated* inside vacuoles, two mutually nonexclusive proposals have been made. The first one considers that the concentration gradient is created by the energy-dependent transport process itself. In this case the problem, as discussed above, is to understand which mechanisms limit and regulate the accumulation.

The second proposal is that, superimposed on the transmembrane translocation, a variety of "trapping" mechanisms occur. The general term of "trapping" will be used in this chapter to designate the reversible intravacuolar events which lower the activity of the translocated metabo-

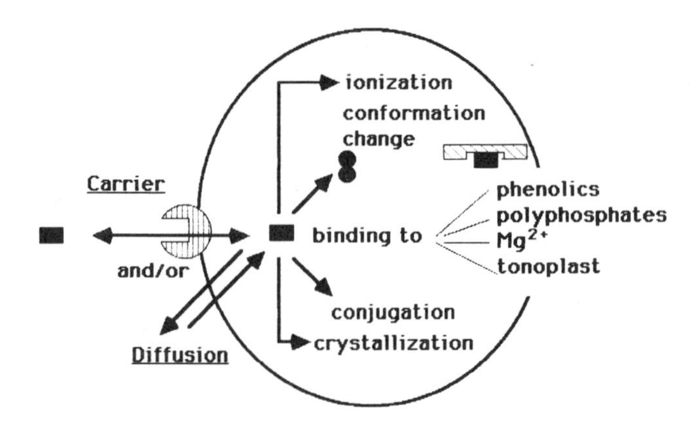

Fig. 2. Diagrammatic representation of several reversible intravacuolar "trapping" mechanisms which displace the equilibrium between uptake and efflux of secondary metabolites and contribute to their vacuolar accumulation.

lite, displace the equilibrium between uptake and efflux, and determine the overall accumulation (Fig. 2). An extreme example of a trapping mechanism concerns some alkaloids, having a charged form with limited diffusion capacity through membranes. Their concentration gradient is not created by the transmembrane translocation (simple diffusion of the neutral form) but corresponds to a secondary effect of the transtonoplast pH gradient with a strong shift of the equilibrium between permeable and non- or poorly permeable forms of solutes (Müller, 1976; Neumann *et al.*, 1983; Renaudin and Guern, 1982). A related mechanism has been proposed by Rataboul *et al.* (1985) for *O*-coumaric glucoside and by Matern *et al.* (1983, 1986) for apigenin malonylglucoside, wherein accumulation is linked to a conformation shift of the absorbed metabolite to give a form with a restricted ability to cross the membrane.

Association of the absorbed metabolites to various vacuolar binding sites such as Mg^{2+} for citrate (Marin and Chréstin, 1985), phenolics for alkaloids (Matile, 1976), polyphosphates for arginine and other basic amino acids (Boller, 1985), or the tonoplast itself for alkaloids (Matile, 1976, and our own unpublished results showing serpentine binding to the tonoplast) has attracted interest as a way to shift the equilibrium created by the transtonoplast translocation processes toward an increased accumulation of various solutes. The extreme case where "trapping" excludes a part of the solutes from the bidirectional exchange through the tonoplast is the occurrence of crystallization, e.g., of salts of berberine (Nakagawa *et al.*, 1984). Unfortunately, the accumulation processes based on "trapping" are generally complex as exemplified by arginine trapping in yeast and *Neurospora* (Boller, 1985; Cramer and Davis, 1984) where binding to polyphosphates is probably not the sole determinant of the vacuolar accumulation.

Another obvious mechanism by which the efflux of solutes absorbed by vacuoles can be restricted is metabolization. A special case is when metabolization is linked to the transmembrane transfer per se, as demonstrated by the group transport of sucrose from the substrate UDPglucose in sugarbeet and sugarcane vacuoles (Thom and Maretzki, 1985; Thom *et al.*, 1986). Many of the products stored in vacuoles are accumulated as glycosides. Glycosylation restricts the diffusion of rather lipophilic aglycones such as those of the phenylpropanoid group or hormone molecules. In that sense, glycosylation is an important determinant of the compartmentation of these metabolites. However, the glycosylation step and the physical compartmentation issued from the transmembrane transfer step are spatially distinct, at least for the storage of *p*-coumaric glucoside in vacuoles of *Melilotus alba* (Rataboul *et al.*, 1985) or the storage of esculine in vacuoles of *Hordeum vulgare* (Werner and Matile, 1985). But such spatial separation may not always be the case since glucosyltransferase activity has been detected in membrane preparations possibly including tonoplast (see review by Werner and Matile, 1985).

Furthermore, metabolization of solutes accumulated inside vacuoles can occur, as illustrated by the enzymatic synthesis of hydroxycinnamic acid esters of malic acid inside vacuoles of *Raphanus sativus* (Sharma and Strack, 1985; Strack and Sharma, 1985) and the synthesis of fructans in cereals (Wagner *et al.*, 1983) and in *Helianthus tuberosus* (Frehner *et al.*, 1984). This calls for more attention to be paid to the intravacuolar metabolization of solutes during uptake experiments as most often what is measured is the uptake and accumulation of radioactivity without checking whether the radioactive label is still associated with the unmodified substrate.

Opposing views have been expressed for the compartmentation of alkaloids. One extreme model is based on a diffusive transmembrane transfer of the neutral form of alkaloids leading to a concentration of the unpermeable ionized form in the acidic vacuole. The overall specificity of uptake, based purely on the permeability coefficients at the tonoplast, the pK_a of the molecules, and the amplitude of the ΔpH, is low (Müller, 1976; Neumann *et al.*, 1983; Renaudin and Guern, 1982). In contrast, another extreme model proposed by Deus-Neumann and Zenk (1984a, 1986) is based on the presence of highly specific carriers at the tonoplast. Strict specificity of uptake of a given alkaloid was observed when comparing homologous and nonhomologous systems, i.e., vacuoles prepared from plants or cell cultures producing or not producing the alkaloid (Deus-Neumann and Zenk, 1984a), or when comparing the uptake of eniantiomeric forms of alkaloids (Deus-Neumann and Zenk, 1986). This has led to the general assumption that "for every alkaloid group, may be even for every single alkaloidal molecular species, there is a highly specific alkaloid carrier, . . ." The extreme structural diversity of binding sites implicit in this hypothesis was compared to that shown by the immunosystem in animals. Such a view has also been expressed for alkaloid sequestration in latex vesicles of *Papaver somniferum* where preliminary evidence indicates that "each alkaloid has its own specific channel protein" (Homeyer and Roberts, 1984).

To reconcile these extreme views of the vacuolar uptake and accumulation of alkaloids is not yet possible as further experiments are needed in order to get a more general view of the problem. Guidelines for future investigation will only be suggested.

First, in most studies of secondary metabolite uptake, including those just cited, the measurement of vacuolar uptake includes not only the transmembrane transfer per se but also any superimposed process, such as binding to vacuolar components, isomerization, and metabolization, leading to solute accumulation. Thus the specificity which can be observed in such a way is at the outset that of the overall process of vacuolar accumulation. This suggests that the saturation kinetics and the specificity of uptake could result not from the uptake process per se (i.e., the transmembrane transfer) but also in some cases from the intravacuolar binding to proteins (including enzymes with a marked specificity for the absorbed alkaloids) or to other binding sites. Particularly significant in this respect are the results described by Matile (1976) showing that the binding of alkaloids to the cell sap of *Chelidonium majus* occurs with distinct specificities. *In vitro* binding to gallic or chelidonic acids is more specific for sanguinarine than for coptisine. Furthermore, typical displacement of berberine bound to tannins by sanguinarine was also observed. Parr *et al.* (1986) have also shown that the efflux of quinoline alkaloids from *Cinchona* ssp. cultured cells was controlled by several mechanisms, of which some were specific. Accordingly, to interpret the results of uptake experiments, the unambiguous distinction between transmembrane transfer and the subsequent intravacuolar events requires attention.

A second point arises from the supposition that several ways of transmembrane translocation could coexist for the same molecule, e.g., a specific carrier and a diffusion process. In such a case, the relative importance of these two components in the overall uptake should differ according to the type of metabolite and the experimental material. As a matter of fact, we observed that vacuoles prepared from *Catharanthus roseus* cells with 0.5 *M* NaCl as an osmoticum (Deus-Neumann and Zenk, 1984a, 1986) were not equivalent to vacuoles isolated in 0.55 *M* sorbitol (Renaudin *et al.*, 1986). Measurement of the vacuolar pH using the ^{31}P-NMR technique revealed that the NaCl treatment strongly decreased the vacuolar acidity (pH 6.35) compared to vacuoles isolated in sorbitol (pH 5.33). This drop of the transtonoplast ΔpH drastically lowers the potential importance of alkaloid accumulation linked to intravacuolar ionization (Barbier-Brygoo *et al.*, 1987). Moreover, at variance with results of Deus-Neumann and Zenk (1984), the vacuoles isolated in sorbitol did accumulate ^{14}C-nicotine ca. ten times more than ^{14}C-ajmalicine, and this higher accumulation was well in agreement, considering an ion-trap mechanism, with the fact that nicotine has a higher acidity constant ($pK_a = 8.0$) than ajmalicine ($pK_a = 6.3$) (Renaudin and Guern, 1987).

Thus, if one admits that a binding component with a high specificity (i.e., a membrane carrier or an intravacuolar binding moiety) coexists with an ion-trapping component associated with a transmembrane diffusion of neutral alkaloids, dramatic changes in the relative importance of the two components should occur according to the experimental conditions. The uptake of an alkaloid with a low pK_a by vacuoles having lost their acidity should mainly reveal the highly specific binding component. Conversely, the uptake of a more basic alkaloid by vacuoles with a high transtonoplast ΔpH should be mainly driven by the diffusive and ion-trapping component.

Our general proposal is that the ion-trapping model and the carrier model are not exclusive of each other and likely coexist. Their relative importance would be dependent on the very large diversity of molecular structures and physicochemical properties (i.e., lipophilicity, basicity, polarity) in the alkaloid group, coupled to the variety of physiological situations in terms of driving forces for vacuolar accumulation (e.g., the value of the transtonoplast pH difference).

B. The Extracellular Compartment: A Key Characteristic of Plant Cell Cultures

1. Relative Size of the Extracellular Compartment

Different metabolites, including small molecules, polysaccharides, and enzymes, are excreted outside the cells or synthesized at the cell surface in the entire plant. This process contributes to the building of extracellular structures such as the cell wall and to the exchange of solutes between cells or organs. The amount of excretion in plant tissues is restricted by the small volume of the apoplast compared to that of the cells and by strong structural limitations to the diffusive exchanges, through a complex pathway, with the external medium at the root level. A reasonable figure is that, in organs, the total cell wall space is equivalent to the cumulated cytoplasmic volume of the cells, namely, 5–20% only of the cumulated total intracellular volume of the cells.

There is a considerable amplification of the importance of the extracellular compartment when plant cells are grown in liquid culture. In this case, the extracellular space immediately available to the cell can be considered as the sum of the cell wall space plus the extracellular medium per se. From cell populations inoculated at low density (about 3000 cells ml^{-1}) to cell populations at stationary phase (about 3×10^6 cells ml^{-1}) the ratio of the extracellular volume to the cumulated volume of the cells varies from approximately 10^4 to 10, or the ratio of the extracellular volume to the cumulated cytoplasmic volume would vary from 10^5 to 10^2. In such conditions, efflux or leakage of secondary metabolites from the cells is amplified by the fact that most often large external concentrations are not built by such processes even if the extracellular amount of metabolites exceeds the intracellular amount.

2. Excretion of Secondary Metabolites in the Extracellular Compartment

A list of some secondary metabolites found in the extracellular compartment is summarized in Table III. It offers only a limited view of the diversity of molecules excreted by plant cells, neglecting, for example, the excretion of enzymes (Olson *et al.*, 1969; Wink, 1984), polysaccharides (Olson *et al.*, 1969), and peptides such as glutathione (Bergmann and Rennenberg, 1978).

Most of the extracellular secondary metabolites are alkaloids. The amounts collected from the culture medium represent the largest fraction if not the totality of the overall production for atropine and scopolamine in *Datura innoxia* cell suspensions (Kibler and Neumann, 1979) and caffeine in *Coffea arabica* calli (Frischknecht and Baumann, 1977; Waller *et al.*, 1983). But other kinds of secondary metabolites have been detected in large relative amounts in the culture medium, including lignin in *Rosa glauca* suspensions (Mollard and Robert, 1984), capsaicin in *Capsicum frutescens* suspensions (Lindsey *et al.*, 1983), and monoterpenoids in *Thuja occidentalis* cell suspensions (Witte *et al.*, 1983).

The real excretion of secondary metabolites by plant cells grown *in vitro* has been a matter of controversy for several years. The presence of metabolites in the medium was first negated (Scott *et al.*, 1979) or attributed to cell lysis (Hinz and Zenk, 1981). It has been also emphasized that secondary metabolites have not been systematically assayed in the culture medium (Pétiard and Courtois, 1983). Careful checks on the kinetics of cell death compared to the kinetics of metabolite excretion has shown that in most cases the excretion of phytochemicals is a real physiological process contributing to the overall dynamics of these metabolites.

The excretion of berberine alkaloids by various Berberidaceae cell suspensions and the excretion of ajmalicine by *Catharanthus roseus* cells provide good examples of the large variability encountered in the intensity of metabolite excretion. Sato and Yamada (1984) found that *Coptis japonica* cell suspensions excreted up to 50% of their berberine production, as opposed to the lack of berberine excretion reported by Yamamoto *et al.* (1986). *Catharanthus roseus* cells were first described as not excreting ajmalicine into the culture medium (Scott *et al.*, 1979; Kurz, personal communication, 1985) as opposed to the results of Mérillon *et al.* (1983) and Renaudin *et al.* (1985) reporting that up to 50% of the ajmalicine produced was recovered in the extracellular medium.

While the reasons for such discrepancies are not totally clear, the argument that the extracellular compartment is fed through cell lysis is no longer tenable. Some reports show that cell strains differ in their ability to excrete secondary metabolites. This is the case for *Catharanthus roseus* cultures (Pétiard and Courtois, 1983) and for *Cinchona ledgeriana* cell suspensions where the degree of excretion varies from 99% of the total production (Table III) to only 10% according to the variety used (Anderson *et al.*, 1982). This is also nicely illustrated by the work of Sato and Yamada (1984) showing a large variability in the total production of berberine and the extent of alkaloid excretion (from 2 to 50% of the total production) among cell lines of *Coptis japonica* isolated by cloning.

The intensity of excretion of secondary metabolites by a cell strain producing several metabolites of the same family appears also to be specific for a given metabolite. This is evident in the different behaviors of (i) cephalotaxine and deoxyharringtonine in *Cephalotaxus harringtonia* calli (Delfel and Rothfus, 1977), (ii) ajmalicine and serpentine in *Catharanthus roseus* cell suspensions (Mérillon *et al.*, 1983; Renaudin *et al.*, 1985), and (iii) menthane monoterpenes and diterpenes in *Thuja occidentalis* cell suspensions (Witte *et al.*, 1983).

Two examples illustrate the peculiarities of the metabolism and storage of secondary compounds in plant cell cultures in relation to the extracellular compartment. In *Lupinus polyphyllus* plants, quinolizidine

TABLE III

Excretion of Secondary Metabolites into the Extracellular Compartment

Metabolite	Plant species and type of culture	Amount in the medium (% of total production)	References
Alkaloids			
Atropine	*Datura innoxia*		
	Cell suspension	50–100%	Kibler and Neumann (1979)
Berberine	*Berberis* sp.		
	Cell suspensions	ϵ%[a]	Hinz and Zenk (1981)
	Coptis japonica		
	Cell suspension	0–30%	Yamada and Sato (1981)
	Cell suspension	24–46%	Sato and Yamada (1984)
	Cell suspension	No excretion	Nakagawa *et al.* (1984)
	Cell suspension	No excretion	Yamamoto *et al.* (1986)
	Thalictrum minus		
	Cell suspension	88%[b]	Nakagawa *et al.* (1984)
	Cell suspension	90%[b]	Nakagawa *et al.* (1986)
	Cell suspension	None	Rueffer (1985)
	Cell suspension	68–95%[c]	Yamamoto *et al.* (1986)
Caffeine	*Coffea arabica*		
	Callus	Up to 90%	Frischknecht and Baumann (1977)
	Callus	60–90%	Waller *et al.* (1983)
Catharanthus alkaloids	*Catharanthus roseus*		
	Cell suspension		
Ajmalicine		No excretion	Scott *et al.* (1979)
		5–50%	Mérillon *et al.* (1983)
		25–50%	Renaudin *et al.* (1985)
Serpentine		No excretion	Scott *et al.* (1979)
		3–15%	Mérillon *et al.* (1983)
		2–6%	Renaudin *et al.* (1985)
Cephalotaxine and harringtonine derivatives	*Cephalotaxus harringtonia*		
	Callus		
Cephalotaxine		Most[d]	Delfel and Rothfus (1977)
Deoxyharringtonine		100%	

Cinchona alkaloids	*Cinchona ledgeriana*		
	Cell suspension		Anderson *et al.* (1982)
Cinchonine		17%	
Cinchonidine		67%	
Quinine		99%	
Quinidine		8%	
Lupanine	*Lupinus polyphyllus*		
	Cell suspension	20–30%	Wink *et al.* (1982)
Nicotine	*Nicotiana tabacum*		
	Callus	12–50%	Ogino *et al.* (1978)
	Cell suspension	5–30%	Mantell *et al.* (1983)
Protopine	*Macleaya cordata*		
Sanguinarine	Callus	25–33%	Böhm and Franke (1982)
	Cell suspension	Most	Franke and Böhm (1982)
Scopolamine	*Datura innoxia*		
	Cell suspension	80–97%	Kibler and Neumann (1979)
Naphthoquinones			
Shikonin	*Lithospermum erythrorhizon*		
	Cell suspension	Most[e]	Tsukada and Tabata (1984)
Phenolics			
Capsaicin	*Capsicum frutescens*		
	Cell suspension	99%	Lindsey *et al.* (1983)
	Immobilized cells	99%	
Terpenoids			
	Thuja occidentalis		
	Cell suspension		
Monoterpenes		Most[f]	Witte *et al.* (1983)
Diterpenes		None	Berlin *et al.* (1984)

[a] Presence in culture medium attributed to cell lysis; 36 species of the genus *Berberis* were checked.
[b] One-third as crystals of berberine nitrate.
[c] According to the cell strain and culture conditions.
[d] According to the age of call.
[e] Accumulation at the cell surface as hydrophobic globules.
[f] Can be trapped with a lipophilic triglyceride phase.

alkaloids are synthesized in the chloroplasts of leaves and translocated to other organs, mainly epidermal and subepidermal cells, where they are concentrated and stored. In cell suspensions of the same species, the alkaloids are excreted into the culture medium, degraded by enzymes present in the medium, and consequently they do not accumulate (Wink, 1985).

In *Macleaya microcarpa* cultivated *in vitro* on solid medium some cells appear specialized in alkaloid accumulation and probably accumulate alkaloids formed by the surrounding cells (Böhm and Franke, 1982). As a consequence only 20–30% of the alkaloids are recovered in the culture medium. When grown in liquid medium, cells excrete nearly the total alkaloid production due to a low number of alkaloid-accumulating cells. However, the differentiation of these cells occurs only in cell aggregates with a size above a critical level, and the mean size of the cell aggregates is under negative control by the auxin concentration (Franke and Böhm, 1982). As a consequence, for high auxin concentrations, the aggregates are small with few alkaloid accumulating cells and nearly all the alkaloids produced are recovered in the extracellular compartment.

3. Possible Mechanisms of Excretion and Means to Modify It

The state of knowledge about the mechanisms by which secondary metabolites are excreted by plant cells is rather poor. The terminology itself is rather ambiguous in the sense that the term excretion means that an active process is involved. A major reason for this confused situation is the lack of real quantitative studies. Simply measuring the relative amounts of metabolites inside the cells and in the extracellular medium is not very informative. The ratio of the intracellular and extracellular concentrations is more significant. It is interesting to note, for example, that the excretion of half the total production of ajmalicine by *Catharanthus roseus* cells still corresponds to an accumulation of this alkaloid within cells against a concentration gradient of 10 (Renaudin *et al.*, 1985).

According to the type of metabolite, the process of excretion could arise by quite different mechanisms. In the case of some nonpolar alkaloids, a simple diffusion process of the neutral form of alkaloids could account for the distribution of alkaloids between the various compartments of the cell suspension including the extracellular compartment (Renaudin and Guern, 1982; Neumann *et al.*, 1983). In agreement with such a scheme, simply lowering the extracellular pH or diluting the medium increases the amount of extracellular alkaloids. In contrast, intracellular vesicles which fuse with the plasmalemma and discharge their contents outside the cells have been described in several cases, and will be discussed in the next section.

Modifying the relative importance and the properties of the extracellular compartment is a topic of high biotechnological interest; most aspects of this will be dealt with in the chapters devoted to continuous culture of plant cells (see Chapter 14, this volume) and cell immobilization (see Chapter 10). A few remarks on various attempts to increase the excretion of secondary metabolites will be made here.

The first line of action has been to permeabilize the cells to provoke leakage of the molecules of interest into the extracellular medium. The problem is to induce the release of the products without affecting cell viability. For this aspect of induced cell decompartmentation, the reader is referred to Brodelius and Nilsson (1983), Parr *et al.* (1984), and Rueffer (1985).

The second line of action is more relevant to our subject. It involves increasing the "trapping properties" of the extracellular medium. A typical example is given by the use of a triglyceride phase to "trap" the sesquiterpene bisabolol excreted by *Matricaria chamomilla* cells (Bisson *et al.*, 1983) and the monoterpenes excreted by *Thuja occidentalis* cells (Berlin *et al.*, 1984). However, other reports show that such a lipophilic trap is not of general use (Rueffer, 1985).

C. Dynamics of the Compartmentation of Secondary Metabolities: The Role of Vesicles

Vacuolar storage and extracellular excretion have been described above as the result of the mere action of transmembrane translocation through the tonoplast and the plasmalemma considered as static membranes. This is an oversimplification of what is in fact the dynamics of the compartmentation of metabolites.

A continuous flow of membrane-bound vesicles has been described: these derive mainly from specialized regions of smooth endoplasmic reticulum, are closely associated with Golgi stacks, and coalesce further with the large vacuole (Marty *et al.*, 1980; Hilling and Amelunxen, 1985). There is some evidence that provacuoles and/or endoplasmic reticulum are the sites of synthesis and/or accumulation of some phytochemicals and are involved in their transport to the vacuoles or to the cell wall (Stafford, 1981). This situation has been well documented in the case of tannins, notably with plant cell cultures (Chafe and Durzan, 1983; Baur and Walkinshaw, 1974; Parham and Kaustinen, 1977). Anthocyanins are also accumulated and perhaps synthesized in anthocyanoplasts which have been located within vacuoles (Pecket and Small, 1980). In *Papaver* species, morphinan alkaloids are mostly associated with vacuolar vesicles in the latex (Fairbairn *et al.*, 1984; Roberts *et al.*, 1983; Kutchan *et al.*, 1985, 1986). These vesicles are able to take up exogenous alkaloids (Roberts *et al.*, 1983; Homeyer and Roberts, 1984). Their role in the biosynthesis of alkaloids remains an open question. They have been referred to as alkaloid vesicles implicated in the storage and transport of alkaloids from the stem to the fruit (Fairbairn *et al.*, 1974).

A cytochrome *P*-450–dependent geraniol hydroxylase in *Catharanthus roseus* seedlings is associated with the membrane of a special class of vesicles (diameter 0.1–0.8 μm, density 1.09/1.10 g cm^{-3}) differing from Golgi apparatus, endoplasmic reticulum, and plasma membrane (Madyastha *et al.*, 1977). The authors postulate the compartmentalization of at least one part of the biosynthetic pathway of the indole alkaloids. Amann *et al.* (1986) have presented interesting new results about vesicles associated with alkaloid metabolism. They have isolated and characterized, from cell suspensions of *Berberis wilsoniae,* vesicles putatively of Golgian origin (0.1–1 μm diameter, density 1.14 g cm^{-3}) containing two oxidases (out of the eight enzymes of berberine synthesis) and also some quaternary alkaloids such as jatrorrhizine, which are the methylated products of the two oxidases. The specific methyltransferase was not present in vesicles which supports the spatial channeling of the intermediates inside and outside the vesicles during biosynthesis. The authors postulate that the quaternary alkaloids were trapped in the vesicles owing to their high polarity, and that they finally accumulated in the large vacuole by fusion of the vesicles with the vacuole. Measurement of the relative concentrations of the quaternary alkaloids in vesicles and in cytosol, together with the amount of alkaloids in vesicles related to that in vacuoles, is required to substantiate such a view.

Vesicles are also implicated in the excretion of phytochemicals into the extracellular medium. The induction of shikonin synthesis in cell cultures of *L. erythrorhizon* begins by the appearance in the cytoplasm of highly electron-dense vesicles deriving from rough endoplasmic reticulum and probably containing shikonin. Electron micrographs show the subsequent fusion of these vesicles with plasma membrane and the final deposition of shikonin granules on the outside of cell walls (Tsukada and Tabata, 1984).

In cultured cells part of the flow of vesicles normally directed toward vacuoles in the whole plant may be directed toward the medium, thus being responsible for the presence of some compounds in this compartment. Wink (1984) has reported that numerous enzymes which are typically associated with vacuoles are present in significant amounts in the culture medium of cells from 12 different species. However, some of these enzymes are also present in cell walls. Capsaicin is exclusively excreted by cultured cells of *Capsicum frutescens* (Lindsey *et al.*, 1983), whereas it is stored in vacuoles and small vesicles in the fruit of *Capsicum annuum* (Fujiwake *et al.*, 1980). In cultured cells of two members of the family Ranunculaceae, the induction of berberine synthesis is associated with the appearance of abundant cytoplasmic vesicles supposed to be involved in the synthesis of alkaloids (Yamamoto *et al.*, 1986). In *Coptis japonica* suspensions, berberine is stored within cells, probably in vacuoles, whereas in *Thalictrum minus* suspensions berberine is mainly present as crystals in the medium (Nakagawa *et al.*, 1984). One likely explanation could be that the berberine-containing vesicles are directed toward the vacuoles in *C. japonica* and toward the medium in *T. minus*. Interestingly, this appears to depend on the culture conditions: in presence of 1 μM 2,4-dichlorophenoxyacetic acid (2,4-D) the production of berberine by *T. minus* cells was low and it was accumulated within cells, whereas with 60 μM naphthaleneacetic acid (NAA) and 10 μM benzylaminopurine (BAP), the production of berberine was strongly stimulated and 90% was excreted (Nakagawa *et al.*, 1986).

This rapidly growing body of evidence for an important role of vesicles in the dynamics of cell compartmentation calls for more effort to be directed toward their isolation and characterization in terms of solute content, enzymes, membrane properties, and origin. The problem of the mechanisms which determine the migration of the vesicles toward the tonoplast or the plasmalemma is also of utmost interest. The vacuolar apparatus is not really fully represented by the large vacuoles issued from the classical procedures of isolation. A more realistic view would be that of large vacuoles with satellite vesicles involved in a dynamic channeling of metabolites toward the vacuoles, the carriers which catalyze the transmembrane transfer of the metabolites being distributed between the membrane of the vesicles and the tonoplast per se.

IV. COMPARTMENTATION RELATED TO GROWTH AND CELL SPECIALIZATION

In the whole plant the expression of secondary metabolism is correlated with development and is typical of specific organs, tissues, and cells (Wiermann, 1981). The developmental aspects of secondary metabolite production are also encountered in cultured cells, but so far most of the studies have only addressed either the net content of the cells or their biosynthetic potential.

A. Compartmentation and Growth

It is recalled that the relative size of the storage compartment represented by the vacuolar apparatus and most likely its membrane properties (i.e., the presence of permeases and the intensity of driving forces for uptake) are related to the growth stage. Meristematic cells do not contain large vacuoles but rather numerous small vacuolar vesicles. Mature cells, on the other hand, have one or several large central vacuoles. In cultured cells, the relative volume of the vacuoles was shown to change from 55 to 90% from the period of rapid cell division to the stationary growth phase (Owens and Poole, 1979). Unfortunately, nothing is known of the relationship between the relative size of vacuoles and their capacity to accumulate secondary metabolites.

Numerous studies have shown that the production of secondary metabolites by cultured plant cells is greatly increased when growth is restricted or stopped (Phillips and Henshaw, 1977). Practically, the maximum production occurs generally at the stationary phase of growth e.g., for ajmalicine in *Catharanthus roseus* suspension culture cells (Fig. 3). We have observed, however, that significant changes in the compartmentation of ajmalicine could be monitored along with the growth cycle (Fig. 3). At the exponential phase of growth, the total production of ajmalicine is low, and about 20–50% of it is present in the extracellular medium, giving a low concentration ratio (intracellular versus extracellular concentrations, C_i/C_e) between 10 and 40. At the beginning of the stationary phase of growth, when the total production is quickly increasing, there is a sharp and reproducible peak of the concentration ratio, which reaches values as high as 800 (less than 1% ajmalicine is in the medium) before falling to values below 50. There is thus in this case a correlation between maximum production and maximum compartmentation of ajmalicine within cells. This indicates that either different compartmentation mechanisms occur at different periods or the relative intensity of particular compartmentation mechanisms changes with time. An explanation of the results of Fig. 3 could be that the cells synthesize phenolics, which could act as traps for alkaloids, mainly at the stationary phase. The rapid decline of the C_i/C_e ratio could be attributed to an increase in membrane permeability and/or the appearance of phenolics in the medium concurrent with the onset of senescence.

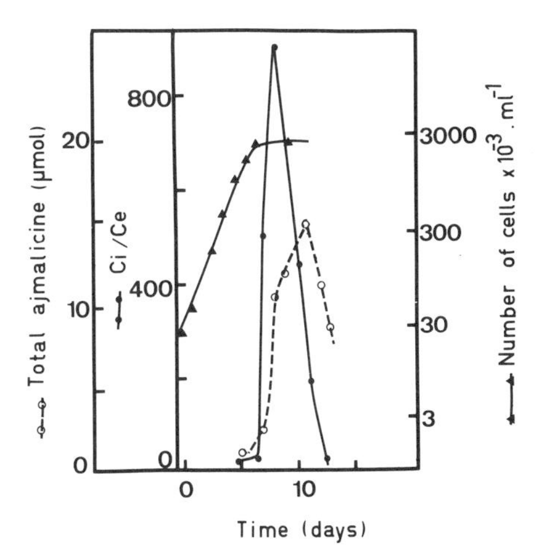

Fig. 3. Kinetics of growth, ajmalicine production, and ajmalicine compartmentation of cultured cells of *Catharanthus roseus*. The culture conditions are described in Renaudin and Guern (1982). Ajmalicine production is expressed as the whole amount of ajmalicine in a 200-ml suspension. Ajmalicine compartmentation is expressed as the ratio of the concentration of ajmalicine in cells, C_i, versus the concentration of ajmalicine in the extracellular medium, C_e.

The relationship between overall production and compartmentation should be examined in each case. Accurate kinetics are required due to the rapidity of the physiological changes occurring at the beginning of the stationary phase and at the onset of senescence.

B. Compartmentation and Cell Specialization

In plants, a specialization of functions between cells has been observed with separate sites of synthesis and sites of storage of secondary metabolites (see, for example, Wink, 1985). Such a compartmentation at the tissue or organ level implies the excretion of the metabolite from the site of synthesis, its transport, and then its reabsorption at the sites of storage. Such situations are difficult to study at the plant level, and the contribution of plant cell suspensions to examination of this problem is quite significant. Populations of cultured cells show a large cell-to-cell variability of content of secondary metabolites. This variability could provide models to study the cooperation of cells with different abilities to synthesize and/or to store secondary metabolites, the common extracellular medium mediating the exchanges between the different cell types.

The concentrations of serpentine and rosmarinic acid in individual cells were found to span over more than a 20-fold range in *Catharanthus roseus* (Neumann *et al.*. 1983; Brown *et al.*, 1984) and *Coleus blumei* (Chaprin and Ellis, 1984) suspensions, respectively. The genetic and/or physiological basis for this quantitative variability remains unclear. However, an analysis by flow cytometry of the fluorescence, due to serpentine, of *Catharanthus roseus* protoplasts has shown that protoplasts with the higher serpentine concentrations also had a more acidic vacuolar pH (Brown *et al.*, 1984). This correlation between the content of serpentine and the vacuolar parameter indicates that, in some cases, the dispersion of individual contents could be linked to a dispersion of the capacity of individual cells to store, not to synthesize, the compound.

An extreme case of cell-to-cell variability is the formation of specialized cells named idioblasts, present both in plants and in cultured cells and observed mostly in the case of alkaloids. They have been attributed a role in the storage rather than the synthesis of the alkaloids (Böhm and Franke, 1982; Franke and Böhm, 1982). The possibility of sorting such cells with good yield by flow cytometry, on the basis of high fluorescence, for instance, should provide insight into the biochemical and physiological peculiarities responsible for their high accumulation capacity.

The selection of high-producing cell lines from individuals with a high content or the selection of cells with a low internal content of secondary metabolites resulting from a high excreting ability are of great potential interest. Up to now, however, no real success has been obtained, apparently because of the instability of the lines selected (Deus-Neumann and Zenk, 1984b).

V. CONCLUSION

Plant cell cultures have been a powerful tool in the study of the expression of secondary metabolism in plants. Most of the recent advances in the knowledge of the compartmentation mechanisms have been obtained by using cultured cells. It is clear that the overall production of secondary metabolites by a plant cell culture is determined by a complex set of factors such as the rates of biosynthesis and degradation, the mechanisms and intensity of the forces determining accumulation in the intracellular storage compartments and determining the excretion out of the cell, and the possibility and intensity of reabsorption of the metabolites from the extracellular compartment. This gives a complex picture, widely disposed to genetic variability and physiological modulation.

Our knowledge of the compartmentation of phytochemicals has now begun to move from an analytical phase to the study of mechanisms. Much has to be done in this direction to identify the molecular mechanisms of cell compartmentation with two main aspects: the membrane components involved in the control of the process (translocators, ATPases, and ionic pumps creating driving forces, multienzyme complexes coupling biosynthesis and membrane translocation) and the properties of the sets of enzymes linked to metabolic pathways. At the cellular level, it is of central importance to obtain a better knowledge of the properties of vesicles and a better understanding of their role in the biosynthesis and intracellular translocation of secondary metabolites. It is also necessary to give more attention to the mechanisms which insure the reversibility of the accumulation of metabolites in the storage compartments.

A rational optimization of the production of secondary metabolites *in vitro* should be based on the knowledge of the relative contribution of the different components regulating that production. This knowledge should offer the possibility to build efficient screening procedures to search for compartmentation mutants, either totally excreting their production or with a high storage capacity in vacuoles. This could prove to be an alternative route toward stable, high yielding strains.

REFERENCES

Abe Imoto, S., and Ohta, Y. (1985). Intracellular localization of lunularic acid and prelunularic acid in suspension-cultured cells of *Marchantia polymorpha*. *Plant Physiol.* **79,** 751–755.

Alibert, G., and Boudet, A. M. (1982). Progrès, problèmes et perspectives dans l'obtention et l'utilisation de vacuoles isolées. *Physiol. Veg.* **20,** 289–302.

Alibert, G., Ranjeva, R., and Boudet, A. M. (1977). Organisation subcellulaire des voies de synthèse des composés phénoliques. *Physiol. Veg.* **15,** 279–301.

Amann, M., Wanner, G., and Zenk, M. H. (1986). Intracellular compartmentation of two enzymes of berberine biosynthesis in plant cell cultures. *Planta* **167,** 310–320.

Anderson, L. A., Keene, A. T., and Philippson, J. D. (1982). Alkaloid production by leaf organ, root organ and cell suspension cultures of *Cinchona ledgeriana*. *Planta Med.* **46,** 25–27.

Banthorpe, D. V., Branch, S. A., Njar, V. C. O., Osborne, M. G., and Watson, D. G. (1986). Ability of plant callus cultures to synthesize and accumulate lower terpenoids. *Phytochemistry* **25,** 629–636.

Barbier-Brygoo, H., Renaudin, J. P., Manigault, P., Mathieu, Y., Kurkdjian, A., and Guern, J. (1987). Properties of vacuoles as a function of the isolation procedure. *In* "Plant Vacuoles: Their Importance in Plant Cell Compartmentation and Their Application in Biotechnology" (B. Marin, ed.). Plenum, New York (in press).

Barz, W., and Köster, J. (1981). Turnover and degradation of secondary (natural) products. *In* "The Biochemistry of Plants" (E. E. Conn, ed.), Vol. 7, pp. 35–84. Academic Press, New York.

Baur, P. S., and Walkinshaw, C. H. (1974). Fine structure of tannin accumulations in callus cultures of *Pinus elliotti* (slash pine). *Can. J. Bot.* **52,** 615–619.

Bergmann, L., and Rennenberg, H. (1978). Efflux und Produktion von Glutathion in Suspensionskulturen von *Nicotiana tabacum. Z. Pflanzenphysiol.* **88,** 175–178.

Berlin, J., Witte, L., Schubert, W., and Wray, V. (1984). Determination and quantification of monoterpenoids secreted into the medium of cell cultures of *Thuja occidentalis. Phytochemistry* **23,** 1277–1279.

Bisson, W., Beiderbeck, R., and Reichling, J. (1983). Die Produktion ätherischer Öle durch Zellsuspensionen der Kamille in einem Zweiphasensystem. *Planta Med.* **47,** 164–168.

Blume, D. E., Jaworski, J. G., and McClure, J. W. (1979). Uridine-diphosphate-glucose: isovitexin 7-*O*-glucosyltransferase from barley protoplasts: Subcellular localization. *Planta* **146,** 199–202.

Böhm, H., and Franke, J. (1982). Accumulation and excretion of alkaloids by *Macleaya microcarpa* cell cultures. I. Experiments on solid medium. *Biochem. Physiol. Pflanz.* **177,** 345–356.

Boller, T. (1985). Intracellular transport of metabolites in protoplasts: Transport between cytosol and vacuole. *In* "The Physiological Properties of Plant Protoplasts" (P. E. Pilet, ed.), pp. 76–86. Springer-Verlag, Berlin and New York.

Boller, T., and Wiemken, A. (1986). Dynamics of vacuolar compartmentation. *Ann. Rev. Plant Physiol.* **37,** 137–164.

Boller, T., Dürr, M., and Wiemken, A. (1975). Characterization of a specific transport system for arginine in isolated yeast vacuoles. *Eur. J. Biochem.* **54,** 81–91.

Boudet, A. M., Alibert, G., and Marigo, G. (1984). Vacuoles and tonoplast in the regulation of cellular metabolism. *In* "Membranes and Compartmentation in the Regulation of Plant Functions" (A. M. Boudet, G. Alibert, G. Marigo, and P. J. Lea, eds.), pp. 29–47. Oxford Univ. Press (Clarendon), London and New York.

Brodelius, P., and Nilsson, K. (1983). Permeabilization of immobilized plant cells, resulting in release of intracellular stored products with preserved cell viability. *Eur. J. Appl. Microbiol. Biotechnol.* **17,** 275–280.

Brown, S. C., Renaudin, J. P., Prévot, C., and Guern, J. (1984). Flow cytometry and sorting of plant protoplasts: Technical problems and physiological results from a study of pH and alkaloids in *Catharanthus* cells. *Physiol. Veg.* **22,** 541–554.

Buser-Suter, C., Wiemken, A., and Matile, P. (1982). A malic acid permease in isolated vacuoles of a crassulacean acid metabolism plant. *Plant Physiol.* **69,** 456–459.

Chafe, S. C., and Durzan, D. J. (1973). Tannin inclusions in cell suspension cultures of white spruce. *Planta* **113,** 251–262.

Chaprin, N., and Ellis, B. E. (1984). Microspectrophotometric evaluation of rosmarinic acid accumulation in single cultured plant cells. *Can. J. Bot.* **62,** 2278–2282.

Charrière-Ladreix, Y. (1977). La secrétion flavonique chez *Populus nigra* L.: Signification fonctionnelle des remaniements ultrastructuraux glandulaires. *Physiol. Veg.* **15,** 619–640.

Conn, E. E. (1984). Compartmentation of secondary compounds. *In* "Membranes and Compartmentation in the Regulation of Plant Functions" (A. M. Boudet, G. Alibert,

G. Marigo, and P. J. Lea, eds.), pp. 1–28. Oxford Univ. Press (Clarendon), London and New York.
Courtois, D., Kurkdjian, A., and Guern, J. (1980). Tryptamine uptake and accumulation by *Catharanthus roseus* cells cultivated in liquid medium. *Plant Sci. Lett.* **18,** 85–96.
Cramer, C. L., and Davis, R. H. (1984). Polyphosphate–cation interaction in the amino acid–containing vacuole of *Neurospora crassa. J. Biol. Chem.* **259,** 5152–5157.
Delfel, N. E., and Rothfus, J. A. (1977). Antitumor alkaloids in callus cultures of *Cephalotaxus harringtonia. Phytochemistry* **16,** 1595–1598.
Deus-Neumann, B., and Zenk, M. H. (1984a). A highly selective alkaloid uptake system in vacuoles of higher plants. *Planta* **162,** 250–260.
Deus-Neumann, B., and Zenk, M. H. (1984b). Instability of indole alkaloid production in *Catharanthus roseus* cell suspension cultures. *Planta Med.* **50,** 427–431.
Deus-Neumann, B., and Zenk, M. H. (1986). Accumulation of alkaloids in plant vacuoles does not involve an ion-trap mechanism. *Planta* **167,** 44–53.
Doll, S., Rodier, F., and Willenbrink, J. (1979). Accumulation of sucrose in vacuoles isolated from red beet tissue. *Planta* **144,** 407–411.
Dougall, D. K. (1981). Tissue culture and the study of secondary (natural) products. *In* "The Biochemistry of Plants" (E. E. Conn, ed.), Vol. 7, pp. 21–34. Academic Press, New York.
Dürr, M., Urech, K., Boller, T., and Wiemken, A. (1979). Sequestration of arginine by polyphosphates in vacuoles of yeast (*Saccharomyces cerevisiae*). *Arch. Microbiol.* **121,** 169–175.
Fairbairn, J. W., Hakim, F., and El Kheir, Y. (1974). Alkaloidal storage, metabolism and translocation in the vesicles of *Papaver somniferum* latex. *Phytochemistry* **13,** 1133–1139.
Franke, J., and Böhm. H. (1982). Accumulation and excretion of alkaloids by *Macleaya microcarpa* cell cultures. II. Experiments in liquid medium. *Biochem. Physiol. Pflanz.* **177,** 501–507.
Frehner, M., Keller, F., and Wiemken, A. (1984). Localization of fructan metabolism in the vacuoles isolated from protoplasts of Jerusalem artichoke tubers (*Helianthus tuberosus* L.). *J. Plant Physiol.* **116,** 197–208.
Frischknecht, P. M., and Baumann, T. W. (1977). Tissue culture of *Coffea arabica.* Growth and caffeine formation. *Planta Med.* **4,** 344–350.
Fry, S. C. (1979). Phenolic components of the primary cell wall and their possible role in the hormonal regulation of growth. *Planta* **146,** 343–351.
Fujiwake, H., Suzuki, T., and Iwai, K. (1980). Intracellular localization of capsaicin and its analogues in *Capsicum* fruit. II. The vacuole as the intracellular accumulation site of capsaicinoid in the protoplast of *Capsicum* fruit. *Plant Cell Physiol.* **21,** 1023–1030.
Grob, K., and Matile, P. (1979). Vacuolar location of glucosinolates in horseradish root cells. *Plant Sci. Lett.* **14,** 327–335.
Guy, M., Reinhold, L., and Michaeli, D. (1979). Direct evidence for a sugar transport mechanism in isolated vacuoles. *Plant Physiol.* **64,** 61–64.
Heinstein, P. F. (1985). Future approaches to the formation of secondary natural products in plant cell suspension cultures. *J. Nat. Prod.* **48,** 1–9.
Hilling, B., and Amelunxen, F. (1985). On the development of the vacuole. II. Further evidence for endoplasmic reticulum origin. *Eur. J. Cell Biol.* **38,** 195–200.
Hinz, H., and Zenk, M. H. (1981). Production of protoberberine alkaloids by cell suspension cultures of *Berberis* species. *Naturwissenschaften* **68,** 620–621.
Hollander-Czytko, H., and Amrhein, N. (1983). Subcellular compartmentation of shikimic acid and phenylalanine in buckwheat cell suspension cultures grown in the presence of shikimate pathway inhibitors. *Plant Sci. Lett.* **29,** 89–96.

Homeyer, B. C., and Roberts, M. F. (1984). Alkaloid sequestration by *Papaver somniferum* latex. *Z. Naturforsch., C: Biosci.* **39C,** 876–881.

Hopp, W., Hinderer, W., Petersen, M., and Seitz, U. (1985). Anthocyanin-containing vacuoles isolated from protoplasts of *Daucus carota* cell cultures. *In* "The Physiological Properties of Plant Protoplasts" (P. E. Pilet, ed.), pp. 116–121. Springer-Verlag, Berlin and New York.

Hrazdina, G., Alscher-Herman, R., and Kish, V. M. (1980). Subcellular localization of flavonoid synthesizing enzymes in *Pisum, Phaseolus, Brassica* and *Spinacia* cultivars. *Phytochemistry* **19,** 1355–1359.

Kaiser, G., and Heber, U. (1984). Sucrose transport into vacuoles isolated from barley mesophyll protoplasts. *Planta* **161,** 562–568.

Keller, F. (1986). Gentiopicroside is located in the vacuoles of root protoplasts of *Gentiana lutea. J. Plant Physiol.* **122,** 473–476.

Kibler, R., and Neumann, D. (1979). Alkaloidgehalte in haploiden und diploiden blättern und zellsuspensionen von *Datura innoxia. Planta Med.* **35,** 354–359.

Kurkdjian, A. (1982). Absorption and accumulation of nicotine by *Acer pseudoplatanus* and *Nicotiana tabacum* cells. *Physiol Veg.* **20,** 73–83.

Kutchan, T. M.. Ayabe, S., Krueger, R. J., Coscia, E. M., and Coscia, C. J. (1983). Cytodifferentiation and alkaloid accumulation in cultured cells of *Papaver bracteatum. Plant Cell Rep.* **2,** 281–284.

Kutchan, T. M., Ayabe, S., and Coscia, C. J. (1985). Cytodifferentiation and *Papaver* alkaloid accumulation. *In* "The Chemistry and Biology of Isoquinoline Alkaloids" (J. D. Phillipson, M. F. Roberts, and M. H. Zenk, eds.), pp. 281–294. Springer-Verlag, Berlin and New York.

Kutchan, T. M., Rush, M., and Coscia, C. J. (1986). Subcellular localization of alkaloids and dopamine in different vacuolar compartments of *Papaver bracteatum. Plant Physiol.* **81,** 161–166.

Leigh, R. A. (1983). Methods, progress and potential for the use of isolated vacuoles in studies of solute transport in higher plant cells. *Physiol. Plant.* **57,** 390–396.

Leigh, R. A., Ap Rees, T., Fuller, W. A., and Banfield, J. (1979). The location of acid invertase activity and sucrose in the vacuoles of storage roots of beetroot (*Beta vulgaris*). *Biochem. J.* **178,** 539–547.

Lindsey, K., Yeoman, M. M., Black, G. M., and Mavituna, F. (1983). A novel method for the immobilization and culture of plant cells. *FEBS Lett.* **155,** 143–149.

Löffelhardt, W., Kopp, B., and Kubelka, W. (1979). Intracellular distribution of cardiac glycosides in leaves of *Convallaria majalis. Phytochemistry* **18,** 1289–1291.

Luckner, M. (1980). Expression and control of secondary metabolism. *In* "Encyclopedia of Plant Physiology, New Series" (E. A. Bell and B. V. Charlwood, eds.), Vol. 8, pp. 23–63. Springer-Verlag, Berlin and New York.

Luckner, M., Diettrich, B., and Lerbs, W. (1980). Cellular compartmentation and channelling of secondary metabolism in microorganisms and higher plants. *Prog. Phytochem.* **6,** 103–142.

Madyastha, K. M., Ridgway, J. E., Dwyer, J. G., and Coscia, C. J. (1977). Subcellular localization of a cytochrome *P*-450 dependent monooxygenase in vesicles of the higher plant *Catharanthus roseus. J. Cell Biol.* **72,** 302–313.

Mantell, S. H., Pearson, D. W., Hazell, L. P., and Smith, H. (1983). The effect of initial phosphate and sucrose levels on nicotine accumulation in batch suspension cultures of *Nicotiana tabacum* L. *Plant Cell Rep.* **2,** 73–77.

Marin, B., and Chréstin, H. (1985). Compartmentation of solutes and the role of the tonoplast ATPase in *Hevea* latex. *In* "Biochemistry and Function of Vacuolar Adenosine-Triphosphatase in Fungi and Plants" (B. P. Marin, ed.), pp. 212–226. Springer-Verlag, Berlin and New York.

Martinoia, E., Flügge, I., Kaiser, G., Heber, U., and Heldt, H. W. (1985). Energy-dependent uptake of malate into vacuoles isolated from barley mesophyll protoplasts. *Biochim. Biophys. Acta* **806,** 311–319.

Marty, F., Branton, D., and Leigh, R. A. (1980). Plant vacuoles. *In* "The Biochemistry of Plants" (P. K. Stumpf and E. E. Conn, eds.), Vol. 1, pp. 625–658. Academic Press, New York.

Matern, U., Heller, W., and Himmelspach, K. (1983). Conformational changes of apigenin 7-*O*-(6-*O*-malonylglucoside), a vacuolar pigment from parsley, with solvent composition and proton concentration. *Eur. J. Biochem.* **133,** 439–448.

Matern, U., Reichenbach, C., and Heller, W. (1986). Efficient uptake of flavonoids into parsley (*Petroselinum hortense*) vacuoles requires acylated glycosides. *Planta* **167,** 183–189.

Matile, P. (1976). Localization of alkaloids and mechanism of their accumulation in vacuoles of *Chelidonium majus* laticifers. *Nova Acta Leopold., Suppl.* **7,** 139–156.

Matile, P. (1978). Biochemistry and function of vacuoles. *Annu. Rev. Plant Physiol.* **29,** 193–213.

Matile, P. (1982). Vacuoles come of age. *Physiol. Veg.* **20,** 303–310.

Matile, P. (1984). Das toxische Kompartiment der Pflanzenzelle. *Naturwissenschaften* **71,** 18–24.

Mérillon, J. M., Chénieux, J. C., and Rideau, M. (1983). Cinétique de croissance, évolution du métabolisme glucido-azoté et accumulation alcaloïdique dans une suspension cellulaire de *Catharanthus roseus*. *Planta Med.* **47,** 169–176.

Mollard, A., and Robert, D. (1984). Etude de la lignine pariétale et extracellulaire des suspensions cellulaires de *Rosa glauca*. *Physiol. Veg.* **22,** 3–17.

Moskowitz, A. H., and Hrazdina, G. (1981). Vacuolar contents of fruit subepidermal cells from *Vitis* species. *Plant Physiol.* **68,** 686–692.

Mothes, K. (1955). Physiology of alkaloids. *Annu. Rev. Plant Physiol.* **6,** 393–432.

Müller, E. (1976). Principles in transport and accumulation of secondary products. *Nova Acta Leopold. Suppl.* **7,** 123–128.

Müller, E., Nelles, A., and Neumann. D. (1971). Beiträge zur Physiologie der Alkaloide. I. Aufnahme von Nikotin in Gewebe von *Nicotiana rustica* L. *Biochem. Physiol. Pflanz.* **162,** 272–294.

Nakagawa, K., Konagai, A., Fukui, H., and Tabata, M. (1984). Release and crystallization of berberine in the liquid medium of *Thalictrum minus* cell suspension cultures. *Plant Cell Rep.* **3,** 254–257.

Nakagawa, K., Fukui, H., and Tabata, M. (1986). Hormonal regulation of berberine production in cell suspension cultures of *Thalictrum minus*. *Plant Cell Rep.* **5,** 69–71.

Neumann, D. (1976). Interrelationship between the morphology of alkaloid storage cells and the possible mechanism of alkaloid storage. *Nova Acta Leopold., Suppl.* **7,** 77–81.

Neumann, D., and Müller, E. (1972). Beiträge zur Physiologie der Alkaloide. III. *Chelidonium majus* L. und *Sanguinaria canadensis* L.: Ultrastruktur der Alkaloidbehalter, Alkaloidaufnahme und -Verteilung. *Biochem. Physiol. Pflanz.* **163,** 375–391.

Neumann, D., Krauss, G., Hieke, M., and Gröger, D. (1983). Indole alkaloid formation and storage in cell suspension cultures of *Catharanthus roseus*. *Planta Med.* **48,** 20–23.

Oba, K., Conn, E. E., Canut, H., and Boudet, A. M. (1981). Subcellular localization of 2-(β-D-glucosyloxy)cinnamic acids and the related β-glucosidase in leaves of *Melilotus alba* Desr. *Plant Physiol.* **68,** 1359–1363.

Ogino, T., Hiraoka, N., and Tabata, M. (1978). Selection of high nicotine-producing cell lines of tobacco callus by single-cell cloning. *Phytochemistry* **17,** 1907–1910.

Ohsumi, Y., and Anraku, Y. (1981). Active transport of basic amino acids driven by a proton motive force in vacuolar membrane vesicles of *Saccharomyces cerevisiae*. *J. Biol. Chem.* **256,** 2079–2082.

Olson, A. C., Evans, J. J., Frederick, D. P., and Jansen, E. (1969). Plant suspension culture media macromolecules—Pectic substances, protein, and peroxidase. *Plant Physiol.* **44,** 1594–1600.

Owens, T., and Poole, R. J. (1979). Regulation of cytoplasmic and vacuolar volumes by plant cells in suspension culture. *Plant Physiol.* **64,** 900–904.

Parham, R. A., and Kaustinen, H. M. (1977). On the site of tannin synthesis in plant cells. *Bot. Gaz. (Chicago)* **138,** 465–467.

Parr, A. J., Robins, R. J., and Rhodes M. J. C. (1984). Permeabilization of *Cinchona ledgeriana* cells by dimethylsulfoxide. Effects on alkaloid release and long-term membrane integrity. *Plant Cell Rep.* **3,** 262–265.

Parr, A. J., Robins, R. J., and Rhodes, M. J. C. (1986). Alkaloid transport in *Cinchona* ssp. cell cultures. *Physiol. Veg.* **24,** 419–429.

Pecket, R. C., and Small, C. J. (1980). Occurrence, location and development of anthocyanoplasts. *Phytochemistry* **19,** 2571–2576.

Pétiard, V., and Courtois, D. (1983). Recent advances in research for novel alkaloids in Apocynaceae tissue cultures. *Physiol. Veg.* **21,** 217–227.

Pfeiffer, B., Roos, W., and Luckner, M. (1982). Accumulation of purpureaglycoside A in vacuoles of *Digitalis lanata* cells cultivated *in vitro. Planta Med.* **45,** 154.

Phillips, R., and Henshaw, G. G. (1977). The regulation of synthesis of phenolics in stationary phase cell cultures of *Acer pseudoplatanus* L. *J. Exp. Bot.* **28,** 785–794.

Rataboul, P., Alibert, G., Boller, T., and Boudet, A. M. (1985). Intracellular transport and vacuolar accumulation of *O*-coumaric acid glucoside in *Melilotus alba* mesophyll cell protoplasts. *Biochim. Biophys. Acta* **816,** 25–36.

Renaudin, J. P. (1981). Uptake and accumulation of an indole alkaloid, ^{14}C-tabernanthine, by cell suspension cultures of *Catharanthus roseus* (L.) G. Don and *Acer pseudoplatanus* L. *Plant Sci. Lett.* **22,** 59–69.

Renaudin, J. P., and Guern, J. (1982). Compartmentation mechanisms of indole alkaloids in cell suspension cultures of *Catharanthus roseus. Physiol. Veg.* **20,** 533–547.

Renaudin, J. P., and Guern, J. (1987). Ajmalicine transport in vacuoles isolated from *Catharanthus roseus* cells. *In* "Plant Vacuoles: Their Importance in Plant Cell Compartmentation and Their Application in Biotechnology" (B. Marin, ed.), pp. 339–348. Plenum, New York.

Renaudin, J. P., Brown, S. C., and Guern, J. (1985). Compartmentation of alkaloids in a cell suspension of *Catharanthus roseus:* A reappraisal of the role of pH gradients. *In* "Primary and Secondary Metabolism of Plant Cell Cultures" (K. H. Neumann, W. Barz, and E. Reinhard, eds.), pp. 124–132. Springer-Verlag, Berlin and New York.

Renaudin, J. P., Brown, S. C., Barbier-Brygoo, H., and Guern, J. (1986). Quantitative characterization of protoplasts and vacuoles from suspension-cultured cells of *Catharanthus roseus. Physiol. Plant.* **68,** 695–703.

Roberts, M. F., McCarthy, D., Kutchan, T., and Coscia, C. J. (1983). Localization of enzymes and alkaloidal metabolites in *Papaver* latex. *Arch. Biochem. Biophys.* **222,** 599–609.

Rueffer, M. (1985). The production of isoquinoline alkaloids by plant cell cultures. *In* "The Chemistry and Biology of Isoquinoline Alkaloids" (J. D. Phillipson, M. F. Roberts, and M. H. Zenk, eds.), pp. 265–280. Springer-Verlag, Berlin and New York.

Sato, F., and Yamada, Y. (1984). High berberine–producing cultures of *Coptis japonica* cells. *Phytochemistry* **23,** 281–285.

Saunders, J. A. (1979). Investigations of vacuoles isolated from tobacco. *Plant Physiol.* **64,** 74–78.
Saunders, J. A., and Conn, E. E. (1978). Presence of the cyanogenic glucoside dhurrin in isolated vacuoles from *Sorghum*. *Plant Physiol.* **61,** 154–157.
Saunders, J. A., and McClure, J. W. (1976). The occurrence and photoregulation of flavonoids in barley plastids. *Phytochemistry* **15,** 805–807.
Schnepf, E. (1976). Morphology and cytology of storage spaces. *Nova Acta Leopold., Suppl.* **7,** 23–44.
Scott, A. I. A., Mizumaki, H., and Lee, S. L. (1979). Characterization of a 5-methyltryptophan resistant strain of *Catharanthus roseus* cultured cells. *Phytochemistry* **18,** 795–798.
Sharma, V., and Strack, D. (1985). Vacuolar localization of 1-sina-polyglucose : L-malate sinapoyltransferase in protoplasts from cotyledons of *Raphanus sativus*. *Planta* **163,** 563–568.
Staba, E. J. (1985). Milestones in plant tissue culture systems for the production of secondary products. *J. Nat. Prod.* **48,** 203–209.
Stafford, H. A. (1981). Compartmentation in natural products biosynthesis by multienzyme complexes. *In* "The Biochemistry of Plants" (E. E. Conn, ed.), Vol. 7, pp. 117–137. Academic Press, New York.
Strack, D., and Sharma, V. (1985). Vacuolar localization of the enzymatic synthesis of hydroxycinnamic acid esters of malic acid in protoplasts from *Raphanus Sativus* leaves. *Physiol. Plant.* **65,** 45–50.
Thom, M., and Komor, E. (1984). H^+–sugar antiport as the mechanism of sugar uptake by sugarcane vacuoles. *FEBS Lett.* **173,** 1–4.
Thom, M., and Maretzki, A. (1985). Group translocation as a mechanism for sucrose transfer into vacuoles from sugarcane cells. *Proc. Natl. Acad. Sci. U.S.A.* **82,** 4697–4701.
Thom, M., Leigh, R. A., and Maretzki, A. (1986). Evidence for the involvement of a UDPglucose-dependent group translocator in sucrose uptake into vacuoles of storage roots of red beet. *Planta* **167,** 410–413.
Tsukada, M., and Tabata, M. (1984). Intracellular localization and secretion of naphtoquinone pigments in cell cultures of *Lithospermum erythrorhizon*. *Planta Med.* **50,** 338–340.
Wagner, G. J. (1979). Content and vacuole/extravacuole distribution of neutral sugars, free amino acids, and anthocyanin in protoplasts. *Plant Physiol.* **64,** 88–93.
Wagner, G. J. (1982). Compartmentation in plant cells: The role of the vacuole. *Recent Adv. Phytochem.* **16,** 1–45.
Wagner, G. J. (1983). Higher plant vacuoles and tonoplast. *In* "Isolation of Membranes and Organelles from Plant Cells" (J. L. Hall and A. L. Moore, eds.), pp. 83–118. Academic Press, London.
Wagner, G. J., Keller, F., and Wiemken, A. (1983). Fructan metabolism in cereals: Induction in leaves and compartmentation in protoplasts and vacuoles. *Z. Pflanzenphysiol.* **112,** 359–372.
Waller, G. R., McVean, C. D., and Suzuki, T. (1983). High production of caffeine and related enzyme activities in callus cultures of *Coffea arabica*. *Plant Cell Rep.* **2,** 109–112.
Weissemböck, G., and Schneider, V. (1974). Untersuchungen zur Lokalisation von Flavonoiden in Plastiden. III. Flavonoidgehalte der Chloroplasten aus *Avena Sativa* L. und das Problem der Kontamination isolierter Plastiden durch Flavonoide. *Z. Pflanzenphysiol.* **72,** 23–35.
Werner, C., and Matile, P. (1985). Accumulation of coumarylglucosides in vacuoles of barley mesophyll protoplasts. *J. Plant Physiol.* **118,** 237–249.
Wiermann, R. (1981). Secondary plant products and cell and tissue differentiation. *In* "The

Biochemistry of Plants" (E. E. Conn, ed.), Vol. 7, pp. 85–116. Academic Press, New York.

Willenbrink, J., and Doll, S. (1979). Characteristics of the sucrose uptake system of vacuoles isolated from red beet tissue. Kinetics and specificity of the sucrose uptake system. *Planta* **147,** 159–162.

Wink, M. (1984). Evidence for an extracellular lytic compartment of plant cell suspension cultures: The cell culture medium. *Naturwissenschaften* **71,** 635–637.

Wink, M. (1985). Metabolism of quinolizidine alkaloids in plants and cell suspension cultures: Induction and degradation. *In* "Primary and Secondary Metabolism of Plant Cell Cultures" (K. H. Neumann, W. Barz, and E. Reinhard, eds.), pp. 107–116. Springer-Verlag, Berlin and New York.

Wink, M., Schiebel, H. M., and Hartmann, T. (1982). Quinolizidine alkaloids from plants and their cell suspension cultures. *Planta Med.* **44,** 15–20.

Witte, L., Berlin, J., Wray, V., Schubert, W., Kohl, W., Höfle, G., and Hammer, J. (1983). Mono- and diterpenes from cell cultures of *Thuja occidentalis. Planta Med.* **49,** 216–221.

Yamada, Y., and Sato, F. (1981). Production of berberine in cultured cells of *Coptis japonica. Phytochemistry* **20,** 545–547.

Yamamoto, H., Nagakawa, K., Fukui, H., and Tabata, M. (1986). Cytological changes associated with alkaloid production in cultured cells of *Coptis japonica* and *Thalictrum minus. Plant Cell Rep.* **5,** 65–68.

Zenk, M. H. (1980). Enzymatic synthesis of ajmalicine and related indole alkaloids. *J. Nat. Prod.* **43,** 438–451.

Zenk, M. H. (1985). Enzymology of benzylisoquinoline alkaloid formation. *In* "The Chemistry and Biology of Isoquinoline Alkaloids" (J. D. Phillipson, M. F. Roberts, and M. H. Zenk, eds.), pp. 240–256. Springer-Verlag, Berlin and New York.

CHAPTER **4**

Regulation of Synthesis of Phenolics

Ragai K. Ibrahim

Plant Biochemistry Laboratory
Department of Biology
Concordia University
Montreal, Quebec, Canada H3G 1M8

I. INTRODUCTION

Plant tissue and cell cultures are a potentially useful source of specific metabolites that may be of pharmaceutical or commercial interest, and they may as well lead to the discovery of new biochemicals. Cultured tissues also provide an efficient system for the study of biosynthesis, productivity, biotransformation, and regulation of metabolite synthesis. The aim of this chapter is to review the recent developments in the regulation of synthesis of phenolic compounds in cultured tissues. These compounds include the phenylpropanoids and their derivatives, flavonoids, naphthoquinones, and anthraquinones. Due to the enormous body of information available no attempt will be made to give a comprehensive qualitative or quantitative coverage of the compounds reported recently. Instead, discretionary examples will be selected to represent the different groups of phenolic compounds and to put in perspective those which have been the subject of regulation studies. It is now evident that the biosynthesis and productivity of phenolic compounds are regulated by a variety of external and biological factors. This review will concern itself with those factors involved in the induction or stimulation of secondary metabolite synthesis. Recent articles which have dealt with this topic include those of Butcher (1976), Hahlbrock (1977), Böhm (1980), Staba (1980), Barz and Ellis (1981), and DiCosmo and Towers (1984).

II. PHENOLIC PRODUCTION IN CULTURED TISSUES

Since higher plants are often held "hostage" of their environment, they adapt to and interact with other organisms in their neighborhood by producing a variety of chemicals, known as secondary metabolites, many of which are of a phenolic nature. Table I lists a variety of phenolic compounds produced by cultured tissues. Whereas this listing is by no means exhaustive, it is intended to represent the different classes of phenolic compounds that will be discussed in this chapter.

A. Phenylpropanoids

Most of the phenylpropanoids found in cultured tissues consist of the common hydroxycinnamic acids: *p*-coumaric, caffeic, ferulic, and, to a much lesser extent, sinapic acids as well as their benzoic acid analogs. The first group usually occurs as the glucose or quinic acid esters, whereas the last occurs mostly as glucosides. Apart from chlorogenic (caffeoylquinic) acid, other unusual depsides, such as rosmarinic acid, lithospermic acid, and verbascoside, have been reported to occur in a limited number of cultures. A few cultured tissues have been reported to produce simple substituted coumarins, both in free and glucoside forms, as well as a variety of furanocoumarins. The formation of two unique phenylisocoumarins, hydrangenol and phyllodulcin, as well as their 8-*O*-glucosides has also been reported. Other less common metabolites have been reported in cultured tissues, such as the stilbenes, resveratrol and its isopentenyl derivative, as well as few lignans. Lignin formed in soybean cell cultures, although not sought as a secondary metabolite, has been shown to have a similar syringyl/guaiacyl ratio as that of the intact tissue (Nimz *et al.*, 1975).

B. Flavonoids

Recent reports on the formation of flavonoid compounds (see also Hinderer and Seitz, in Volume 5, this treatise) in cultured tissues indicate that the cultures can be induced, under appropriate conditions, to produce almost all classes of flavonoids (Table I). These include the unusual retrochalcones, licodione, and its methyl ether, echinatin,

TABLE I

Phenolic Compounds in Cultured Tissues

Compound	Plant species[a]	Reference
Phenylpropanoids		
Hydroxycinnamic acids	*Daucus carota* (s)	Sugano *et al.* (1975)
Caffeoyl esters	*Perilla ocymoides* (s)	Ibrahim and Edgar (1976)
Hydroxycinnamoylquinic and glucose esters	*Pyrus malus* (s)	Koumba and Macheix (1981)
Hydroxycinnamoyl-putrescines	*Nicotiana tabacum* (c)	Mizusaki *et al.* (1971)
	Nicotiana tabacum (s)	Knobloch *et al.* (1982)
Rosmarinic acid	*Coleus blumei* (s)	Zenk *et al.* (1977)
	Anchusa officinalis(c)	De-Eknamkul and Ellis (1984)
Lithospermic acid	*Lithospermum erythrorhizon* (s)	Fukui *et al.* (1984)
Verbascoside	*Syringa vulgaris* (s)	Ellis (1983)
Coumarin derivatives	*Hydrangea macrophylla* (s)	Suzuki *et al.* (1977)
Scopolin and scopoletin	*Nicotiana tabacum* (s)	Ibrahim and Boulay (1980)
Hydrangenol and phyllodulcin	*Hydrangea macrophylla* (s)	Suzuki *et al.* (1977)
Stilbenes	*Arachis hypogea* (c)	Fritzmeier *et al.* (1983)
Podophyllotoxin (lignan)	*Podophyllum peltatum* (c)	Kadkade (1982)
Lignins	*Glycine max* (s)	Nimz *et al.* (1975)
Flavonoids		
Licodione and echinatin	*Glycyrrhiza echinata* (s)	Ayabe *et al.* (1980)
Flavanone glucosides	*Prunus avium*	Treutter *et al.* (1985)
Flavone glucosides	*Petroselinum hortense* (s)	Kreuzaler and Hahlbrock (1973)
Flavone glycoside-malonyl esters	*Petroselinum hortense* (s)	Matern *et al.* (1981)
Polymethoxy flavones	*Citrus mitis* (c)	Brunet and Ibrahim (1973)
Isoflavones	*Vigna angularis* (s)	Hattori and Ohata (1985)
	Glycyrrhiza echinata (c)	Ayabe *et al.* (1980)
Quercetin glucouronide	*Anethum graveolens* (s)	Moehle *et al.* (1985)
Anthocyanins and acylated derivatives	*Vitis* sp. (s)	Yamakawa *et al.* (1983)
Cyanin and peonin	*Strobilanthes dyeriana* (c)	Smith *et al.* (1981)
Petunidin and malvidin	*Petunia hybrida* (c)	Coljin *et al.* (1981)
Condensed tannins	*Acer pseudoplatanus* (s)	Westcott and Henshaw (1976)
	Crataegus monogyna (c)	Schrall and Becker (1977)
	Paul's Scarlet rose (s)	Muhitch and Fletcher (1985)
	Cryptomeria japonica (c)	Samejima *et al.* (1982)
Naphthoquinones and anthraquinones		
Shikonin derivatives	*Lithospermum erythrorhizon* (s)	Tabata *et al.* (1974)
Plumbagin	*Plumbago zeylanica* (s)	Heble *et al.* (1974)
Anthraquinone derivatives	*Cassia tora* (c)	Tabata *et al.* (1975)
	Morinda citrifolia (s)	Zenk *et al.* (1975)
	Galium mollugo (s)	Bauch and Leistner (1978)
	Rumex alpinum (c)	Van den Berg and Labadie (1981)
	Cinchona ledgeriana (c)	Mulder-Kreiger *et al.* (1982)
	Rubia cordifolia (c)	Suzuki *et al.* (1984)

[a] (c), Callus culture; (s), suspension culture.

which are not natural constituents of the intact tissue, as well as a variety of flavanone, flavone, isoflavone, and flavonol derivatives, both as free and glycoside forms. The formation of anthocyanins has frequently been reported for cultures derived from both cyanic and acyanic species. Whereas cyanidin and its glycosides appear to be the most common pigments formed, few cultures produce the O-methylated derivatives, peonidin, petunidin and malvidin. A few cultured tissues have also been reported to produce condensed tannins (catechins and proanthocyanidins).

C. Naphthoquinones and Anthraquinones

No other group of phenolic compounds exhibits such a variety of metabolite formation *in vitro* as the anthraquinones (see also Chapters by Tabata and Koblitz, Volume 5, this treatise). Although reported for only six genera, some cultured tissues may produce up to 12 compounds, both as free or bound forms. Whereas the anthraquinones of *Galium* suspension cultures differed in pattern from those of the intact plant (Inoue *et al.*, 1984), those of *Cinchona* are considered phytoalexins since they are not natural constituents of the intact tissues (Wijnsma *et al.*, 1984). In contrast to anthraquinones, reports on the production of naphthoquinones cover a narrow range of compounds only, such as plumbagin and shikonin derivatives.

III. REGULATION OF PHENOLIC SYNTHESIS

A. Nutrient Media

It is now evident that the composition of the most commonly used nutrient media (White's, Heller's MS, LS, B5; for details see Ozias-Akins and Vasil, Volume 2, this treatise) has a marked effect on the production of secondary metabolites. However, nutrient-rich media, such as MS, LS, or B5, usually support prolific cell or tissue growth. It is not surprising, therefore, that most cultured cells fail to produce their normal pattern of secondary metabolites, whereas others may produce novel compounds in culture (e.g., Ayabe *et al.*, 1980; Mulder-Krieger *et al.*, 1982). As a result of intensive work on the effect of different components of culture media on the productivity of phenolic metabolites, a few generalizations can now be made.

1. Carbon Source

The increase of sucrose concentration in the nutrient medium, above the 2–3% normally used, has been reported to stimulate the production of polyphenols (Westcott and Henshaw, 1976; Chandler and Dodds, 1983), phenolic depsides (Knobloch and Berlin, 1980; De-Eknamkul and Ellis, 1985a), coumarins and phenylisocoumarins (Suzuki *et al.*, 1981), anthocyanins (Knobloch *et al.*, 1982; Yamakawa *et al.*, 1983), naphthoquinone derivatives (Fujita *et al.*, 1981a), and anthraquinones (Schulte *et al.*, 1984; Suzuki *et al.*, 1984; Harkes *et al.*, 1985; Khouri *et al.*, 1986). In most cases, the sucrose concentration used was moderately high (4–6%), whereas in others it varied between 7 and 14%. Such high sucrose levels may create a state of osmotic stress in some cultured tissues, thus resulting in the inhibition of cell growth. On the other hand, Okazaki *et al.* (1982) reported that optimum coumarin production in tobacco cultures occurred at low sucrose levels. Whereas the general consensus is that high sucrose concentrations (above 3%) increase phenolic production, it is not known, in most reported cases, whether this can be traced to stimulation of cell growth, to altered phenolic metabolism, or to inconsistency in the method of reporting metabolite yields (Zenk, 1978).

In contrast to the case of microorganisms which exhibit catabolite repression of fermentation products (Drew and Demain, 1977), there is an evident lack of knowledge of the process of carbon assimilation (glycolysis, pentose phosphate pathway, and Krebs cycle) and its relation to the productivity of secondary metabolites in general and phenolic compounds in particular. It should also be noted that the effect of carbon source on phenolic synthesis is closely related to nitrate (Jessup and Fowler, 1976) and phosphate (Schiel *et al.*, 1984). Such relationships may shift the balance of carbon flow from the predominant pathway of glycolysis to the pentose phosphate pathway. The latter supplies part of the carbon skeleton of phenolic compounds and the NADPH required for their biosynthesis (Pryke and Ap Rees, 1977).

2. Mineral Nutrients

In view of the competition by two biochemical processes (growth and secondary metabolite synthesis) for some common precursors (aromatic and indole amino acids), it is reasonable to assume that conditions which are favorable for one pathway may suppress the other. As a result of recent studies it has become evident that in order to increase the productivity of most groups of secondary metabolites, cultured plant cells have to be maintained under limiting conditions of certain nutrients, especially nitrogen and inorganic phosphate. This led to the formulation of nutrient media for optimum productivity of secondary metabolites (Zenk *et al.*, 1975; Fujita *et al.*, 1981a; Knobloch and Berlin, 1981; Schulte *et al.*, 1984; Wijnsma *et al.*, 1984; Khouri *et al.*, 1986).

a. Nitrogen. Most nutrient media contain nitrogen in the form of a mixture of NO_3^- and NH_4^+ in different ratios. Recent studies have shown that both the type and amount of the nitrogen source affect the pattern and yield, if any, of the phenolics formed. The production of shikonin derivatives in *Lithospermum* cells could only be achieved with NO_3^- as the sole nitrogen source and was inhibited by NH_4^+ (Fujita *et al.*, 1981a). These cells ceased to yield naphthoquinones when grown on LS medium and produced, instead, both rosmarinic and lithospermic acids (Fukui *et al.*, 1984). In contrast to naphthoquinones, anthraquinone production was best attained on MS medium (Suzuki *et al.*, 1984), or B5 medium supplemented with 0.2% NZ amine Type A (Bauch and Leistner, 1978). However, the majority of studies seem to indicate that decreasing the nitrogen level stimulated the production of polyphenols (Westcott and Henshaw, 1976; Chandler and Dodds, 1983), coumarins (Okazaki *et al.*, 1982), depsides (Knobloch and Berlin, 1980; De-Eknamkul and Ellis, 1985a), anthocyanins (Knobloch *et al.*, 1982; Yamakawa *et al.*, 1983), and anthraquinones (Zenk *et al.*, 1975; Suzuki *et al.*, 1984; Harkes *et al.*, 1985; Khouri *et al.*, 1986). The addition of other organic supplements, such as casein hydrolysate, yeast extract, or peptone, has been reported to either inhibit (Mizukami *et al.*, 1977; Okazaki *et al.*, 1982) or stimulate (Zenk *et al.*. 1977; Bauch and Leistner, 1978) the production of different phenolic metabolites.

b. Phosphorus. In contrast to the stimulation of anthraquinone production (Zenk *et al.*, 1975), high phosphate levels have been reported to decrease the accumulation of coumarins (Okazaki *et al.*, 1982), polyphenols (Suzuki *et al.*, 1981; Chandler and Dodds, 1983; Schiel *et al.*, 1984), rosmarinic acid (De-Eknamkul and Ellis, 1985a), and anthocyanins (Dougall and Weyrauch, 1980; Knobloch *et al.*, 1982; Yamakawa *et al.*, 1983). Since most secondary metabolites are synthesized via phosphorylated intermediates, the inhibition by phosphate of phenolic production may be due to inhibition of the phosphatases responsible for product release. Recently, Wray *et al.* (1984) reported the use of high-field ^{31}P-NMR spectroscopy as a means of monitoring the levels of inorganic and organic phosphorus compounds in cell cultures and used them as criteria for the transfer of cultured cells to induction media for optimum metabolite productivity.

c. Other Elements. Whereas the role of mineral nutrients other than nitrogen and phosphorus in various metabolic processes has been extensively studied in intact plants (Rains, 1976), there have seldom been parallel studies in cultured cells, especially in relation to the production of secondary metabolites. Except for the reported effects of Ca^{2+} on lignification (Lipetz, 1962) and the production of rosmarinic acid (De-Eknamkul and Ellis, 1985a), of Ca^{2+}, SO_4^{2-}, and Cu^{2+} on naphthoquinones (Mizukami *et al.*, 1977; Fujita *et al.*, 1981b), and of micronutrients on anthraquinones (Zenk *et al.*, 1975), there is a conspicuous gap in our knowledge regarding the role of some minor elements in the production of phytochemicals. For example, S-containing compounds are essential for the formation of the methyl group donor, *S*-adenosyl-L-methionine, as well as the sulfate group donor, 3′-phosphoadenosine-5′-phosphosulfate, both of which are involved in the biosynthesis of methylated and sulfated flavonoids.

B. Phytohormones

There is sufficient evidence to indicate that the type and concentration of auxins and cytokinins, as well as their relative ratios in the culture medium, control the biosynthesis and accumulation of secondary metabolites. Whether the response to growth regulators is the cause or the consequence of cell differentiation or organization is debatable. It is indeed difficult to evaluate the effects of individual auxins and cytokinins on phenolic production independently of the nutritional status of the cultured tissue. However, the following discussion will provide an overview of the effect of phytohormones, singly or in combination, on the productivity of different classes of phenolic compounds.

The accumulation of simple phenolics, coumarins, depsides, and lignans, has been reported to be stimulated in the presence of low auxin levels, especially naphthaleneacetic acid (NAA) (King, 1976; De-Eknamkul and Ellis, 1985b), or 2,4-dichlorophenoxyacetic acid (2,4-D) (Sugano *et al.*, 1975; Kadkade, 1982). However, increasing the auxin concentration either stimulated (Zenk *et al.*, 1977; Okazaki *et al.*, 1982) or inhibited (Ibrahim and Edgar, 1976; Sahai and Shuler, 1984; De-Eknamkul and Ellis, 1985b) phenolic production. On the other hand, the removal of 2,4-D from the culture medium resulted in a significant production of phenolics in stationary phase tobacco cells (Sahai and Shuler, 1984) and in the induction of anthocyanin formation in carrot cells (Ozeki and Komamine, 1981). Increasing the level of cytokinin, in the presence of auxin, stimulated the formation of caffeoyl esters (Ibrahim and Edgar, 1976), coumarin derivatives (Okazaki *et al.*, 1982), and lignins (Kuboi and Yamada, 1976). There was no effect of added auxin or cytokinin on the production of coumarins or phenylisocoumarins in *Hydrangea* cell cultures (Suzuki *et al.*, 1984). The production of polyphenols was stimulated by increasing both 2,4-D and kinetin (K) (Shah *et al.*, 1976), whereas that of leucoanthocyanidins was suppressed by increasing 2,4-D but was stimulated by cytokinin and gibberellic acid (GA_3) (Samejima *et al.*, 1982). Abscisic acid (ABA) was reported to inhibit both growth and accumulation of coumarins, chlorogenic acid, and lignin, whereas GA_3 (0.1–10 ppm) resulted in an increase in the last two constituents (Li *et al.*, 1970), presumably by counteracting ABA inhibition.

The induction of anthocyanin synthesis in carrot cells was reported to occur in the dark on removal of 2,4-D from the culture medium; however, pigment accumulation was inhibited by low levels of 2,4-D but was stimulated by 10^{-6}–10^{-8} M of either K, isopentenyladenine (IPA), benzyladenine (BA), or zeatin (Ozeki and Komamine, 1981). Nevertheless, most studies seem to indicate that anthocyanin production can be achieved under conditions of low auxin, especially indoleacetic acid (IAA) or NAA (Kinnersley and Dougall, 1980; Coljin *et al.*, 1981; Smith *et al.*, 1981; Nishimaki and Nozue, 1985) and may be stimulated (Yamakawa *et al.*, 1983) or inhibited by increasing the auxin level (Lam and Street, 1977; Dougall and Weyrauch, 1980; Coljin *et al.*, 1981; Smith *et al.*, 1981). A combination of auxin and cytokinin increased (Nishimaki and Nozue, 1985), had no effect (Coljin *et al.*, 1981), or inhibited (Kinnersley and Dougall, 1980) pigment production in carrot cultures. GA_3 was reported to inhibit anthocyanin formation by blocking its synthesis at the level of chalcone synthase (Hinderer *et al.*, 1984).

Similar to simple phenolics and anthocyanins, optimum production of naphthoquinones and anthraquinones was achieved in low auxin media (Zenk *et al.*, 1975; Bauch and Leistner, 1978; Harkes *et al.*, 1985) and was stimulated by increasing the level of IAA (Tabata *et al.*, 1975; Khouri *et al.*, 1986), NAA (Zenk *et al.*, 1977; Suzuki *et al.*, 1984; Khouri *et al.*, 1986), or 2,4-D (Tabata *et al.*, 1975; Van den Berg and Labadie, 1981), but was suppressed by high levels of auxin or as little as 10^{-7} M GA_3 (Yoshikawa *et al.*, 1986). Recently, Zenk *et al.* (1984) have studied the effect of 40 different phenoxyacetic acids on the growth and product accumulation in *Morinda citrifolia* cell cultures. The results of this study confirmed the view that subtle differences in the substitution pattern of synthetic auxins play an important role in the induction or suppression of the biosynthetic pathways leading to anthraquinone formation. The effect of cytokinin was found to either inhibit naphthoquinone formation (Fujita *et al.*, 1981b) or not affect anthraquinone production (Harkes *et al.*, 1985; Khouri *et al.*, 1986).

C. Light

It is now evident that phenolic production in cultured tissues is influenced by both the quality and intensity of light. Except for carrot cell strains which produce anthocyanin pigments in the dark (Ozeki and Komamine, 1981), white light has been reported to induce or stimulate the formation of flavone derivatives (Brunet and Ibrahim, 1973; Kreuzaler and Hahlbrock, 1973) and anthocyanins (Coljin *et al.*, 1981; Knobloch *et al.*, 1982), although it was shown to be inhibitory for polyphenol (Ibrahim and Edgar, 1976) and anthraquinone (Suzuki *et al.*, 1984; Harkes *et al.*, 1985) production. Blue light was reported to be most effective in the formation of polyphenols (Suzuki *et al.*, 1981), but it inhibited the formation of podophyllotoxin (Kadkade, 1982) and naphthoquinones (Tabata *et al.*, 1974). Light stimulation of phenolic synthesis can be related to its effect on the regulation of the enzymes of the phenylpropanoid pathway (Hahlbrock, 1977), as well as other phytochrome-regulated enzymes (Schopfer, 1977). The induction of flavonoid synthesis by UV-B light (280–320 nm), which has recently been reported in parsley cell culture, is believed to involve the participation of a blue light photoreceptor which required P_{fr} for maximum effect (Duell-Pfaff and Wellmann, 1982). UV irradiation also triggered the formation of quercetin-3-*O*-glucuronide in dill cell culture (Moehle *et al.*, 1985). On the other hand, stilbene formation in peanut cell culture (Fritzmeier *et al.*, 1983) was selectively induced by UV-C light (260–270 nm) and was not controlled by phytochrome.

D. Precursors

Attempts to induce or increase the production of metabolite formation in cultured tissues by supplying precursors or intermediate compounds have produced encouraging results. L-Phenylalanine or L-tyrosine, administered to *Coleus blumei* cell culture, was transformed to rosmarinic acid (Razzaque and Ellis, 1977; Zenk *et al.*, 1977). The addition of phenylalanine has also been reported to increase the accumulation of hydroxycinnamoyl esters in apple cell culture (Koumba and Macheix, 1981). of polyphenols in tobacco cells (Sahai and Shuler, 1984), and of naphthoquinones in *Lithospermum* cell culture (Mizukami *et al.*, 1977). Flavonolignans were produced in cell cultures of *Silybum marianum* when administered luteolin and coniferyl alcohol (Schrall and Becker, 1977). Anthocyanin synthesis in carrot cell culture, which was blocked by GA_3 at the chalcone synthase level, was restored by the addition of naringenin, eriodictyol, or dihydroquercetin (Hinderer *et al.*, 1984), all of which are known precursors of anthocyanin synthesis (Fritsch and Grisebach, 1975). Furthermore, the addition of dihydroquercetin to carrot cell cultures also restored pigment formation which was formerly repressed under the influence of 2,4-D (Ozeki and Komamine, 1985). Increased production of anthraquinones in *Galium mollugo* cell culture was achieved by the addition of *o*-succinyl benzoic acid, especially to media which supported low product yield (Bauch and Leistner, 1978) as well as those under phosphate-limiting conditions (Wilson and Balagué, 1985).

Unsuccessful attempts to induce or increase product yield may be due to our lack of knowledge concerning the timing of addition of such compounds, their uptake, and their compartmentation in relation to the enzymes involved in their utilization. Further experimentation is required in this vital area of research.

E. Enzymes of Major Pathways

The admirable studies by the Freiburg group, over the last decade, have contributed extensively to our knowledge of the regulation of phenolic synthesis in cultured tissues (Hahlbrock, 1977). Using soybean and parsely cell suspension cultures, alternatively, these studies demonstrated the highly coordinated changes in activities of the enzymes of the general phenylpropanoid pathway, the Group I enzymes [phenylalanine ammonia-lyase (PAL), cinnamate-4-hydroxylase (C4H), and 4-hydroxycinnamate CoA ligase (4HCL)] by various exogenous and endogenous factors (Hahlbrock, 1977). Significant changes in enzyme activities were observed on continuous irradiation of parsley cell culture with light (Hahlbrock *et al.*, 1976) or dilution of the culture (Hahlbrock and Schröder, 1975). Simultaneous induction by dilution and irradiation resulted in synergistically increased enzyme activities. Group I enzymes are considered, therefore, to represent a distinct metabolic sequence leading to both lignin and/or flavonoid pathways.

Light induction of PAL activity has also been reported in several cultured tissues and coincided with maximum production of hydroxycinnamic acids (Ibrahim and Edgar, 1976; Koumba and Macheix, 1981), flavone derivatives (Brunet and Ibrahim, 1973), and anthocyanins (Stark *et al.*, 1976; Ozeki and Komamine, 1985). Induction or increase in PAL activity by factors other than light was also reported on increasing the sucrose level and/or depletion of nitrate or phosphate from the culture medium (Westcott and Henshaw, 1976; Knobloch *at al.*, 1982), removal of 2,4-D (Ozeki and Komamine, 1981) or GA_3 (Heinzmann and Seitz, 1977), addition of PAL inhibitors (Westcott, 1976; Noe and Seitz, 1982), as well as the addition of abiotic (Hattori and Ohata, 1985) or biotic (Ebel *et al.*, 1976; Dixon and Bendall, 1978) elicitors, with the occasional production of multiple forms of PAL (Bolwell *et al.*, 1985).

The first and last of the Group I enzymes were highly purified, and the latter was shown to occur as two distinct isoenzymes, 4HCL1 and 4HCL2 in soybean cell cultures, whereas parsley cultures contained only 4HCL2. The differences observed in substrate specificities and kinetic constants of the two isoenzymes indicated tight regulation, by both isoenzymes and their products, of lignin and flavonoid biosynthesis in soybean and parsley cultures, respectively (Knobloch and Hahlbrock, 1975). Two enzymes involved in lignin synthesis, cinnamoyl-CoA : NADPH reductase (Wengenmayer *et al.*, 1976) and cinnamyl alcohol : NADP oxidoreductase (Wyrambick and Grisebach, 1975), were purified from soybean cell cultures and their properties studied. O-Methylation of lignin precursors has also been reported in soybean (Poulton *et al.*, 1976) and tobacco (Kuboi and Yamada, 1976; Tsang and Ibrahim, 1979a,b) cell cultures, as well as the O-glucosylation of coniferyl alcohol (Ibrahim and Grisebach, 1976), by substrate-specific enzymes. Coniferin β-glucosidase activity has been shown to coincide with lignification in a number of cell cultures (Hösel *et al.*, 1982). None of the enzymes of the lignin pathway was reported to be stimulated by light irradiation (Hahlbrock, 1977).

In contrast with Group I enzymes, those of Group II, enzymes of the flavonoid pathway, were not light induced, although they exhibited coordinated enzyme activities similar to those of Group I, but with differences in the lag periods, times of peak activities, and apparent half-lives. These included acetyl-CoA carboxylase (Ebel and Hahlbrock, 1977), chalcone synthase (Hrazdina *et al.*, 1976; Saleh *et al.*, 1978), flavone/flavonol 3′-*O*-methyltransferase (Poulton *et al.*, 1977), isoflavone 4′-*O*-methyltransferase (Wengenmayer *et al.*, 1974), flavone/flavonol

3-*O*-glucosyltransferase (Poulton and Kauer, 1977), UDP-apiose transferase (Wellmann and Baron, 1974) and the flavone/flavonol-3-*O*-glucoside and -7-*O*-glucoside malonyltransferases (Matern *et al.*, 1981, 1983). These studies indicate that phenolic synthesis in cultured cells is regulated at the enzyme level by the high substrate specificities of individual enzymes of the three different pathways, the existence of isoenzymes, and the coordinated changes in enzyme activities by various exogenous and endogenous factors (Hahlbrock, 1977).

Several enzymes involved in the biosynthesis of different groups of phenolic compounds have also been reported in cultured tissues. These include the stilbene synthase from peanut cell culture (Fritzmeier *et al.*, 1983), 4HCL (Heinzmann *et al.*, 1977) and chalcone synthase (Hinderer and Seitz, 1985; Ozeki *et al.*, 1985) from carrot cell cultures, a dihydroxycoumarin 7-*O*-glucosyltransferase from tobacco cell culture (Ibrahim and Boulay, 1980), and several position-specific flavonol *O*-methyltransferases from calamondin orange peel callus (Brunet and Ibrahim, 1980), apple fruit (Macheix and Ibrahim, 1984), and tobacco cell (Tsang and Ibrahim, 1979a,b) cultures.

Despite the intensive studies on the biosynthesis of the anthraquinone skeleton (Zenk and Leistner, 1968; Leistner, 1973; Bentley, 1975; Inoue *et al.*, 1984), there has been an evident lack of knowledge regarding the enzymes involved in the later steps of anthraquinone formation, such as prenylation, hydroxylation, O-methylation, and/or O-glucosylation. Recently, however, five anthraquinone-specific glucosyltransferases have been characterized from *Cinchona succirubra* cell cultures and shown to be involved in the biosynthesis of anthraquinone glucosides of this tissue (Khouri and Ibrahim, 1987). Moreover, further investigation calls for a study of the localization of the enzymes of secondary metabolism and the compartmentation of final products in order to complete our understanding of the regulation of their synthesis.

IV. CONCLUSION

The recent developments in the regulation of phenolic synthesis and accumulation in cultured tissues have been impressive, even though most of these have been achieved by the use of conventional approaches. Noteworthy by their absence are studies of the effect of the natural growth inhibitors, abscisic acid and ethylene, on extending the stationary phase of culture growth in relation to metabolite production. Further knowledge at the molecular level is urgently needed for understanding the role of growth substances in the induction and suppression of secondary metabolite synthesis. No explanation for the striking differences in their effects will be attained until auxin receptors are isolated and characterized.

Limitation of space prevented the coverage of such recently applied techniques as screening for variant cell lines, single cell cloning, and immobilization of cultured cells for improving metabolite production; also omitted was the use of specific inhibitors as probes of metabolic blocks that may be responsible for the lack of expression of secondary metabolite synthesis. These topics are reviewed and discussed in other chapters of this volume.

A very promising approach to the regulation of metabolite synthesis would be the application of the techniques of genetic engineering in modifying cell lines which have been induced for high activity of one or more of the key enzymes of secondary metabolism. A system combining *in vivo* labeling of such enzymes, their translation in a wheat germ cell-free system, and immunoprecipitation of their specific mRNAs has recently been proposed (Berlin *et al.*, 1985). This would be followed by the construction of gene banks and their screening for positive clones. Notwithstanding the difficulties involved in such techniques, it may be possible, in the near future, to realize transformed cells with high biosynthetic potential for specific secondary metabolites.

ACKNOWLEDGMENT

The author is extremely grateful to the many researchers who provided reprints, prepublication copies of manuscripts, or personal communications. Research work cited from the author's laboratory was supported in part by grants from the Natural Sciences and Engineering Research Council of Canada and the Department of Higher Education, Government of Quebec.

REFERENCES

Ayabe, S., Kobayashi, M., Hikishi, M., Matsumoto, K., and Furuya, T. (1980). Flavonoids from the cultured cells of *Glycyrrhiza echinata*. *Phytochemistry* **19**, 2179–2183.

Barz, W., and Ellis, B. (1981). Plant cell cultures and their biotechnological potential. *Ber. Dtsch. Bot. Ges.* **94**, 1–26.

Bauch, H., and Leistner, E. (1978). Aromatic metabolites in cell suspension cultures of *Galium mollugo*. *Planta Med.* **33,** 105–123.

Bentley, R. (1975). Biosynthesis of quinones. *Biosynthesis* **3,** 181–246.

Berlin, J., Beiber, H., Fecker, L., Forche, E., Noe, N., Sasse, F., and Schiel, O. (1985). Conventional and new approaches to increase alkaloid production of plant cell cultures. *In* "Primary and Secondary Metabolism of Plant Cell Cultures" (K.-H. Neumann, W. Barz, and E. Reinhard, eds.), pp. 273–280. Springer-Verlag, Berlin and New York.

Böhm, H. (1980). The formation of secondary metabolites in plant tissue and cell cultures. *Int. Rev. Cytol., Suppl.* **11B,** 183–208.

Bolwell, G., Bell, J., Cramer, C., Schuch, W., Lamb, C., and Dixon, R. (1985). PAL from *Phaseolus vulgaris:* Characterization and differential induction of multiple forms from elicitor-treated cell suspension cultures. *Eur. J. Biochem.* **149,** 411–419.

Brunet, G., and Ibrahim, R. (1973). Tissue culture of citrus peel and its potential for flavonoid synthesis. *Z. Pflanzenphysiol.* **69,** 152–162.

Brunet, G., and Ibrahim, R. (1980). O-Methylation of flavonoids by cell-free extracts of calamondin orange. *Phytochemistry* **19,** 741–746.

Butcher, D. (1976). Secondary products in tissue cultures. *In* "Applied and Fundamental Aspects of Plant Cell, Tissue and Organ Culture" (J. Reinert and Y. Bajaj, eds.), pp. 668–693. Springer-Verlag, Berlin and New York.

Chandler, S., and Dodds, J. (1983). The effect of phosphate, nitrogen and sucrose on the production of phenolics and solasidine in callus cultures of *Solanum laciniatum*. *Plant Cell Rep.* **2,** 205–208.

Coljin, C., Jonsson, L., Schram, A.. and Kool, A. (1981). Synthesis of malvidin and petunidin in tissue cultures of *Petunia hybrida*. *Protoplasma* **107,** 63–68.

De-Eknamkul, W., and Ellis, B. (1984). Rosmarinic acid production and growth characteristics of *Anchusa officinalis* suspension cultures. *Planta Med.* **51,** 346–350.

De-Eknamkul, W., and Ellis, B. (1985a). Effects of macronutrients on growth and rosmarinic acid formation in cell suspension cultures of *Anchusa officinalis*. *Plant Cell Rep.* **4,** 46–49.

De-Eknamkul, W., and Ellis, B. (1985b). Effects of auxins and cytokinins on growth and rosmarinic acid formation in cell suspension cultures of *Anchusa officinalis*. *Plant Cell Rep.* **4,** 50–53.

DiCosmo, F., and Towers, G. (1984). Stress and secondary metabolism in plant cells. *In* "Phytochemical Adaptations to Stress" (B. Timmermann, C. Steelink, and F. Loewus, eds.), pp. 95–175. Plenum, New York.

Dixon, R., and Bendall, D. (1978). Changes in the levels of enzymes of phenylpropanoid and isoflavonoid synthesis during phaseollin production in cell suspension cultures of *Phaseolus vulgaris*. *Physiol. Plant Pathol.* **13,** 295–306.

Dougall, D., and Weyrauch, K. (1980). Growth and anthocyanin production by carrot suspension cultures grown under chemostat conditions with phosphate as the limiting nutrient. *Biotechnol. Bioeng.* **22,** 337–352.

Drew, S., and Demain, A. (1977). Effects of primary metabolites on secondary metabolism. *Annu. Rev. Microbiol.* **31,** 343–356.

Duell-Pfaff, N., and Wellmann, E. (1982). Involvement of phytochrome and a blue light photoreceptor in UV-B induced flavonoid synthesis in parsley cell suspension cultures. *Planta* **156,** 213–217.

Ebel, J., and Hahlbrock, K. (1977). Enzymes of flavone and flavonol glycoside biosynthesis: Coordinated and selective induction in cell suspension cultures of *Petroselinum hortense*. *Eur. J. Biochem.* **75,** 201–209.

Ebel, J., Ayers, A., and Albersheim, P. (1976). Responses of suspension cultured soybean cells to the elicitor isolated from *Phytophthora megasperma* var. *sojae*, a fungal pathogen of soybean. *Plant Physiol.* **57,** 775–779.

Ellis, B. (1983). Production of hydroxyphenylethanol glycosides in suspension cultures of *Syringa vulgaris. Phytochemistry* **22,** 1941–1943.

Fritsch, H., and Grisebach, H. (1975). Biosynthesis of cyanidin in cell cultures of *Haplopappus gracilis. Phytochemistry* **14,** 2437–2442.

Fritzmeier, K., Rolfs, C., Pfau, J., and Kindl, H. (1983). Action of ultraviolet C on stilbene formation in callus of *Arachis hypogea. Planta* **159,** 25–29.

Fujita, Y., Hara, Y., Ogino, T., and Suga, C. (1981a). Production of shikonin derivatives by cell suspension cultures of *Lithospermum erythrorhizon*. I. Effects of nitrogen sources on production of shikonin derivatives. *Plant Cell Rep.* **1,** 59–60.

Fujita, Y., Hara, Y., Suga, C., and Morimoto, T. (1981b). Production of shikonin derivatives by cell suspension cultures of *Lithospermum erythrorhizon*. II. A new medium for the production of shikonin derivatives. *Plant Cell Rep.* **1,** 61–63.

Fukui, H., Yazaki, K., and Tabata, M. (1984). Two phenolic acids from *Lithospermum erythrorhizon* cell suspension cultures. *Phytochemistry* **23,** 2398–2399.

Hahlbrock, K. (1977). Regulatory aspects of phenylpropanoid biosynthesis in cell cultures. *In* "Plant Tissue Culture and Its Biotechnological Application" (W. Barz, E. Reinhard, and M. Zenk, eds.), pp. 95–111. Springer-Verlag, Berlin and New York.

Hahlbrock, K., and Schröder. J. (1975). Specific effects of enzyme activities upon dilution of *Petroselinum hortense* into water. *Arch. Biochem. Biophys.* **171,** 500–506.

Hahlbrock, K., Knobloch, K., Kreuzaler, F., Potts, R., and Wellmann, E. (1976). Coordinated induction and subsequent activity changes of two groups of metabolically interrelated enzymes. Light-induced synthesis of flavonoid glycosides in cell suspension cultures of *Petroselinum hortense. Eur. J. Biochem.* **61,** 199–206.

Harkes, P., Krijbolder, L., Libbenga, K., Wijnsma, R., Nsengiyaremge, T., and Verpoorte, R. (1985). Influence of various media constituents on the growth of *Cinchona ledgeriana* tissue cultures and the production of alkaloids and anthraquinones. *Plant Cell, Tissue Organ Cult.* **4,** 199–214.

Hattori, T., and Ohata, Y. (1985). Induction of PAL activation and isoflavone glucoside accumulation in suspension-cultured cells of red bean, *Vigna angularis*, by phytoalexin elicitors, vanadate and elevation of medium pH. *Plant Cell Physiol.* **26,** 1101–1110.

Heble, M., Narayanaswamy, S., and Chadha, M. (1974). Tissue differentiation and plumbagin synthesis in variant cell strains of *Plumbago zeylanica. Plant Sci. Lett.* **2,** 405–409.

Heinzmann, U., and Seitz, U. (1977). Synthesis of PAL in anthocyanin-containing and anthocyanin-free callus cells of *Daucus carota* L. *Planta* **135,** 63–67.

Heinzmann, U., Seitz, U., and Seitz, U. (1977). Purification and substrate specificities of hydroxycinnamate:CoA ligase from anthocyanin-containing and anthocyanin-free carrot cells. *Planta* **135,** 313–318.

Hinderer, W., and Seitz, H. (1985). Chalcone synthase from cell suspension cultures of *Daucus carota. Arch. Biochem. Biophys.* **240,** 265–272.

Hinderer, W., Petersen, M., and Seitz, H. (1984). Inhibition of flavonoid biosynthesis by gibberellic acid in cell suspension cultures of *Daucus carota*. L. *Planta* **160,** 544–549.

Hösel, W., Fiedler-Preiss, A., and Borgmann, E. (1982). Regulation of coniferin β-glucosidase to lignification in various plant cell suspension cultures. *Plant Cell, Tissue Organ Cult.* **1,** 137–148.

Hrazdina, G., Kreuzaler, F., Hahlbrock, K., and Grisebach, H. (1976). Substrate specificity of flavanone synthase from cell suspension cultures of parsley and structure of release products *in vitro. Arch. Biochem. Biophys.* **175,** 392–399.

Ibrahim, R., and Boulay, B. (1980). Purification and some properties of UDPglucose : *o*-dihydroxycoumarin 7-*O*-glucosyltransferase from tobacco cell cultures. *Plant Sci. Lett.* **18,** 177–184.

Ibrahim, R., and Edgar, D. (1976). Phenolic synthesis in *Perilla* cell suspension cultures. *Phytochemistry* **15,** 129–131.

Ibrahim, R., and Grisebach, H. (1976). Purification and properties of UDPglucose : coniferyl alcohol glucosyltransferase from suspension cultures of Paul's Scarlet rose. *Arch. Biochem. Biophys.* **176,** 700–706.

Inoue, K., Shiobara, Y., Nayeshiro, H., Inouye, H., Wilson, G., and Zenk, M. (1984). Biosynthesis of anthraquinones and related compounds in *Galium mollugo* cell suspension cultures. *Phytochemistry* **23,** 307–311.

Jessup, W., and Fowler, M. (1976). Interrelationship between carbohydrate metabolism and nitrogen assimilation in cultured plant cells. I. Effects of glutamate and nitrate as alternative nitrogen sources on cell growth. *Planta* **132,** 119–123.

Kadkade, P. (1982). Growth and podophyllotoxin production in callus tissues of *Podophyllum peltatum*. *Plant Sci. Lett.* **25,** 107–115.

Khouri, H., and Ibrahim, R. K. (1987). Purification and some properties of five anthraquinone-specific glucosyltransferases from *Cinchona succirubra* cell suspension cultures. *Phytochemistry* **26** (in press).

Khouri, H., Ibrahim, R. K., and Rideau, M. (1986). Effects of nutritional and hormonal factors on growth and production of anthraquinone glucosides in cell suspension cultures of *Cinchona succirubra*. *Plant Cell Rep.* **5,** 423–426.

King, P. (1976). Utilization of 2,4-D by steady-state cell cultures of *Acer pseudoplatanus*. *J. Exp. Bot.* **27,** 1053–1072.

Kinnersley, A., and Dougall, D. (1980). Increase in anthocyanin yield from wild carrot cell cultures by a selection method based on cell-aggregate size. *Planta* **149,** 200–204.

Knobloch, K., and Berlin, J. (1980). Influence of medium composition on formation of secondary compounds in cell suspension cultures of *Catharanthus roseus* *Z. Naturforsch., C: Biosci.* **35C,** 551–556.

Knobloch, K., and Berlin, J. (1981). Phosphate-mediated regulation of cinnamoyl putrescine biosynthesis in cell suspension cultures of *Nicotiana tabacum*. *Planta Med.* **42,** 167–172.

Knobloch, K., and Hahlbrock, K. (1975). Isoenzymes of *p*-coumarate : CoA ligase from cell suspension of *Glycine max*. *Eur. J. Biochem.* **52,** 311–320.

Knobloch, K., Bast, G., and Berlin, J. (1982). Medium-induced and light-induced formation of serpentine and anthocyanins in cell suspension cultures of *Catharanthus roseus*. *Phytochemistry* **21,** 591–594.

Koumba, D., and Macheix, J. (1981). Biosynthesis of hydroxycinnamic derivatives in apple fruit cell suspension culture. *Physiol. Veg.* **20,** 137–142.

Kreuzaler, F., and Hahlbrock, K. (1973). Flavonoid glycosides from illuminated cell suspension cultures of *Petroselinum hortense*. *Phytochemistry* **12,** 1149–1152.

Kuboi, T., and Yamada, Y. (1976). Caffeic acid *O*-methyltransferase in a suspension of cell aggregates of tobacco. *Phytochemistry* **15,** 397–400.

Lam, T., and Street, H. (1977). The effects of selected aryloxyalkanecarboxylic acids on growth and levels of soluble phenols in cultured cells of *Rosa damascena*. *Z. Pflanzenphysiol.* **84,** 121–128.

Leistner, E. (1973). Biosynthesis of morindone and alizarin in intact plants and cell suspension cultures of *Morinda citrifolia*. *Phytochemistry* **12,** 1669–1674.

Li, H., Rice, E., Rohrbaugh, L., and Wender, S. (1970). Effects of abscisic acid on phenolic content and lignin biosynthesis in tobacco tissue culture. *Physiol. Plant.* **23,** 928–936.

Lipetz, J. (1962). Calcium and the control of lignification in tissue cultures. *Am. J. Bot.* **49,** 460–464.
Macheix, J., and Ibrahim. R. (1984). The *O*-methyltransferase system of apple fruit cell suspension culture. *Biochem. Physiol. Pflanz.* **179,** 659–664.
Matern, U., Potts, R., and Hahlbrock, K. (1981). Partial purification and some properties of malonyl CoA : flavone/flavonol 7-*O*-glycoside malonyltransferase and malonyl CoA : flavonol 3-*O*-glucoside malonyltransferase from cell suspension cultures of *Petroselinum hortense. Arch. Biochem. Biophys.* **208,** 233–241.
Matern, U., Feser, C., and Hammer, D. (1983). Further characterization and regulation of malonyl CoA : flavonoid glucoside malonyltransferase from parsley cell suspension cultures. *Arch. Biochem. Biophys.* **226,** 206–217.
Mizukami, H., Konoshima, M., and Tabata, M. (1977). Effect of nutritional factors on shikonin derivative formation in *Lithospermum* callus cultures. *Phytochemistry* **16,** 1183–1186.
Mizusaki, S., Tanabe, T., Noguchi, M., and Tamaki, E. (1971). *p*-Coumaroyl-putrescine, caffeoyl-putrescine and feruloyl-putrescine from callus cultures of *Nicotiana tabacum. Phytochemistry* **10,** 1347–1350.
Moehle, B., Heller, W., and Wellmann, E. (1985). UV-induced biosynthesis of quercetin-3-glucuronide in dill (*Anethum graveolens*). *Phytochemistry* **24.** 465–468.
Muhitch, M. J., and Fletcher, J. S. (1985). Influence of culture age and spermidine treatment on the accumulation of phenolic compounds in suspension cultures. *Plant Physiol.* **78,** 25–28.
Mulder-Krieger, T., Verpoorte, R., Water, A., Gessel, M., Overten, B., and Barheim, S. (1982). Identification of the alkaloids and anthraquinones in *Cinchona ledgeriana* callus cultures. *Planta Med.* **46,** 19–24.
Nimz, H., Ebel, J., and Grisebach, H. (1975). On the structure of lignin from soybean suspension cultures. *Z. Naturforsch., C: Biosci.* **30C,** 442–445.
Nishimaki, T., and Nozue, M. (1985). Isolation and culture of protoplasts from high anthocyanin-producing callus of sweet potato. *Plant Cell Rep.* **4,** 248–251.
Noe, W., and Seitz, H. (1982). Induction of *de novo* synthesis of PAL by α-aminooxy-β-phenylpropionic acid in suspension cultures of *Daucus carota. Planta* **154,** 454–458.
Okazaki, M., Hino, F., Nagasawa, K., and Miura, Y. (1982). Effects of nutritional factors on formation of scopoletin and scopolin in tobacco tissue cultures. *Agric. Biol. Chem.* **46,** 601–607.
Ozeki, Y., and Komamine, A. (1981). Induction of anthocyanin synthesis in relation to embryogenesis in a carrot suspension culture. *Physiol. Plant.* **53,** 570–577.
Ozeki, Y., and Komamine, A. (1985). Changes in activities of enzymes involved in general phenylpropanoid metabolism during induction of anthocyanin synthesis in carrot suspension culture as regulated by 2,4-D. *Plant Cell Physiol.* **26,** 903–911.
Ozeki, Y., Sakano, K., Komamine, A., Tanaka, Y., Noguchi, H., Sanakawa, U., and Suzuki, T. (1985). Purification and some properties of chalcone synthase from carrot suspension culture induced for anthocyanin synthesis and preparation of its specific antiserum. *J. Biochem. (Tokyo)* **98,** 9–17.
Poulton, J., and Kauer, M. (1977). Identification of an UDPglucose : flavonol 3-*O*-glucosyltransferase from cell suspension cultures of soybean (*Glycine max*). *Planta* **136,** 53–59.
Poulton, J., Hahlbrock, K., and Grisebach, H. (1976). Purification and properties of SAM : caffeic acid 3-*O*-methyltransferase from soybean cell suspension cultures. *Arch. Biochem. Biophys.* **176,** 449–456.
Poulton, J., Hahlbrock, K., and Grisebach, H. (1977). O-Methylation of flavonoid sub-

strates by a partially purified enzyme from soybean cell suspension cultures. *Arch. Biochem. Biophys.* **180,** 543–549.

Pryke, J., and Ap Rees, T. (1977). The pentose phosphate pathway as a source of NADPH for lignin biosynthesis. *Phytochemistry* **16,** 557–560.

Rains, D. (1976). Mineral metabolism. *In* "Plant Biochemistry" (J. Bonner and E. Varner, eds.), 3rd ed., pp. 561–597. Academic Press, New York.

Razzaque, A., and Ellis, B. (1977). Rosmarinic acid production in *Coleus* cell cultures. *Planta* **137,** 287–291.

Sahai, O., and Shuler, M. (1984). Environmental parameters influencing phenolic production by batch cultures of *Nicotiana tabacum. Biotechnol. Bioeng.* **26,** 111–120.

Saleh, N., Fritsch, H., Kreuzaler, F., and Grisebach, H. (1978). Flavanone synthase from cell suspension cultures of *Haplopappus gracilis* and comparison with the synthase from parsley. *Phytochemistry* **17,** 183–186.

Samejima, M., Yamaguchi, T., Fukuzumi, T., and Yoshimoto, T. (1982). Effects of phytohormones on accumulation of flavanols in callus cultures of woody plants *In* "Plant Tissue Culture 1982" (A. Fujiwara, ed.), pp. 353–354. Maruzen, Tokyo.

Schiel, O.. Jarchow-Redecker, K., Piehl, G., Lehmann, J., and Berlin, J. (1984). Increased formation of cinnamoyl putrescines by fedbatch fermentation of cell suspension cultures of *Nicotiana tabacum. Plant Cell Rep.* **3,** 18–20.

Schopfer, P. (1977). Phytochrome control of enzymes. *Annu. Rev. Plant Physiol.* **28,** 233–252.

Schrall, R., and Becker, H. (1977). Tissue and suspension cultures of *Silybum marianum:* Formation of flavonolignans by feeding suspension cultures with flavonoids and coniferyl alcohol. *Planta Med.* **32,** 27–32.

Schulte, U., El-Shagi, H., and Zenk, M. (1984). Optimization of 19 Rubiaceae species in cell culture for the production of anthraquinones. *Plant Cell Rep.* **3,** 51–54.

Shah, R., Subbaiah, K., and Mehta, A. (1976). Hormonal effect on polyphenol accumulation in *Cassia* tissue cultured *in vitro. Can. J. Bot.* **54,** 1240–1245.

Smith, S., Slywka, G., and Kreuger, G. (1981). Anthocyanins of *Strobilanthes dyeriana* and their production in callus culture. *J. Nat. Prod.* **44,** 609–610.

Staba, E. J., ed. (1980). "Plant Tissue Culture as a Source of Biochemicals." CRC Press, Boca Raton, Florida.

Stark, V., Alfermann, A., and Reinhard, E. (1976). PAL activity and biosynthesis of anthocyanins and chlorogenic acid in tissue cultures of *Daucus carota. Planta Med.* **30,** 104–117.

Sugano, N., Iwata, R., and Nishi, A. (1975). Formation of phenolic acids in carrot cell suspension culture. *Phytochemistry* **14,** 1205–1207.

Suzuki, H., Matsumoto, T., Kisaki, T., and Noguchi, M. (1981). Influence of cultural conditions on polyphenol formation and growth of Amacha cells (*Hydrangea macrophylla*). *Agric. Biol. Chem.* **45,** 1067–1077.

Suzuki, H., Matsumoto, T., and Mikami, Y. (1984). Effects of nutritional factors on the formation of anthraquinones by *Rubia cordifolia* cells in suspension culture. *Agric. Biol. Chem.* **48,** 603–610.

Tabata, M., Mizukami, H., Hiraoka, N., and Konoshima, M. (1974). Pigment formation in callus cultures of *Lithospermum erythrorhizon. Phytochemistry* **13,** 927–932.

Tabata, M., Hiraoka, N., Ikenoue, M., Sano, Y., and Konoshima, M. (1975). The production of anthraquinones in callus cultures of *Cassia tora. Lloydia* **38,** 131–134.

Treutter, D., Galensa, R.. Feucht, W., and Schmid, P. (1985). Flavanone glucosides of *Prunus avium:* Identification and stimulation of their synthesis. *Physiol. Plant.* **65,** 95–101.

Tsang, Y., and Ibrahim, R. (1979a). Two forms of *O*-methyltransferase in tobacco cell suspension culture. *Phytochemistry* **18,** 1131–1136.

Tsang, Y., and Ibrahim, R. (1979b). Evidence for the existence of meta and para directing *O*-methyltransferases in tobacco cell cultures. *Z. Naturforsch., C: Biosci.* **34C,** 46–50.

Van den Berg, A., and Labadie, R. (1981). The production of acetate-derived hydroxyanthraquinones, -dianthrones, -naphthalenes, and -benzenes in tissue cultures of *Rumex alpinum*. *Planta Med.* **41,** 169–173.

Wellmann, E., and Baron, D. (1974). Durch Phytochrom Kontrollierte Enzyme der Flavonoidsynthese in Zellsuspensionkulturen von *Petroselinum hortense*. *Planta* **119,** 161–164.

Wengenmayer, H., Ebel, J., Grisebach, H. (1974). Purification and properties of *S*-adenosylmethionine : isoflavone 4'-*O*-methyltransferase from cell suspension cultures of *Cicer arietinum* L. *Eur. J. Biochem.* **50,** 135–143.

Wengenmayer, H., Ebel, J., and Grisebach, H. (1976). Enzymic synthesis of lignin precursors. Purification and properties of a cinnamoyl CoA : NADPH reductase from cell suspension cultures of soybean (*Glycine max*). *Eur. J. Biochem.* **65,** 529–536.

Westcott, R. (1976). Changes in phenolic metabolism of suspension cultures of *Acer pseudoplatanus* caused by the addition of 2-(chloroethyl)phosphonic acid. *Planta* **131,** 209–210.

Westcott, R., and Henshaw, G. (1976). Phenolic synthesis and PAL activity in suspension cultures of *Acer pseudoplatanus*. *Planta* **131,** 67–73.

Wijnsma, R., Verpoorte, R., Mulder-Krieger, T., and Baerheim, S. (1984). Anthraquinones in callus cultures of *Cinchona ledgeriana*. *Phytochemistry* **23,** 2307–2311.

Wilson, G., and Balagué, C. (1985). Biosynthesis of anthraquinones by cells of *Galium mollugo* with limiting sucrose or phosphate. *J. Exp. Bot.* **36,** 485–493.

Wray, V., Schiel, O., and Berlin, J. (1984). High field ^{31}P-NMR investigation of the phosphate metabolites in cell suspension cultures of *Nicotiana tabacum*. *J. Plant Physiol.* **112,** 215–220.

Wyrambik, D., and Grisebach, H. (1975). Purification and properties of isozymes of cinnamyl alcohol dehydrogenase from soybean cell suspension cultures. *Eur. J. Biochem.* **59,** 9–15.

Yamakawa, T., Kato, S., Ishida, K., Kodama, T., and Minoda, Y. (1983). Production of anthocyanins by *Vitis* cells in suspension culture. *Agric. Biol. Chem.* **47,** 2185–2191.

Yoshikawa, N., Fukui, H., and Tabata, M. (1986). Effect of GA_3 on shikonin production in *Lithospermum* callus cultures. *Phytochemistry* **25,** 621–622.

Zenk, M. (1978). The impact of plant tissue culture on industry. *In* "Frontiers of Plant Tissue Culture 1978" (T. A. Thorpe, ed.), pp. 1–13. Univ. of Calgary, Calgary, Canada.

Zenk, M., and Leistner, E. (1968). Biosynthesis of quinones. *Lloydia* **31,** 275–292.

Zenk, M., El-Shagi, H., and Schulte, U. (1975). Anthraquinone production by cell suspension cultures of *Morinda citrifolia*. *Planta Med., Suppl.,* pp. 79–101.

Zenk, M., El-Shagi, H., and Ulbrich, B. (1977). Production of rosmarinic acid by cell suspension cultures of *Coleus blumei*. *Naturwissenschaften* **64,** 585–586.

Zenk, M., Schulte, U., and El-Shagi, H. (1984). Regulation of anthraquinone formation by phenoxyacetic acids in *Morinda* cell cultures. *Naturwissenschaften* **71,** 266.

CHAPTER 5

Cell Growth and Accumulation of Secondary Metabolites

Masaaki Sakuta[1]
Atsushi Komamine

Biological Institute
Faculty of Science
Tohoku University
Sendai 980, Japan

I. INTRODUCTION

It is well known that secondary metabolism is expressed in specific tissues and cells at specific stages of growth in higher plants, implying that the expression of secondary metabolism in higher plants is closely correlated with the growth and morphological differentiation of cells. Growth may be divided into two phases: increase in cell number (cell division) and increase in cell volume (cell expansion or cell elongation). In most cases of growth in intact plants these two phenomena occur simultaneously. Furthermore, growth patterns and growth rates are different in different tissues. It is therefore difficult to analyze the relationship between growth and secondary metabolism in intact plants.

In recent years, suspension cultures of plant cells have been used as model systems for studying the regulatory mechanisms of secondary metabolism in higher plants. Suspension cultures are suitable for studying the regulation of secondary metabolism in relation to growth, since cells cultured in this manner are relatively homogeneous and environmental conditions can be easily controlled. Moreover, it is possible to analyze the relationship between secondary metabolite production and growth in terms of cell number and cell morphology. The aim of this

[1]Present address: Advanced Research Laboratory, Hitachi Ltd., Kokubunji, Tokyo 185, Japan.

chapter is to outline how and to what extent the method of cell culture has advanced our knowledge of the relationship between growth and the accumulation of secondary metabolites and to describe the contribution of knowledge obtained to the applied aspects of secondary metabolite production.

II. FACTORS CONTROLLING GROWTH IN PLANT CELL CULTURES

A. Growth Patterns and Characteristics of Growth Phases in Batch-Grown Suspension Cultures

In recent years, suspension cultures have been used more frequently than callus cultures for metabolic studies on cultured plant cells for reasons of homogeneity and superior growth rate. Fresh weight, dry weight, packed cell volume, and cell number are widely used as growth parameters for plant cell suspension cultures. Cell number is usually estimated by counting protoplasts and/or cells using a hemocytometer after enzymatic maceration using pectinase and cellulase or chemical maceration with a mixture of 1% HCl and 1% chromic acid. The growth kinetics of suspension culture plant cells are essentially the same as those of microorganisms. Batch-grown suspension cultures of plant cells exhibit sigmoidal growth in three phases, i.e., the lag, logarithmic (log, exponential), and stationary phases. In general, suspension cultures in the logarithmic phase consist of small, round cells or cell clusters which develop to form larger cell aggregates as a consequence of active cell division, while most cells in cell clusters in the lag or stationary phases are composed of expanded or elongated large cells.

Cells undergo not only morphological changes during culture but also intracellular metabolic changes. Respiratory activity per cell reaches a maximum in the late lag phase in suspension cultures of sycamore (*Acer pseudoplatanus*) (Shimizu *et al.*, 1977). The maximum activity of phosphofructokinase (PFK), a key regulatory enzyme in glycolysis (Givan, 1968), also occurs in the late lag stage in sycamore suspension cultures (Fowler, 1971). However, the maxima of other glycolytic enzymes are not always observed in the same phase as PFK. In contrast, the activities of all enzymes involved in the pentose phosphate pathway attain maxima during the early logarithmic phase (Fowler, 1971). Maximal production of NADPH coincides with maximum pentose phosphate pathway activity. Protein content per cell reaches a maximum during the logarithmic phase. The lag phase can be regarded as an energy-producing phase and the logarithmic phase as a biosynthetic phase (Shimizu *et al.*, 1977).

The pattern of growth and the length of each phase largely depend on subculture interval and inoculum size. When cells in the stationary phase are subcultured, the culture passes through lag and logarithmic phases and then enters a stationary phase. When cells in the logarithmic phase are transferred to fresh medium, however, they immediately begin exponential growth without an intervening lag phase. The growth of suspension cultures is also affected by inoculum size (Nash and Davies, 1972; Ozeki and Komamine, 1985). When a relatively small number of cells are inoculated into fresh medium, the lag phase becomes longer and the tangent at logarithmic phase on the growth curve is diminished. In contrast, high cell density at subculture brings about a limited number of cell divisions before the stationary phase is entered. Optimal inoculum size differs between cultures and depends on the species and culture conditions.

B. Effect of Medium Composition on Growth

It is important to improve the growth rates of plant cell cultures, since these are much lower than those of microorganisms, leading to low productivity in plant culture systems. A variety of culture media have been reported to support the growth of cultured cells. Suspension culture media usually contain sucrose as a carbon source, NO_3^- and/or NH_4^+ as a nitrogen source, PO_4^{3-} as a phosphate source, SO_4^{2-} as a sulfur source, K^+, Ca^{2+}, and Mg^{2+} as macronutrients, micronutrients essential for growth, and plant growth regulators. Among these various components, plant growth regulators play an important role in the growth of suspension culture cells. In general, auxin and/or cytokinin are added to culture media. The addition of auxin is essential to induce and maintain growth in cultured cells, except for crown gall and habituated cells. 2,4-Dichlorophenoxyacetic acid (2,4-D) and naphthaleneacetic acid (NAA), both synthetic auxins, are used widely since they are more stable than indoleacetic acid (IAA), a naturally occurring auxin which is easily degraded by oxidase (Schneider and Wightman, 1974). The requirement for an exogenous supply of cytokinin depends on the plant species. The effects of plant growth regulators on growth were reviewed by Dougall (1980).

Generally, suspension culture plant cells do not posess photosynthetic ability except for photoautotrophic cultures (Chandler *et al.*, 1972; Berlyn and Zelitch, 1975; Hüsemann and Barz, 1977). Therefore, the addition of a carbon source is essential for growth. Sucrose and its component monosaccharides, glucose and fructose, are commonly used as carbon sources. The sugar concentration giving maximal growth differs among species, but growth rate and final biomass are affected by the initial sucrose concentration in the medium. Sucrose can be a factor limiting growth in the stationary phase of normal plant cell cultures.

Nitrate and ammonium are used as nitrogen sources in plant cell culture media. Ammonium alone gives poor growth compared with nitrate alone (Dougall, 1980). Media containing both nitrate and ammonium are widely used, since the addition of both nitrogen sources gives superior growth (Takayama *et al.*, 1977). The effects of media with a wide range of ammonium to nitrate ratios have been reported. In the case of media containing both nitrogen sources, ammonium is preferentially used in the early stages of growth, and when exhausted nitrate utilization begins (Wilson and Marron, 1978; De-Eknamkul and Ellis, 1984).

Phosphorus is an important nutritional element for plant cells since it is required for nucleic acid metabolism and energy metabolism as a component of UDP, ADP, ATP, etc. Accordingly, phosphate concentration can also be a factor limiting growth in plant cell cultures. In fact, a linear relationship between initial phosphate concentration (0–2.5 m*M*) and cell number in the stationary phase was observed in suspension cultures of *Catharanthus roseus* (Amino *et al.*, 1983). In suspension cultures of *Agrostemma githago,* phosphate also stimulated growth in the range 2.5–5 m*M* (Takayama *et al.*, 1977).

C. Synchronous Suspension Cultures

Synchronous growth systems are useful tools not only for understanding the mechanism of cell division but also for characterizing the dynamics of metabolism in relation to growth. Consequently, the establishment of highly synchronous suspension cultures has been attempted by several workers (see also Chapter 21, Volume 1, this treatise). The induction of synchrony has been achieved by treatment with DNA synthesis inhibitors such as aphidicolin (Nagata *et al.*, 1982; Sala *et al.*, 1983), by the removal and readdition of plant growth regulators (Constabel *et al.*, 1977; Nishi *et al.*, 1977; Everett *et al.*, 1981), or by the withdrawal of some nutrients which regulate growth (King *et al.*, 1974; Gould and Street, 1975). Gould *et al.* (1981) showed that phosphate starvation prolonged the viability of cells, compared to sucrose or nitrogen starvation, in suspension cultures of *Acer pseudoplatanus.*

Amino *et al.* (1983) reported a highly synchronized cell division system for suspension cultures of *Catharanthus roseus* using the double phosphate starvation method. This is based on the observation that initial phosphate concentration in the medium limits maximum growth. The rapid incorporation of phosphate into cells on readdition of phosphate contributes to the high synchronization of this system. Synchronous culture systems will be useful for investigation of the regulatory mechanisms of secondary metabolite accumulation in relation to cell growth.

III. PRODUCTION OF SECONDARY METABOLITES AND CONTROL OF CELL GROWTH

A. Plant Growth Regulators

Secondary metabolism in higher plants is strongly influenced by environmental factors. In cell suspension cultures, plant growth regulators and nutritional factors affect the production of secondary metabolites as well as growth. Different types of growth regulators, i.e., auxin, cytokinin, gibberellin, abscisic acid, are known to show different effects on growth and secondary metabolism. Various effects of auxins on secondary metabolism have been reported; in many cases, synthetic auxins, especially 2,4-D, inhibit the production of secondary metabolites. For instance, 2,4-D inhibited nicotine synthesis in *Nicotiana tabacum* cultures (Furuya *et al.*, 1971; Tabata *et al.*, 1971), shikonin derivatives in *Lithospermum erythrorhizon* cultures (Tabata *et al.*, 1974), anthraquinones in *Morinda citrifolia* cultures (Zenk *et al.*, 1975), phenolic compounds in *Acer pseudoplatanus* cultures (Phillips and Henshaw, 1977), L-dihydroxyphenylalanine (L-dopa) in *Stizolobium hassjoo* cultures (Obata-Sasamoto and Komamine, 1983), and ferruginol in *Salvia miltiorrhiza* cultures (Miyasaka *et al.*, 1985).

Ozeki and Komamine (1981) described the induction of anthocyanin synthesis in carrot suspension cultures in which small cell aggregates (31–81 μm) were selected by sieving through nylon screens. The elimination of auxin from the medium was essential for anthocyanin induction. When cells were transfered from a medium containing 2,4-D to one lacking 2,4-D, suppression of cell division and initiation of cell elongation coincided with induction of anthocyanin formation. The readdition of 2,4-D simultaneously caused elongated cells to divide and anthocyanin to disappear. This suggests that anthocyanin formation correlates with cell growth and cell morphology. In contrast, stimulation of secondary metabolite synthesis by 2,4-D was reported for carotenoid synthesis in *Daucus carota* cultures (Mok *et al.*, 1976). Berberine production in suspension cultures of *Thalictrum minus* is also enhanced by 2,4-D and NAA (Nakagawa *et al.*, 1986). In some cases, the synthetic auxin NAA and the natural auxin IAA either have no effect or positive effects on secondary metabolite synthesis whereas 2,4-D has a negative effect. Endress (1976) showed that betacyanin accumulation is stimulated by NAA in callus cultures of *Portulaca.* It has also been shown that NAA and IAA stimulate anthocyanin formation in suspension cultures of *Populus* (Matsumoto *et al.*, 1973).

Cytokinins have different effects depending on the type of metabolite and the species concerned. For example, kinetin stimulated L-dopa accumulation in callus culture of *Stizolobium hassjoo* (Obata-Sasamoto and Komamine, 1983). Berberine production in suspension cultures of *Thalictrum minus* (Nakagawa *et al.*, 1984) and anthocyanin accumulation in suspension cultures of *Happlopappus gracilis* (Constabel *et al.*, 1971) are also promoted by 6-benzyladenine (BA) and kinetin. In contrast, kinetin inhibits the formation of anthocyanin in suspension cultures of *Populus* (Matsumoto *et al.*, 1973) and carotenoid synthesis in carrot cultures (Mok *et al.*, 1976).

Gibberellin inhibits shikonin production in callus cultures of *Lithospermum* but does not affect growth (Yoshikawa *et al.*, 1986). Gibberellic acid and abscisic acid suppressed the accumulation of anthocyanin in carrot suspension cultures (Ozeki and Komamine, 1986) and betacyanin accumulation in *Phytolacca americana* (M. Sakuta and A. Komamine, unpublished data). In both cases, gibberellic acid had no effect on growth while abscisic acid inhibited growth.

As described above, the effects of plant growth regulators on the production of secondary metabolites are complex and rather contradictory. For example, kinetin stimulated the formation of anthocyanin in *Haplopappus gracilis* (Constabel *et al.*, 1971) and carrot suspension cultures (Ozeki and Komamine, 1981) but inhibited its formation in *Pop-*

ulus (Matsumoto *et al.*, 1973). These contradictory effects are difficult to explain, but the endogenous levels of plant growth regulators in the cultured cells must be considered. It has been reported that larger cell aggregates contain higher levels of endogenous cytokinin than smaller cell aggregates (Szweykowska, 1975). Relationships between cell aggregate size and secondary metabolite accumulation have been demonstrated (Kinnersley and Dougall, 1980; Ozeki and Komamine, 1981; Franke and Böhm, 1982; Sakuta *et al.*, 1986). It has been pointed out that cell-to-cell interaction and gradients of substances such as plant growth regulators may play important roles in morphological and metabolic differentiation as well as growth (Ozeki and Komamine, 1981; Lindsey and Yeoman, 1983).

B. Nutrition

Nutritional factors play an important role in the regulation of secondary metabolism in plants. It is well known that sucrose as a carbon source can support growth in callus and suspension cultures of plant cells. Dougall (1980) showed that sucrose concentration in the medium affects not only growth but also the production of secondary metabolites. The formation of shikonin derivatives in callus cultures of *Lithospermum erythrorhizon* increased with increasing sucrose concentration in the range 1–5% (Mizukami *et al.*, 1977). The accumulation levels of shikonin derivatives remained almost constant for sucrose concentrations in the range 7–10%. while maximum growth occurred at 5%. The accumulation of indole alkaloids (serpentine, ajmalicine) was enhanced by increasing the sucrose concentration in suspension cultures of *Catharanthus roseus* (Zenk *et al.*, 1977). The enhancement of production of other secondary metabolites by sucrose has also been observed: anthocyanin formation in *Populus* (Matsumoto *et al.*, 1973), polyphenol synthesis in Paul's Scarlet Rose (Davies, 1972), diosgenin production in suspension cultures of *Dioscorea deltoidea* (Tal and Goldberg, 1982), anthraquinone synthesis in continuous chemostat cultures of *Galium mollugo* (Wilson and Balagué, 1985), and anthocyanin synthesis in carrot suspension cultures (Ozeki and Komamine, 1985). On the other hand, increased sucrose levels inhibited the formation of ubiquinone in suspension cultures of *Nicotiana tabacum* (Ikeda *et al.*, 1976).

Concerning the mechanism behind these effects of sucrose on secondary production, increased osmotic pressure should be considered in addition to the role of sucrose as a carbon source. Kimball *et al.* (1975)

showed that cell size decreases with increasing sucrose, glucose, mannitol, or sorbitol concentration in soybean cell cultures. In suspension cultures of *Phytolacca americana*, experiments in which mannitol replaced sucrose indicated that sucrose itself caused an increase in cell number, but cell size was decreased by the effects of both sucrose concentration and osmotic pressure. Betacyanin accumulation in *Phytolacca* cultures decreased with increasing osmotic pressure (M. Sakuta, T. Takagi, and A. Komamine, unpublished data). In general, cell size decreases with increasing osmotic pressure (Kimball *et al.*, 1975; Takayama *et al.*, 1977). However, definitive data indicating a clear-cut effect of osmotic pressure on secondary metabolite accumulation are lacking.

Yamakawa *et al.* (1983) showed that anthocyanin production in suspension cultures of *Vitis* was influenced by the inorganic nitrogen/sucrose ratio of the culture medium. Tal *et al.* (1982) also demonstrated that the carbon-to-nitrogen ratio of the medium was an important factor influencing diosgenin production in *Dioscorea deltoidea*. Knobloch and Berlin (1981) indicated that a nitrogen source was necessary for optimal formation of cinnamoyl putrescines in *Nicotiana tabacum*. The formation of shikonin derivatives in callus cultures of *Lithospermum erythrorhizon* increased with increasing total nitrogen concentration in the range 67–104 m*M* and decreased for nitrogen concentrations higher than 134 m*M* (Mizukami *et al.*, 1977). Davies (1972) showed that polyphenol accumulation at 20 m*M* nitrate was less than at 10 m*M* nitrate in suspension cultures of Paul's Scarlet Rose. Leucoanthocyanin accumulation in the same suspension cultures also decreased with increasing nitrate concentration in media containing 0.2 m*M* glucose (Amorin *et al.*, 1977).

In *Dioscorea deltoidea*, both ammonium and nitrate were required for growth and diosgenin production, which was maximal at higher amounts of both nitrogen sources (Tal and Goldberg, 1982; Tal *et al.*, 1982). In contrast to these results, good growth and ubiquinone formation occurred in the media with nitrate as the sole nitrogen source (Ikeda *et al.*, 1977). The authors also found that the ratio of ammonium to nitrate affected ubiquinone formation in tobacco cell suspension cultures; higher ratios of ammonium to nitrate increased ubiquinone accumulation and decreased growth. Nakagawa *et al.* (1984) showed that the highest yield of berberine in *Thalictrum minus* was obtained when the ratio of nitrate to ammonium was 1 : 2, while maximal growth was observed at a ratio of 5 : 1 (nitrate : ammonium). On the other hand, Fujita *et al.* (1981a) showed that decreasing the ammonium concentration increased the yield of shikonin derivatives while nitrate stimulated it in *Lithospermum erythrorhizon*. On the basis of this finding, they established a new medium suitable for shikonin production containing nitrate as a sole nitrogen source (Fujita *et al.*, 1981b).

As shown in synchronous cultures, the phosphate concentration in the medium can be an important factor controlling the growth of plant cell cultures. Phosphate also affects the production of secondary metabolites. Obata-Sasamoto and Komamine (1983) showed that media in which growth was limited by low phosphate were favorable for the production of L-dopa in callus cultures of *Stizolobium hassjoo*. Knobloch and Berlin (1983) also demonstrated that the accumulation of secondary products (alkaloids and phenolics) could be stimulated by growth-limiting culture media lacking phosphate. In contrast to these results, increased phosphate levels stimulated the formation of digitoxin in *Digitalis purpurea* (Hagimori *et al.*, 1982). Betacyanin accumulation in *Phytolacca americana* was also promoted by phosphate. Betacyanin accumulation was closely related to cell division, and maximum betacyanin accumulation was observed during the logarithmic phase of growth (Sakuta *et al.*, 1986).

C. Two-Phase Culture

In recent years, two-phase culture and immobilized cell culture have been developed as methods for the production of secondary metabolites. An inverse relationship between growth and the accumulation of secondary metabolites has been reported for many plant cell cultures. Two-phase culture is based on the principle that a high yield of secondary products can be expected when cells grown in a medium suitable for growth are transferred to a medium suitable for secondary metabolite production. Fujita *et al.* (1981a,b) showed that media promoting growth differed from those promoting the production of shikonin derivatives in suspension cultures of *Lithospermum erythrorhizon;* growth decreased under culture conditions elevating the production of shikonin derivatives and vice versa. On the basis of this finding, they established a medium suitable for the production of shikonin derivatives (M9), in which ammonium was omitted and the amounts of nitrate, phosphate, sulfate, copper, and sucrose of White's medium (1939) were modified. The yield of shikonin derivatives was elevated to 1400 mg/liter by transferring cells from growth medium to production medium (M9). Shikonin mass-produced using a two-phase culture system, in which the first stage increases biomass and the second stage promotes shikonin production, is now marketed as a purple cosmetic pigment. Recently, other two-phase culture systems giving increased yields of valuable secondary metabolites have been established (e.g., Zenk *et al.*, 1977; Miyasaka *et al.*, 1986).

D. Immobilized Cells

The immobilization of plant cells has also been recently developed as a method for secondary metabolite production. The principle of this method is to manipulate the physical environment of cells by entrapping them within a matrix in which cell growth is suppressed. A variety of methods have been established, e.g., *Mucuna pruriens* cells were entrapped in calcium–alginate gel (Wichers *et al.*, 1983), *Mentha* cells in cross-linked polyacrylamide–hydrazide (Galun *et al.*, 1983), and *Capsicum frutescens* cells in a matrix of reticulate polyethane foam (Lindsey and Yeoman, 1984; Lindsey, 1985).

Under these conditions, metabolism in cells changes markedly. Wichers *et al.* (1983) reported that cells of *Mucuna pruriens* immobilized in calcium–alginate gels were able to transform exogenously supplied tyrosine to L-dopa, 90% of which was released into the medium. Lindsey and Yeoman (1984) also demonstrated that cells of *Capsicum frutescens* immobilized in a matrix of reticulate polyethane foam produce levels of capsaicin higher by two or three orders of magnitude than freely suspended cells. Moreover, supplementing the medium with precursors of capsaicin, such as phenylalanine and isocapric acid, increased the capsaicin production of immobilized cells. Increased secondary metabolite production in immobilized cells seems to be caused by the suppression of growth. A reciprocal relationship between protein synthesis and capsaicin synthesis was observed: the rate of [^{14}C]phenylalanine incorporation into capsaicin increased, while the rate of incorporation into soluble protein decreased when cells were immobilized. Moreover, starvation of nutritional factors and omitting growth regulators from the medium brought about a further stimulation of capsaicin production in immobilized cells of *Capsicum frutescens* (Lindsey and Yeoman, 1984; Lindsey, 1985). These findings indicate that the metabolic flow shifts from primary metabolism supporting growth to secondary metabolism under the conditions of limited growth in immobilized cells.

IV. RELATIONSHIP BETWEEN GROWTH AND ACCUMULATION OF SECONDARY METABOLITES

Relatively little attention has been focused on the relationship between cell growth and the accumulation of secondary metabolites. For understanding the regulatory mechanisms of secondary metabolism, however, it is essential to study the kinetics of cell growth and the

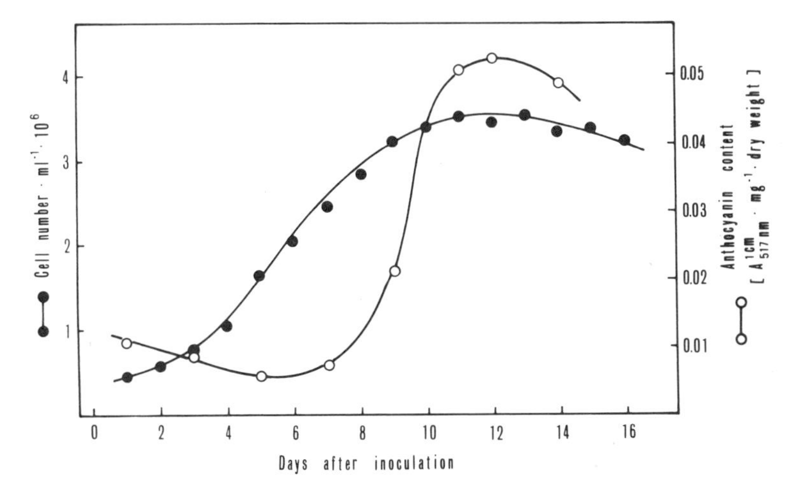

Fig. 1. Kinetics of growth and anthocyanin accumulation in carrot (*Daucus carota*) suspension cultures. (After Noé *et al.*, 1980.)

formation of secondary metabolites. In recent years, in order to elevate the yield of useful secondary metabolites from plant tissue cultures, analytical studies on the relationship between cell growth and the formation of secondary metabolites have been performed.

Patterns of secondary metabolite accumulation in relation to growth fall in two categories: (1) accumulation during the stationary phase with an inverse relationship between growth and accumulation, as seen in anthocyanin accumulation in *Daucus carota* cultures (Fig. 1); (2) accumulation during the logarithmic phase, such as betacyanin accumulation in *Phytolacca americana* shown in Fig. 2, in which accumulation is associated with cell division. The accumulation of most secondary products reaches a maximum during the stationary phase of culture, e.g., ajmalicine and serpentine in *Catharanthus roseus* (Neumann *et al.*, 1983; Mérillon *et al.*, 1984), berberine in *Thalictrum minus* (Nakagawa *et al.*, 1986), diosgenin in *Dioscorea deltoidea* (Tal *et al.*, 1984), anthocyanin in *Daucus carota* (Noé *et al.*, 1980) and *Vitis* (Yamakawa *et al.*, 1983), phenolics in *Acer pseudoplatanus* (Phillips and Henshaw, 1977), rosmarinic acid in *Anchusa officinalis* (De-Eknamkul and Ellis, 1984), and reserpine in *Rauwolfia serpentina* (Yamamoto and Yamada, 1986) cultures.

The increase and decrease in the accumulation of precursors and in the activities of enzymes involved in the synthesis of secondary metabolites precede those of secondary metabolite accumulation, which is considered reasonable. Tal *et al.* (1984) suggested that intermediates in the biosynthetic pathway of diosgenin were formed in early stages of

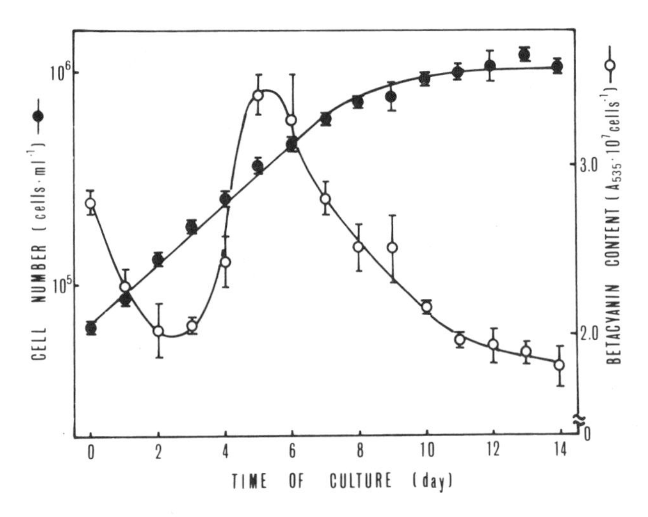

Fig. 2. Kinetics of growth and betacyanin accumulation in suspension cultures of *Phytolacca americana*. Vertical lines indicate SD ($n = 3$). (Sakuta *et al.*, 1986.)

growth and then transformed to diosgenin in the stationary phase in suspension cultures of *Dioscorea deltoidea*. The activity of berberine bridge enzyme increases during the late log phase and peaks in the early stationary phase (Steffens *et al.*, 1985).

Lindsey and Yeomann (1983) showed that fast growing and friable calli usually accumulated low levels of alkaloids, whereas relatively slowly growing calli of the same species, in which the cells were more compactly associated or even organized into morphologically recognizable structures, accumulated higher levels of alkaloids. On the basis of these findings, they referred to an inverse relationship between growth and alkaloid accumulation. Tabata *et al.* (1972) showed that alkaloid production was associated with root organization in *Scopolia parviflora* cultures. Ramawat *et al.* (1985) have also observed a correlation between alkaloid accumulation and morphological differentiation in *Ruta graveolens*. Ozeki and Komamine (1981) indicated that anthocyanin synthesis was induced after cell division had ceased and was closely correlated with somatic embryogenesis in carrot suspension cultures.

Contrasting with the pattern of most secondary metabolites accumulation in relation to growth, maximum accumulation of betacyanin was observed during the logarithmic phase of growth in *Phytolacca americana* (Fig. 2). The production and accumulation of betacyanin may be associated with cell division or may occur just after cell division (Sakuta *et al.*, 1986). Corduan and Reinhard (1972) also demonstrated that maximum accumulation of volatile oil occurred during the logarithmic phase of growth in callus cultures of *Ruta graveolens*.

Betacyanin is a characteristic constituent of most species of Centrospermae (Piattelli, 1976; Mabry, 1980), and its localization in tissues of intact plants differs from that of other secondary metabolites, such as anthocyanin. Anthocyanin is usually accumulated in specific tissues such as the subepidermal layers of cortex, whereas in most cases betacyanin is not localized in specific tissues, e.g., in beet roots.

Figure 3 shows a working hypothesis for the regulation of secondary metabolite accumulation in relation to growth and primary metabolism. The primary metabolic flow branches at point A into two pathways: one is the pathway to proliferation or undifferentiated growth (Pathway I), and the other is the pathway to differentiation and the production of most secondary metabolites, such as anthocyanin (Pathway II). From the results obtained with anthocyanin synthesis in carrot suspension cultures (Ozeki and Komamine, 1981), it is presumed that switching of the metabolic flow between Pathways I and II occurs at point A, since anthocyanin synthesis occurs only when cell division has ceased. It is induced by the same trigger as induction of differentiation (embryogenesis). Furthermore, 2,4-D induces cell division but inhibits strongly both anthocyanin synthesis and embryogenesis.

On the other hand, betacyanin accumulation in suspension cultures of *Phytolacca americana* reveals that the flow to betacyanin may branch off to

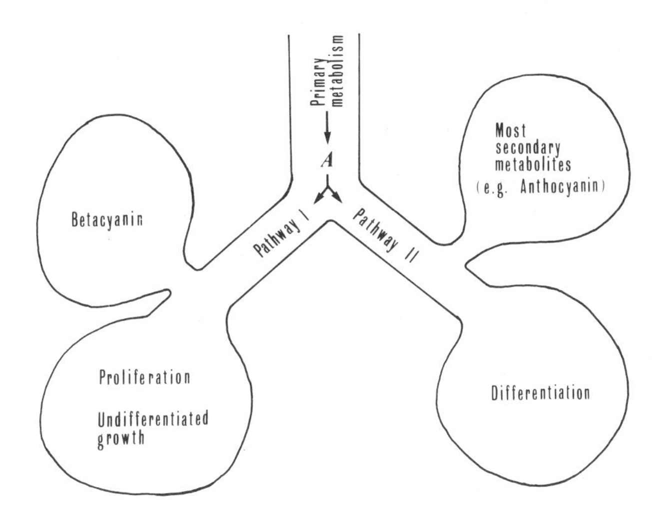

Fig. 3. Possible scheme of metabolic flow in relation to secondary metabolite accumulation.

proliferation or undifferentiated growth (Pathway I), since the peaks of both betacyanin accumulation and growth rate were observed in the same stage of the growth cycle (Sakuta *et al.*, 1986) and 2,4-D promoted cell division as well as betacyanin accumulation. Environments favorable for proliferation also promote betacyanin accumulation (M. Sakuta, T. Takagi, and A. Komamine, unpublished data). Thus, a physiological situation in which proliferation occurs is considered essential for betacyanin synthesis. However, the minimal metabolic flow to support proliferation is not sufficient for betacyanin accumulation, indicating that proliferation occurs predominantly and that metabolites which flow over from the proliferation pathway are used for betacyanin synthesis.

The working hypothesis described above is applicable at least to betacyanin accumulation in suspension cultures of *Phytolacca americana* (Sakuta *et al.*, 1986) and anthocyanin synthesis in carrot suspension cultures (Ozeki and Komamine, 1981). Further investigation will reveal whether or not it can be generalized to cover other aspects of secondary metabolism.

REFERENCES

Amino, S., Fujimura, T., and Komamine, A. (1983). Synchrony induced by double phosphate starvation in a suspension culture of *Catharanthus roseus*. *Physiol. Plant.* **59,** 393–396.

Amorin, H. V., Dougall, D. K., and Sharp, W. R. (1977). The effects of carbohydrate and nitrogen concentration on phenol synthesis in Paul's Scarlet Rose cells grown in tissue culture. *Physiol. Plant.* **39,** 91–95.

Berlyn, M. B., and Zelitch, I. (1975). Photoautotrophic growth and photosynthesis in tobacco callus cells. *Plant Physiol.* **56,** 752–756.

Chandler, M. T., Tandeau de Marsac, N., and de Kouchkovsky, Y. (1972). Photosynthetic growth of tobacco cells in liquid suspension. *Can. J. Bot.* **50,** 2265–2270.

Constabel, F., Shyluk, J. P., and Gamborg, O. L. (1971). The effect of hormones on anthocyanin accumulation in cell cultures of *Haplopappus gracilis*. *Planta* **96,** 306–316.

Constabel, F., Kurz, W. G. W., Chatson, K. B., and Kirkpatrick, J. W. (1977). Partial synchrony in soybean cell suspension cultures induced by ethylene. *Exp. Cell Res.* **105,** 263–268.

Corduan, G., and Reinhard, E. (1972). Synthesis of volatile oils in tissue cultures of *Ruta graveolens*. *Phytochemistry* **11,** 917–922.

Davies, M. E. (1972). Polyphenol synthesis in cell suspension cultures of Paul's Scarlet Rose. *Planta* **104,** 50–65.

De-Eknamkul, W., and Ellis, B. E. (1984). Rosmarinic acid production and growth characteristics of *Anchusa officinalis* cell suspension cultures. *Planta Med.* **50,** 346–350.

Dougall, D. K. (1980). Nutrition and metabolism. *In* "Plant Tissue Culture as a Source of Biochemicals" (E. J. Staba, ed.), pp. 21–58. CRC Press, Boca Raton, Florida.

Endress, R. (1976). Betacyan-Akkumulation in Kallus von *Portulaca grandiflora* var. JR unter dem Einfluss von Phytohormonen und Cu^{2+}-Ionen auf unterschiedlichem Grundmedien. *Biochem. Physiol. Pflanz.* **169,** 87–98.

Everett, N. P., Wang, T. L., Gould, A. R., and Street, H. E. (1981). Studies on the control of the cell cycle in cultured plant cells. II. Effects of 2,4-dichlorophenoxyacetic acid (2,4-D). *Protoplasma* **106,** 15–22.

Fowler, M. W. (1971). Growth in cultures of plant cells. XIV. Carbohydrate oxidation during the growth of *Acer pseudoplatanus* cells in suspension culture. *J. Exp. Bot.* **22,** 715–724.

Franke, J., and Böhm, H. (1982). Accumulation and excretion of alkaloids by *Macleaya microcarpa* cell cultures. II. Experiments in liquid medium. *Biochem. Physiol. Pflanz.* **177,** 501–507.

Fujita, Y., Hara, Y., Ogino, T., and Suga, C. (1981a). Production of shikonin derivatives by cell suspension cultures of *Lithospermum erythrorhizon.* I. Effects of nitrogen sources on the production of shikonin derivatives. *Plant Cell Rep.* **1,** 59–60.

Fujita, Y., Hara, Y., Suga, C., and Morimoto, T. (1981b). Production of shikonin derivatives by cell suspension cultures of *Lithospermum erythrorhizon.* II. A new medium for the production of shikonin derivatives. *Plant Cell Rep.* **1,** 61–63.

Furuya, T., Kojima, H., and Syono, K. (1971). Regulation of nicotine biosynthesis by auxins in tobacco callus tissues. *Phytochemistry* **10,** 1529–1532.

Galun, E., Aviv, D., Dantes, A., and Freeman, A. (1983). Biotransformation by plant cells immobilized in cross-linked polyacrylamide–hydrazide. *Planta Med.* **49,** 9–13.

Givan, V. C. (1968). Short-term changes in hexose phosphates and ATP in intact cells of *Acer pseudoplatanus* L. subjected to anoxia. *Physiol. Plant.* **43,** 948–952.

Gould, A. R., and Street, H. E. (1975). Kinetic aspects of synchrony in suspension cultures of *Acer pseudoplatanus* L. *J. Cell Sci.* **17,** 337–348.

Gould, A. R., Everett, N. P., Wang, T. L., and Street, H. E. (1981). Studies on the control of the cell cycle in cultured plant cells. I. Effects of nutrient limitation and nutrient starvation. *Protoplasma* **106,** 1–13.

Hagimori, M., Matsumoto, T., and Obi, Y. (1982). Studies on the production of *Digitalis* cardenolides by plant tissue cultures. III. Effects of nutrients on digitoxin formation by shoot-forming cultures of *Digitalis purpurea* L. grown in liquid media. *Plant Cell Physiol.* **23,** 1205–1211.

Hüsemann, W., and Barz, W. (1977). Photoautotrophic growth and photosynthesis in cell suspension cultures of *Chenopodium rubrum. Physiol. Plant.* **40,** 77–81.

Ikeda, T., Matsumoto, T., and Noguchi, M. (1976). Effects of nutritional factors on the formation of ubiquinone by tobacco plant cells in suspension culture. *Agric. Biol. Chem.* **40,** 1765–1770.

Ikeda, T., Matsumoto, T., and Noguchi, M. (1977). Effects of inorganic nitrogen sources and physical factors on the formation of ubiquinone by tobacco plant cells in suspension culture. *Agric. Biol. Chem.* **41,** 1197–1201.

Kimball, S. L., Beversdorf, W. D., and Bingham, E. T. (1975). Influence of osmotic potential on the growth and development of soybean tissue cultures. *Crop Sci.* **15,** 750–752.

Kinnersley, A. M., and Dougall, D. K. (1980). Increase in anthocyanin yield from wild-carrot cell cultures by a selection system based on cell-aggregate size. *Planta* **149,** 200–204.

King, P. J., Cox, B. J., Fowler, M. W., and Street, H. E. (1974). Metabolic events in synchronised cell cultures of *Acer pseudoplatanus* L. *Planta* **117,** 109–122.

Knobloch, K. H., and Berlin, J. (1981). Phosphate mediated regulation of cinnamoyl putrescine biosynthesis in cell suspension cultures of *Nicotiana tabacum. Planta Med.* **42,** 167–172.

Knobloch, K. H., and Berlin, J. (1983). Influence of phosphate on the formation of the indole alkaloids and phenolic compounds in cell suspension cultures of *Catharanthus roseus*. I. Comparison of enzyme activities and product accumulation. *Plant Cell, Tissue Organ Cult.* **2,** 333–340.

Lindsey, K. (1985). Manipulation, by nutrient limitation of the biosynthetic activity of immobilized cells of *Capsicum frutescens* Mill. cv. annuum. *Planta* **165,** 126–133.

Lindsey, K., and Yeoman, M. M. (1983). The relationship between growth rate, differentiation and alkaloid accumulation in cell cultures. *J. Exp. Bot.* **34,** 1055–1065.

Lindsey, K., and Yeoman, M. M. (1984). The synthetic potential of immobilised cells of *Capsicum frutescens* Mill. cv. annuum. *Planta* **162,** 495–501.

Mabry, T. J. (1980). Betalains. *In* "Secondary Plant Products. Encyclopedia of Plant Physiology, New Series" (E. A. Bell and B. V. Charlwood, eds.), Vol. 8, pp. 513–533. Springer-Verlag, Berlin and New York.

Matsumoto, T., Nishida, K., Noguchi, M., and Tamaki, E. (1973). Some factors affecting the anthocyanin formation by *Populus* cells in suspension culture. *Agric. Biol. Chem.* **37,** 561–567.

Mérillon, J. M., Rideau, M., and Chénieux, J. C. (1984). Influence of sucrose on levels of ajmalicine, serpentine, and tryptamine in *Catharanthus roseus* cells *in vitro*. *Planta Med.* **50,** 497–501.

Miyasaka, H., Nasu, M., Yamamoto, T., and Yoneda, K. (1985). Production of ferruginol by cell suspension cultures of *Salvia miltiorrhiza*. *Phytochemistry* **24,** 1931–1933.

Miyasaka, H., Nasu, M., Yamamoto, T., Endo, Y., and Yoneda, K. (1986). Regulation of ferruginol and cryptotanshinone biosynthesis in cell suspension cultures of *Salvia miltiorrhiza*. *Phytochemistry* **25,** 637–640.

Mizukami, H., Konoshima, M., and Tabata, M. (1977). Effect of nutritional factors on shikonin derivative formation in *Lithospermum* callus cultures. *Phytochemistry* **16,** 1183–1186.

Mok, M. C., Gabelman, W. H., and Skoog, F. (1976). Carotenoid synthesis in tissue cultures of *Daucus carota* L. *J. Am. Soc. Hortic. Sci.* **101,** 442–449.

Nagata, T., Okada, K., and Takebe, I. (1982). Mitotic protoplasts and their infection with tobacco mosaic virus RNA encapsulated in liposomes. *Plant Cell Rep.* **1,** 250–252.

Nakagawa, K., Konagai, A., Fukui, H., and Tabata, M. (1984). Release and crystallization of berberine in the liquid medium of *Thalictrum minus* cell suspension cultures. *Plant Cell Rep.* **3,** 254–257.

Nakagawa, K., Fukui, H., and Tabata, M. (1986). Hormonal regulation of berberine production in cell suspension cultures of *Thalictrum minus*. *Plant Cell Rep.* **5,** 69–71.

Nash, T. D., and Davies, M. E. (1972). Some aspects of growth and metabolism of Paul's Scarlet rose cell suspensions. *J. Exp. Bot.* **23,** 75–91.

Neumann, D., Krauss, G., Hieke, M., and Gröger, D. (1983). Indole alkaloid formation and storage in cell suspension cultures of *Catharanthus roseus*. *Planta Med.* **48,** 20–23.

Nishi, A., Kato, K., Takahashi, M., and Yoshida, R. (1977). Partial synchronization of carrot cell culture by auxin deprivation. *Physiol. Plant.* **39,** 9–12.

Noé, W., Langebartels, C., and Seitz, H. U. (1980). Anthocyanin accumulation and PAL activity in a suspension culture of *Daucus carota* L. *Planta* **149,** 283–287.

Obata-Sasamoto, H., and Komamine, A. (1983). Effect of culture conditions on dopa accumulation in a callus culture of *Stizolobium hassjoo*. *Planta Med.* **49,** 120–123.

Ozeki, Y., and Komamine, A. (1981). Induction of anthocyanin synthesis in relation to embryogenesis in a carrot suspension culture: Correlation of metabolic differentiation with morphological differentiation. *Physiol. Plant.* **53,** 570–577.

Ozeki, Y., and Komamine, A. (1985). Effects of inoculum density, zeatin and sucrose on

anthocyanin accumulation in a carrot suspension culture. *Plant Cell, Tissue Organ Cult.* **5,** 45–53.
Ozeki, Y., and Komamine, A. (1986). Effects of growth regulators on the induction of anthocyanin synthesis in carrot suspension cultures. Plant Cell Physiol. **27,** 1361–1368.
Phillips, R., and Henshaw, G. G. (1977). The regulation of synthesis of phenolics in stationary phase cell cultures of *Acer pseudoplatanus* L. *J. Exp. Bot.* **28,** 785–794.
Piattelli, M. (1976). Betalains. *In* "Chemistry and Biochemistry of Plant Pigments" (T. W. Goodwin, ed.), 2nd ed., Vol. 1, pp. 569–596. Academic Press, New York.
Ramawat, K. G., Rideau, M., and Chénieux, J. (1985). Growth and quaternary alkaloid production in differentiating and nondifferentiating strains of *Ruta graveolens. Phytochemistry* **24,** 441–445.
Sakuta, M., Takagi, T., and Komamine, A. (1986). Growth related accumulation of betacyanin in suspension cultures of *Phytolacca americana* L. *J. Plant Physiol.* **125,** 337–343.
Sala, F., Galli, M. G., Nielsen, E., Magnien, E., Devreux, M., Pedrali-Noy, G., and Spadari, S. (1983). Synchronization of nuclear DNA synthesis in cultured *Daucus carota* L. cells by aphidicolin. *FEBS Lett.* **153,** 204–208.
Schneider, E. A., and Wightman, F. (1974). Metabolism of auxin in higher plants. *Annu. Rev. Plant Physiol.* **25,** 487–513.
Shimizu, T., Clifton, A., Komamine, A., and Fowler, M. W. (1977). Changes in metabolite levels during growth of *Acer pseudoplatanus* (sycamore) cells in batch suspension culture. *Physiol. Plant.* **40,** 125–129.
Steffens, P., Nagakura, N., and Zenk, M. H. (1985). Purification and characterzation of the berberine bridge enzyme from *Berberis beaniana* cell cultures. *Phytochemistry* **24,** 2577–2583.
Szweykowska, A. (1975). The role of cytokinin in the control of cell growth and differentiation in culture. *In* "Tissue Culture and Plant Science 1974" (H. E. Street, ed.), pp. 416–477. Academic Press, New York.
Tabata, M., Yamamoto, H., Hiraoka, N., Matsumoto, Y., and Konoshima, M. (1971). Regulation of nicotine production in tobacco tissue culture by plant growth regulators. *Phytochemistry* **10,** 723–729.
Tabata, M., Yamamoto, H., Hiraoka, N., and Konoshima, N. (1972). Organization and alkaloid production in tissue cultures of *Scopolia parviflora. Phytochemistry* **11,** 949–955.
Tabata, M., Mizukami, H., Hiraoka, N., and Konoshima, M. (1974). Pigment formation in callus cultures of *Lithospermum erythrorhizon. Phytochemistry* **13,** 927–932.
Takayama, S., Misawa, M., Ko, K., and Misato, T. (1977). Effect of cultural conditions on the growth of *Agrostemma githago* cells in suspension culture and the concomitant production of an anti-plant virus substance. *Physiol. Plant.* **41,** 313–320.
Tal, B., and Goldberg, I. (1982). Growth and diosgenin production by *Dioscorea deltoidea* cells in batch and continuous cultures. *Planta Med.* **44,** 107–110.
Tal, B., Gressel, J., and Goldberg, I. (1982). The effect of medium constituents on growth and diosgenin production by *Dioscorea deltoidea* cells grown in batch cultures. *Planta Med.* **44,** 111–115.
Tal, B., Rokem, J. S., and Goldberg, I. (1984). Timing of diosgenin appearance in suspension cultures of *Dioscorea deltoidea. Planta Med.* **50,** 239–241.
White, P. R. (1939). Potentially unlimited growth of excised plant callus in an artificial medium. *Am. J. Bot.* **26,** 59–64.
Wichers, H. J., Malingré, T. M., and Huizing, H. J. (1983). The effect of some environmental factors on the production of L-dopa by alginate-entrapped cells of *Mucuna pruriens. Planta* **158,** 482–486.

Wilson, G., and Balagué, C. (1985). Biosynthesis of authraquinone by cells of *Galium mallugo* L. grown in a chemostat with limiting sucrose or phosphate. *J. Exp. Bot.* **36,** 485–493.

Wilson, G., and Marron, P. (1978). Growth and anthraquinone biosynthesis by *Galium mollugo* L. cells in batch and chemostat culture. *J. Exp. Bot.* **29,** 837–851.

Yamakawa, T., Kato, S., Ishida, K., Kodama, T., and Minoda, Y. (1983). Production of anthocyanins by *Vitis* cells in suspension culture. *Agric. Biol. Chem.* **47,** 2185–2191.

Yamamoto, O., and Yamada, Y. (1986). Production of reserpine and its optimization in cultured *Rauwolfia serpentina* Benth. cells. *Plant Cell Rep.* **5,** 50–53.

Yoshikawa, N., Fukui, H., and Tabata, M. (1986). Effect of gibberellin A on shikonin production in *Lithospermum* callus cultures. *Phytochemistry* **25,** 621–622.

Zenk, M. H., El-Shagi, H., and Schulte, U. (1975). Anthraquinone production by cell suspension cultures of *Morinda citrifolia. Planta Med., Suppl.,* pp. 79–101.

Zenk, M. H., El-Shagi, H., Arens, H., Stöckigt, J., Weiler, E. W., and Deus, B. (1977). Formation of the indole alkaloids serpentine and ajimalicine in cell suspension cultures of *Catharanthus roseus. In* "Plant Tissue Culture and Its Biotechnological Application" (W. Barz, E. Reinhard, and M. H. Zenk, eds.), pp. 27–43. Springer-Verlag, Berlin and New York.

PART **III**

Special Techniques

CHAPTER **6**

Cell Cloning and the Selection of High Yielding Strains

Donald K. Dougall

Department of Botany
The University of Tennessee
Knoxville, Tennessee 37996–1100

I. INTRODUCTION

The selection of high yielding strains from plant tissue cultures has the practical objective of increasing the yields or productivity of phytochemicals. That high yielding strains can be selected from plant tissue cultures implies that the cultures being considered are heterogeneous. The larger fundamental scientific questions then are what is the source or sources of this heterogeneity and what mechanisms are involved. While high yielding strains can be selected from plant tissue cultures and heterogeneity in cultures for phytochemical yield is well established, the sources of variability and mechanisms involved in cellular diversity in cultures are not. Here, the role of cell cloning as a tool with which to explore the heterogeneity within cultures will be examined.

II. CLONES AND CLONING

The term *clone* is commonly used in four ways. These are (1) plants propagated vegetatively, (2) plants regenerated from tissue cultures, (3) the progeny derived by mitosis from a single cell, and (4) the isolation and replication of a piece of DNA. The third of these ways is the one which will be used here.

Defining a clone as the progeny derived by mitosis from a single cell

CELL CULTURE AND SOMATIC CELL
GENETICS OF PLANTS, VOL. 4

does not require that all cells in that clone be identical. Events such as spontaneous mutations or other changes may occur in individuals in that progeny, which make them different from other individuals. The occurrence of such changes can therefore be detected by taking a single cell, allowing it to grow into a population, taking single cells from that population, allowing them to grow into populations (recloning), and determining whether this new series of populations (subclones) all show the same phenotype made under standard conditions. If some of these populations of the new series (subclones) are different from others, then some event occurred in the progeny of the original single cell which led to the observed phenotypic difference. If instead of taking only one single cell from the starting culture, one takes a series of single cells, grows each up into a population, and finds phenotypic difference between those populations, then one must consider the possibility that the original population was a mixture of different cells. Whether this mixture of different cells was present in the explant from which the culture was initiated or arose after the culture was initiated cannot be determined from the first cloning of a culture but can be distinguished by recloning.

The frequency of subclones exhibiting altered phenotype cannot be related to the rate of change in the clone. This can be easily perceived if the event leading to the altered phenotype is random. By definition there is an equal probability of the event occurring in the original single cell as there is of it occurring in any of the cells in the population just prior to recloning. There is, however, a major difference in the frequency of subclones with the altered phenotype which will be obtained if the event occurs early or late in the growth of the clone.

Other factors which also prevent the frequency of appearance of subclones with a changed phenotype from being relatable to the rate of change of phenotype include differences in the growth rates of the two phenotypes and cross-feeding of one cell type by another allowing the disadvantaged phenotype to survive and grow, albeit at low frequency, in a population for long periods of time. The experimental approaches to measuring the rate of change of phenotype are available and have been discussed by Meins (1983).

III. HETEROGENEITY IN CELL CULTURES

Differences in yield of phytochemicals in cells of a plant cell culture have been documented by several authors. By screening through sieves of definite mesh size followed by density gradient centrifugation, Ozeki and Komamine (1981, 1985) obtained two populations of cells from carrot suspension cultures, the heavier of which gave embryos in the absence of auxin and the lighter of which accumulated anthocyanin under the same conditions. Within the lighter population a maximum of 50% of the cells accumulated anthocyanin.

With *Catharanthus roseus* cultures, Hall and Yeoman (1986) have shown by microspectrophotometry that up to 10% of the cells in a culture accumulate anthocyanin and that cells differ in their absorbance spectra. They discussed changes in vacuolar pH as a possible explanation of these differences and showed that at least a 40-fold range of anthocyanin content and concentration was present in cells in the population. Ellis (1985) and Chaprin and Ellis (1984) have shown wide variation in rosmarinic acid content of single cells in cultures of *Anchusa officinalis* and *Coleus blumei* using UV absorbance and microspectrophotometry.

In addition to these direct observations, differences in phytochemical production capacity between cells in cultures have been inferred from cloning experiments and from selection from callus tissue. Heterogeneity in yield of phytochemicals has been demonstrated by cloning in the following cases:

Anthocyanin in *Daucus carota* cultures (Dougall *et al.*. 1980)
Shikonin (naphthoquinones) in *Lithospermum erythrorhizon* (Fujita *et al.*, 1985)
Nicotine in *Nicotiana tabacum* (Ogino *et al.*, 1978)
Nicotine in *Nicotiana rustica* (Tabata and Hiraoka, 1976)
Serpentine with and without ajmalicine in *Catharanthus roseus* (Deus-Neuman and Zenk, 1984)
Rosmarinic acid in *Anchusa officinalis* (Ellis, 1985)
Berberine in *Coptis japonica* (Sato and Yamada, 1984)
Biotin in *Lavandula vera* (Watanabe and Yamada, 1982; Watanabe *et al.*, 1982)
Ubiquinone-10 in *Nicotiana tabacum* (Matsumoto *et al.*, 1980)

In addition, heterogeneity in cultures has been indirectly demonstrated by observing variation in distribution of pigments in callus cultures, selecting and subculturing the heavily pigmented portions, and showing that pigment yield of these sublines was increased (Mizukami *et al.*, 1978; Yamamoto and Mizuguchi, 1982).

In many of the cases cited above, the clones are not derived from explicitly documented single cells. The populations used for plating consists of units which vary from one to a number of cells in small aggregates. The maximum number of cells per aggregate depends on the mesh size of the screen used as well as the characteristics of the culture. A proportion of the colonies obtained from plating such a population will be derived from the aggregates present if the single cells grow in the plates. If the single cells do not grow, all the colonies obtained will arise from aggregates of cells. The justification for regarding colonies derived from aggregates as true clones arises from the way plant cells divide. Because initially the new cell wall is deposited within the existing cell at cytokinesis, the daughter cells remain joined together. Only later, as a result of cell enlargement and stresses such as shear forces applied to the cells, do they separate. Thus in aggregates where there are clearly common cell walls between cells, the cells are assumed to be the progeny of a single cell and thus clonal. With increasing numbers of cells in an aggregate, the probability of a change in one or more cells in the aggregate increases.

The use of mixtures of single cells and aggregates for producing colonies arises for technical reasons. First, because of variations in the size of single cells it is rarely possible to obtain populations consisting only of single cells by screening. Second, growth of single cells into visible colonies appears to be more demanding than the growth of aggregates and to require more time than the growth of aggregates. The use of aggregates for the generation of clones does not appear to invalidate the concept of cloning, though it does alter some details. Murphy (1982) has examined this problem and concluded from the behavior of models, based on some assumptions about the behavior of aggregates and their growth, that the expected distribution of mutants was modified but that mutants were detectable by fluctuation tests.

IV. STABILITY OF CLONES AND SUBCLONES

In view of the demonstrated heterogeneity with respect to phytochemical yields in cells in culture, questions about the stability of clones or subclones selected for high yield are significant. Clones selected for high yield of phytochemicals have been shown (Ogino *et al.*, 1978; Dougall *et al.*, 1980; Matsumoto *et al.*, 1981; Deus-Neuman and Zenk, 1984; Ellis, 1985) to rapidly become populations of cells whose yields of phytochemicals differ. Despite this diversity of yields which develops, high yielding clones can be routinely obtained from these populations by recloning. As a result of the development of the cellular diversity of yields, the average yield of the population declines. The extent of decline of anthocyanin production was different over six passages in different subclones of carrot and was different in two different media (Dougall *et al.*, 1980). The time courses of the decreases in ajmalicine, serpentine, and secologanin yields in cloned *Catharanthus roseus* cultures have been reported by Deus-Neuman and Zenk (1984).

While the yield of phytochemicals in these populations declines, it appears not to decline to zero, i.e., the ability of the cells to produce phytochemicals does not seem to be completely lost. This suggests that the clonal populations become equilibrium mixtures of cells with different yields and that these equilibrium mixtures may be different in different media. The details of changes in cells which lead to the differences in yields are not known. In addition one may reasonably presume that sublines which have been obtained from a culture by selection and propagation of some unique portion of a culture (e.g., pigmented areas) will also display at least the same general properties as clones. The yields in these cultures will be the average of the yields of the individual cells in the populations.

The stability of these populations on serial passage is required for reproducible yields from cultures. Data on the stability of yield of phytochemicals in the cultures have been assembled by Deus-Neuman and Zenk (1984) and Yamada and Hashimoto (1984). Both list cultures in which yield appears to be stable and also cases where yield appears to be unstable. In their approach to high yields of rosmarinic acid in *Coleus blumei* cultures, Ulbrich *et al.* (1985) identify "strictly controlled constant culture conditions" as critical to the maintenance of stability of yield. Kinnersley and Dougall (1980) showed that the yields of anthocyanin in carrot cultures could be systematically increased or decreased by selection of small or large aggregates from cultures. We have noted that when the person who routinely maintains stock cultures is changed, the average but relatively stable yield of phytochemicals by the stock culture may also change.

These observations are consistent with the idea that the populations in cell cultures are neither stable nor unstable but are a metastable equilibrium which is sensitive to minor and generally difficult to identify changes in the ways the cultures are maintained. These changes may be relatable to the fact that the cells of the cultures exist in aggregates of different sizes which settle at different rates, thus leading to differences in the populations transferred to fresh medium at subculture and the subsequent behavior of the culture (Dougall, 1985). The deliberate or inadvertent selection of large or small aggregates for initiation of new stock cultures could represent a selective pressure applied to the populations which leads to shifts in the population distribution and thus to overall yields of the cultures.

An analog of this situation was described by Bayliss (1977a,b), who examined factors which affected the frequency of tetraploid carrot cells in diploid populations in culture. The frequency of tetraploids was reduced to very low levels when the subculture interval was 7 days, while frequency of diploid cells was very low when the subculture interval was 14 days. Further, he showed that the tetraploid cells had a competitive advantage over the diploid cells when the cultures experienced a period of phosphate limitation (e.g., in the latter half of the 14-day culture period). In contrast, neither type had a competitive advantage when the nitrogen in the medium was limited over the 14-day growth period. Bayliss (1977b) concluded that the competitive advantage of the tetraploid cells was operational only at the time at which phosphate became limiting and that the advantage was not the result of higher growth rates of the tetraploid cells.

V. HIGH YIELDING STRAINS

Despite the potential implication of the above discussion of stability of strains and clones, namely, that populations of cells from a species in the same environment will tend to the same equilibrium mixture and thus the same specific yield, i.e., productivity, there are reports where selection has led to stable increased yields of phytochemicals in cultures (Ogino *et al.*, 1978; Yamamoto and Mizuguchi, 1982; Sato and Yamada, 1984; Ellis, 1985; Fujita *et al.*, 1985; see also Chapter 70, Volume 1, this treatise). These increases in yield suggest that clones and subclones can differ in ways which lead to differences in the equilibrium populations and thus population yields. These differences are clearly over and above the mechanism(s) by which the populations become equilibrium mixtures because in each case the various subclones and clones obtained were evaluated in a constant culture regimen. The constant culture regimen used in these cases also rules out the possibility that yield is differing because the cells are responding physiologically to different culture media and environments. Physiological responses of yield to culture media, etc., are well known and reversible (Meins, 1983). Here different clones and subclones are responding differently to the same culture regimen. In addition to obtaining high yielding subclones or clones by explicit isolation, examination, and propagation under nonselective conditions, there is the possibility that the conditions of maintenance of stock cultures is inadvertently applying selection pressure to them for the maintenance or enhancement of intrinsically high yielding cells in the population.

Yields of phytochemicals have been increased by cloning and selection of high yielding subclones and sublines. The yields obtained in batch cultures are, however, less than the potential yield because of the heterogeneity which develops on serial passage. This implies the potential for increasing yield in batch culture is great either by changes in culture conditions which prevent or greatly decrease the rate of progress to equilibrium populations or by knowledge of the mechanism of the changes in yields and rational interruptions of the mechanism by which yield declines.

REFERENCES

Bayliss, M. W. (1977a). Factors affecting the frequency of tetraploid cells in a predominantly diploid suspension culture of *Daucus carota*. *Protoplasma* **92,** 109–115.

Bayliss, M. W. (1977b). The causes of competition between two cell lines of *Daucus carota* in mixed culture. *Protoplasma* 92, 117–127.

Chaprin, N., and Ellis, B. E. (1984). Microspectrophotometric evaluation of rosmarinic acid accumulation in single cultured plant cells. *Can. J. Bot.* **62,** 2278–2282.

Deus-Neumann, B., and Zenk, M. H. (1984). Instability of indole alkaloid production in *Catharanthus roseus* cell suspension cultures. *Planta Med.*, 427–431.

Dougall, D. K. (1985). Variability in plant cell cultures and its implications for process scale-up and development. *In* "Research Needs in Non-Conventional Bioprocesses" (D. J. Fink, L. M. Curran, and B. R. Allen, eds.), pp. 115–119. Battelle Press, Columbus, Richland.

Dougall, D. K., Johnson, J. M., and Whitten, G. H. (1980). A clonal analysis of anthocyanin accumulation by cell cultures of wild carrot. *Planta* **149,** 292–297.

Ellis, B. I. (1985). Characterization of clonal cultures of *Anchusa officinalis* derived from single cells of known productivity. *J. Plant Physiol.* **19,** 149–158.

Fujita, Y., Takahashi, S., and Yamada, Y. (1985). Selection of cell lines with high productivity of shikonin derivatives by protoplast culture of *Lithospermum erythrorhizon* cells. *Agric. Biol. Chem.* **49,** 1755–1759.

Hall, R. D., and Yeoman, M. M. (1986). Temporal and spatial heterogeneity in the accumulation of anthocyanins in cell cultures of *Catharanthus roseus* (L.) G. Don. *J. Exp. Bot.* **37,** 48–60.

Kinnersley, A. M., and Dougall, D. K. (1980). Increase in anthocyanin yield from wild-carrot cell cultures by a selection system based on cell-aggregate size. *Planta* **149,** 200–204.

Matsumoto, T., Ikeda, T., Kanno, N., Kisaki, T., and Noguchi, M. (1980). Selection of high ubiquinone 10-producing strain of tobacco cultured cells by cell cloning technique. *Agric. Biol. Chem.* **44,** 967–969.

Matsumoto, T., Kanno, N., Ikeda, T., Obi, Y., Kisaki, T., and Noguchi, M. (1981). Selection of cultures tobacco cell strains producing high levels of ubiquinone 10 by a cell cloning technique. *Agric. Biol. Chem.* **45,** 1627–1633.

Meins, F., Jr. (1983). Heritable variation in plant cell culture. *Annu. Rev. Plant Physiol.* **34,** 327–346.

Mizukami, H., Konoshima, M., and Tabata, M. (1978). Variation in pigment production in *Lithospermum erythrorhizon* callus cultures. *Phytochemistry* **17,** 95–97.

Murphy, T. M. (1982). Analysis of distributions of mutants in clones of plant-cell aggregates. *Theor. Appl. Genet.* **61,** 367–372.

Ogino, T., Hiraoka, N., and Tabata, M. (1978). Selection of high nicotine-producing cell lines of tobacco callus by single-cell cloning. *Phytochemistry* **17,** 1907–1910.

Ozeki, Y., and Komamine, A. (1981). Induction of anthocyanin synthesis in relation to embryogenesis in a carrot suspension culture: Correlation of metabolic differentiation with morphological differentiation. *Physiol. Plant.* **53,** 570–577.

Ozeki, Y., and Komamine, A. (1985). Induction of anthocyanin synthesis in relation to embryogenesis in a carrot suspension culture—A model system for the study of expression and repression of secondary metabolism. *In* "Primary and Secondary Metabolism of Plant Cell Cultures" (K.-H. Neuman, W. Barz, and E. Reinhard, eds.), pp. 100–106. Springer-Verlag, Berlin and New York.

Sato, F., and Yamada, Y. (1984). High berberine-producing cultures of *Coptis japonica* cells. *Phytochemistry* **23,** 281–285.

Tabata, M., and Hiraoka, N. (1976). Variation of alkaloid production in *Nicotiana rustica* callus cultures. *Physiol. Plant.* **38,** 19–23.

Ulbrich, B., Wiesner, W., and Arens, H. (1985). Large-scale production of rosmarinic acid from plant cell cultures of *Coleus blumei* Benth. *In* "Primary and Secondary Metabolism of Plant Cell Cultures" (K.-H. Neumann, W. Barz, and E. Reinhard, eds.), pp. 293–303. Springer-Verlag, Berlin and New York.

Watanabe, K., and Yamada, Y. (1982). Selections of variants with high levels of biotin from cultured green *Lavandula vera* cells irradiated with gamma rays. *Plant Cell Physiol.* **23,** 1453–1456.

Watanabe, K., Yano, S.-I., and Yamada, Y. (1982). The selection of cultured plant cell lines producing high levels of biotin. *Phytochemistry* **21,** 513–516.

Yamada, Y., and Hashimoto, T. (1984). Secondary products in tissue culture. *In* "Applications of Genetic Engineering to Crop Improvement" (G. B. Collins and J. G. Petolino, eds.),pp. 561–604. Martinus Nijhoff/Dr. W. Junk Publishers, Dordrecht, The Netherlands.

Yamamoto, Y., and Mizuguchi, R. (1982). Selection of a high and stable pigment-producing strain in cultured *Euphorbia millii* cells. *Theor. Appl. Genet.* **61,** 113–116.

CHAPTER 7

Selection of Mutants which Accumulate Desirable Secondary Compounds

Jack M. Widholm

University of Illinois
Department of Agronomy
Urbana, Illinois 61801

I. INTRODUCTION

Most attempts to obtain plant tissue cultures which produce desirable secondary compounds have involved altering the growth conditions or explant source to "turn on" the synthesis of the compounds. In most cases this procedure has not induced compound levels comparable to that of the source plant and contrasts greatly with the methods employed by microbiologists to obtain compounds like antibiotics and amino acids. The microbiologist generally screens a large number of individuals in a mass screening system either to select mutants which are actually producing high levels of the desired compound or to select specific alterations in pathways which are likely to affect the synthesis of the desired compound.

These selection procedures have not been utilized in many plant tissue culture studies as yet, so the possibilities would seem to have potential if adequately pursued. It must be noted, however, that the few attempts made so far have been only moderately successful, and plant cells are not as easily manipulated, nor as well studied, as microorganisms in general. This lack of basic understanding of biosynthetic pathways is a real obstacle since certain selection techniques to be proposed here require this understanding so that the pathways can be manipulated. Other selection procedures do not need knowledge of the pathway but rely on sensitive analytical procedures which also must be available.

In general one would recommend that in order to carry out the mutant selection work proposed here, cultures should be initiated from high producing individual plants of a species known to produce the correct compound (Zenk *et al.*, 1977). While differentiation is generally disregarded in this discussion it is possible that certain organelles like chloroplasts may be needed for compound synthesis, so cell greening for example, might be necessary in some cases for expression of compound production.

Due to the brevity of this chapter and to the previous review of this research area (Widholm, 1980), the ideas and data will be summarized briefly, and subjects like culture systems and mutagenesis will not be covered. Information on these subjects and selection in general can be found elsewhere (e.g., Widholm, 1980; Gonzales and Widholm, 1985; Berlin and Sasse, 1985).

II. SELECTION SYSTEMS

A. Visual or Chemical Analysis

The selection procedures described in this section involve the straightforward approach of visually or chemically measuring the approximate levels of the desired secondary compound or compounds in individual cells or cell clones to identify the high producers. The most common methods involve visual or microscopic estimations, if the compounds are colored, or extraction and analysis, if the compounds are not visable. Visual analysis has the advantage of being nondestructive while most chemical analysis procedures result in cell death. The cells to be used in these selection systems, as well as those to be used for resistance selection in the next section, need not be haploid, but can be diploid or even of higher ploidy since overproduction usually arises because of increased enzyme levels or altered feedback control. Both mechanisms would be expected to be expressed dominantly or semidominantly.

As an example of visual selection, Eichenberger (1951) visually selected carrot callus with high levels of β-carotene which was maintained in the selected strain for more than 10 years (Naef and Turian, 1963). More recently, strains of *Lithospermum erythrorhizon* which produce high amounts of the commercially important pigment shikonin have been produced by cloning methods (reviewed by Tabata and Fujita, 1985). Such strains have been used commercially to produce shikonin in Japan.

Deus and Zenk (1982) used fluorescence microscopy to identify *Catharanthus roseus* colonies with high fluorescence due to serpentine. Analysis of the selected highly fluorescent colonies by a quantitative radioimmunoassay for serpentine showed that many high yielding strains could be selected by the rapid, nondestructive method. Ultraviolet microspectrophotometry was also used by Ellis (1985) to measure the rosmarinic acid content of individual *Anchusa officinalis* culture cells which were then cultured to form clonal suspension cultures. Unfortunately the rosmarinic levels of the clones were not related to that of the initial cell, but strains with characteristic, relatively stable levels were obtained.

Chemical analysis selection procedures, where a portion of a colony grown from a single cell is analyzed for the desired compound, ideally utilize analytical techniques which are rapid, inexpensive, specific, and very sensitive. An early example of such a system was described by Zenk *et al.* (1977) where portions of colonies derived from single cell clones from *Catharanthus roseus* suspension cultures were analyzed for serpentine and ajmalicine using a sensitive and specific radioimmunoassay. More than 200 samples could be analyzed each day with this procedure. Strains selected with these techniques were high producers although instability was noted.

A massive chemical analysis effort was undertaken by Matsumoto *et al.* (1980) to use recurrent selection with repeatedly cloned tobacco strains to obtain high ubiquinone-10 production (~15 times the wild-type level). Since ubiquinone-10 is a mitochondrial component and correlates with mitochondrial volume of the selected cells (Ikeda *et al.*, 1981), one would wonder if selection for resistance using respiratory inhibitors like antimycin A, piericidin, rotenone, or amytal might also produce the same end result (see Section II,B).

A simpler but less sensitive variation of the last two analytical systems was the "cell squash method" of Ogino *et al.* (1978), where portions of tobacco cell clones were squashed on filter paper and a modified Dragendorff's reagent sprayed on the reverse side of the paper to visualize alkaloids, including nicotine. Strains with increased nicotine levels were selected.

Since both the alkaloid-producing *Caltharanthus roseus* cells and the rosmarinic acid-producing *Anchusa officinalis* cells fluoresce owing to the presence of the compounds, one would expect that these selection techniques could be made more efficient if they were automated by using flow cytometric sorting systems. Brown *et al.* (1984) have sorted *C. roseus* protoplasts according to blue fluorescence apparently caused by their serpentine content as determined by the fluorescence spectrum. These cell sorting instruments can utilize light scattering, cell volume, or fluorescence to sort individual cells or protoplasts at rates of approximately 1,000 units per second. Thus if the desired compound fluoresces, large numbers of cells or protoplasts can be screened and sorted into categories nondestructively using these instruments.

Beside measuring the native fluorescence or color, one might devise staining procedures where an added compound forms a complex with the desired compound which can then be detected. Another possibility might be a fluorescent antibody which would detect the desired compound on the cell surface. These methods, if applied to individual cells or protoplasts, must be nondestructive and permit the cells or protoplasts to grow further.

Destructive methods using stains or antibodies could be used if replica plating procedures were available to obtain replicas of colonies formed from single cells or protoplasts. One replica could be analyzed destructively to detect the high producing colonies while the other would be reserved to allow the rescue of the high producing clones after identification. The first problem with this idea is that no satisfactory replica plating procedure has been developed as yet.

Another colony measuring procedure could be a blotting procedure whereby colonies on a plate are blotted either from above by pressing a paper onto them or from below by removing the colonies in place on some material like filter paper (Horsch *et al.*, 1980). Then a spray or dip staining method specific for the compound could be applied to visualize the high producing colonies. Catt (1981) blotted filter paper onto *Datura innoxia* colonies, measured alkaline phosphatase, β-glucosidase, and peroxidase activity by dipping the paper into the proper reaction mixtures, and did identify stable activity level variants in the case of the first two enzymes. Knoop and Beiderbeck (1985) plated suspension cultures on a cellophane sheet on activated carbon-coated filter paper which was on a nutrient medium. Compounds released by the cells were absorbed in localized areas of the charcoal, and, following removal of the colonies on the cellophane, the replicas were visualized on the filter paper with UV light.

Compounds excreted by plated colonies may also be measured directly using a method like that of Dorn (1965) with *Aspergillus nidulans* where excreted phosphatase produced a red-colored product from the colorless substrate, α-naphthylphosphate, which was incorporated into the medium. High or low phosphatase-excreting colonies can then be identified directly. Some experiments with this system and plant cells have been carried out by Patrick King (personal communication).

I attempted to devise a similar procedure to identify specific amino acid-overproducing plant cell strains by detecting the cross-feeding of amino acid-requiring *Escherichia coli* (J. M. Widholm, unpublished). Small calli were placed on agar-solidified medium containing the auxotrophic bacteria. Wild-type and tryptophan-, phenylalanine-, or proline-overproducing carrot lines and wild-type and tryptophan-overproducing tobacco lines were placed on medium containing the appropriate bacterial mutants. Within a few days a halo caused by bacterial growth formed around each callus piece, but the halo size could not be clearly correlated with amino acid overproduction. This was unexpected since previous studies had shown that the liquid medium surrounding these overproducing cell lines contained increases in the corresponding free amino acids which were approximately proportional to that found within the cells (Widholm, 1977a, and unpublished).

B. Resistance

Selection for resistance to toxic compounds can produce mutants which accumulate certain compounds or detoxify the added toxic compound. Both of these mechanisms can lead to secondary compound accumulation as described below.

Resistance to toxic analogs of a required metabolite can be caused by overproduction of the required metabolite which dilutes the analog thus preventing the toxicity. Such mutants can be selected from large populations by their ability to grow in the presence of the inhibitor. Direct selection for the overproduction of secondary compounds by using secondary compound analogs will probably not be possible since secondary compounds are not, by definition, necessary for cell growth and therefore disruption of the pathway or compound action by an analog will probably not be toxic to the cell. However, overproduction of a primary metabolite which is a precursor of a secondary compound can sometimes increase the synthesis of the secondary compound as described below.

Resistance to toxic analogs can also be due to detoxification of the analog. If this detoxification is carried out by an enzyme which can affect secondary compound production, then selection for resistance can lead to an increase in enzyme levels or to enzyme kinetic characteristics which could cause increased secondary compound production.

There are a number of findings from this laboratory which indicate that resistance selection may produce mutant cell strains capable of accumulating desired compounds. Selection for resistance to the tryptophan analog, 5-methyltryptophan (5MT), with tissue cultures of a number of species, in most cases produces strains which overproduce tryptophan due to the presence of a feedback-altered tryptophan biosynthetic control enzyme, anthranilate synthase (AS) (e.g., Widholm, 1972a,b; Ranch *et al.*, 1983). The free tryptophan concentrations are usually increased about 30-fold in the 5MT-resistant strains. When ten 5MT-resistant carrot and one 5MT-resistant potato strain were tested for auxin autotrophy, five of the carrot and the one potato strain were indeed auxin autotrophic (Widholm, 1977b). The wild-type strains still required auxin for growth. Two 5MT-resistant tobacco (Widholm, 1977b) and 79 5MT-resistant *D. innoxia* strains (Ranch *et al.*, 1983) did not express auxin autotrophy, however. These findings indicate that in some cases with carrot and potato cells, high levels of free tryptophan induces the synthesis of the natural auxin, indoleacetic acid (IAA), which is derived from tryptophan. This was confirmed by Sung (1979) who showed that a wild carrot strain selected as 5MT resistant accumulated free tryptophan, was auxin autotrophic, and also contained from 10 to 40 times the normal IAA levels.

Another amino acid analog, *p*-fluorophenylalanine (PFP), has been used to select a number of resistant cell lines including carrot and tobacco (Palmer and Widholm, 1975; Berlin and Widholm, 1977), *Acer pseudoplatanus* (Gathercole and Street, 1976), and tobacco and *Nicotiana glauca* (Berlin, 1980). The resistance mechanisms appear to include (1) overproduction of phenylalanine which would compete with PFP for incorporation into protein since PFP toxicity is probably due to its incorporation in place of phenylalanine, (2) decreased PFP uptake and, (3) increased activity of phenylalanine ammonia-lyase (PAL) which can detoxify PFP by converting it to *p*-fluorocinnamic acid (Berlin *et al.*, 1982). In the cases where PAL activity was increased, increased synthesis (2- to 12-fold) of phenolics from phenylalanine was also observed.

In more recent work from this laboratory, tobacco (Gonzales *et al.*, 1984) and *Asparagus officinalis* (C. D. Curtiss and J. M. Widholm, unpublished) suspension cultures resistant to the methionine analog, ethionine (ETH), have been selected. In all cases free methionine is increased which in turn leads to the synthesis of *S*-methylmethionine which is undetectable in wild-type tobacco cells. In the *A. officinalis* cells *S*-methylmethionine, as well as *S*-methylcysteine, levels are increased by about 4- and 80-fold, respectively, in the ETH-resistant cells.

These successful accounts of selecting for the overproduction of primary compounds (tryptophan and methionine) or for the metabolism of a toxic analog (PFP) which then leads to secondary compound accumulation indicate that resistance selection procedures may give strains which accumulate desirable compounds. There are also several reports showing that the addition of precursors to the culture medium can increase the levels of secondary compounds.

Several direct tests of these selection ideas have been carried out. Schallenberg and Berlin (1979) selected 10 *Catharanthus roseus* suspension cultures resistant to 5MT. Free tryptophan was increased in all strains up to 30-fold, yet the content of the indole alkaloid, serpentine, which is synthesized from tryptophan, was not increased in the four 5MT resistant strains analyzed. Serpentine synthesis from tryptophan was apparently not increased in these strains since no increase in the levels of tryptophan decarboxylase (TDC) and its product tryptamine, which is an intermediate in the serpentine biosynthetic pathway, were noted in the 5MT-resistant lines. Scott *et al.* (1979) also selected 5MT-resistant *C. roseus* cell strains, and analysis of one showed increased free tryptophan but not increased tryptamine or ajmalicine.

Several tryptophan analogs (4-methyl-, 4-fluoro-, 5-fluoro-, and 5-hydroxytryptophan) were found to be good substrates for TDC while α-methyltryptophan was a competitive inhibitor (Sasse *et al.*, 1983a). It might be possible to select for increased levels of TDC by selecting for resistance to the analogs which are utilized by this enzyme. In order for this selection to be effective, decarboxylation of the analog must make the analog less toxic to the cells. If this occurs then cells with high TDC activity should survive the treatment and the wild-type cells would not. In an attempt to use this technique to increase the TDC activity of *C. roseus* cells, Sasse *et al.* (1983b) selected for resistance to 4-methyltryptophan (4MT) and found that the eight 4MT-resistant cell strains showed 3- to 10-fold increases in TDC activity and that two of these accumulated increased levels of the desired indole alkaloid, ajmalicine. Selection for 4MT resistance with *Peganum* cells has produced strains with around 100-fold higher levels of serotonin than the wild type (J. Berlin, personal communication). These studies demonstrate that in some cases, the selection for increases in TDC activity can produce strains with increased levels of the desired secondary compounds.

C. Auxotrophy

The selection of auxotrophs, i.e., mutants which cannot synthesize a compound needed for growth, might be useful for increasing secondary compound production since a block in a necessary pathway would usually result in the buildup of intermediates before the block. If these intermediates are precursors of secondary compounds or somehow have a positive effect on secondary compound production, then greater secondary compound synthesis may occur. Auxotrophs must be grown in medium with the deficient compound added or growth will not occur, and selection must be done with haploid cells since the traits are generally recessive.

While a number of auxotrophic plants and cultured cells have been selected, the techniques are not routine or efficient except in certain cases, e.g., when chlorate is used to select for nitrate reductase-deficient mutants. The chlorate-resistance selection system is an example of a specific directed "suicide" selection scheme where wild-type cells which contain nitrate reductase activity convert chlorate to the more toxic compound chlorite. Thus these cells are killed while any which lack nitrate reductase survive and grow if a source of reduced nitrogen is present. Müller and Gräfe (1978), for example, used chlorate to select tobacco cell cultures which lacked nitrate reductase. Other directed systems would include allyl alcohol for alcohol dehydrogenase deficiency (Schwartz and Osterman, 1976) and indole analogs for tryptophan synthase deficiency (Widholm, 1981).

More general "suicide" schemes include 5-bromodeoxyuridine (BrdU) which is incorporated into DNA and arsenate which acts as a phosphate analog. Growing cells can utilize both compounds more rapidly than non-growing cells, so are killed on minimal medium. Auxotrophs which do not grow on minimal medium, and so are not killed by the suicide agents, can then be rescued by the addition of the deficient compound. Carlson (1970) applied the BrdU suicide technique to predominantly haploid *Nicotiana tabacum* suspension culture cells. A total of 119 calli formed from more than 1.75×10^6 mutagenized cells plated on enriched medium following BrdU incorporation. The growth of six of these was stimulated by the addition of hypoxanthine, biotin, *p*-aminobenzoic acid, arginine, lysine, or proline. The lesions were incomplete since the strains grew from 17 to 51% as well in minimal medium as with supplementation.

Polacco (1979) used arsenate to kill soybean cells capable of growing with urea as the sole nitrogen source and rescued one strain which would grow only with bovine serum albumin, casein hydrolysate, or soybean culture extracellular proteins as the source of nitrogen. The BrdU and arsenate selection methods have been used by other workers to select some auxotrophs, but the total numbers are very small so these methods cannot be considered to be routine with plant systems.

The auxotrophic selection system which has produced the widest variety of mutants is the total or "brute force" method where cells are individually tested for inability to grow under minimal conditions. In one example, selection was performed with colonies derived from mutagenized *Hyoscyamus muticus* haploid protoplasts grown up on complete medium (containing 20 amino acids, nine vitamins, four nucleic acid bases, inositol, and choline) (Gebhardt *et al.*, 1981). The small colonies were transferred individually to minimal medium and incubated at 32°C. Colonies which showed no growth after 1–2 weeks were rescued by placing on complete medium at 26°C. From 28,872 clones tested, 12 showed auxotrophy or temperature sensitivity. Two clones required histidine. one tryptophan, three nicotinamide, and two required undefined amino acids for growth. Two other clones, which required amino acids for growth, were also chlorate resistant, indicating nitrate reductase deficiency. The two remaining clones either stopped growing or underwent chlorosis and brown pigment accumulation at the restrictive temperature of 32°C.

III. CONCLUSION

I have summarized a series of ideas describing possible ways to select mutant plant cells which produce increased levels of some desired compounds. There could be many other possible schemes, and, since most of those presented are general, specific details will have to be worked out for each desired compound. The recurring problem area, the lack of basic biochemical knowledge, which runs through these discussions can only be remedied by a continued research effort. Once we understand the limiting step in a biosynthetic pathway, we may be able to manipulate levels of this enzyme and/or be able to isolate the key gene and use this gene to transform cells in order to bypass the block.

Another unconventional possibility would be to fuse protoplasts from two different species or strains so that the biosynthetic pathways from each can complement the other to produce the desired end products. This may have been demonstrated in the case of the sexual hybrid between *Baptisia leucantha* and *B. sphaerocarpa* where the leaves contained most of the flavonoids of each parent, as well as four new compounds (Alston and Hempel, 1964).

Generally the selection efforts so far have been plagued by the instability of the selected strains. This indicates that true mutants were not recovered but that cells in a certain differentiated state were selected. This situation might be overcome if the original population were to be mutagenized to produce "stable" mutants where the biochemical alteration would be stabilized. Larger numbers of strains should also be isolated and tested.

Several directed selection studies have not produced strains which synthesize the desired secondary compound, e.g., the tryptophan-accumulating *Catharanthus roseus* strains do not produce more serpentine or ajmalicine (Schallenberg and Berlin, 1979; Scott *et al.*, 1979). In these studies only 11 selected strains were analyzed, and no additional mutagenesis and selection for alkaloid production were carried out. However, in the related selection work where increased TDC activity was selected for (Sasse *et al.*, 1983b; J. Berlin, personal communication), increased levels of ajmalicine and serotonin, respectively, were found.

A very direct problem which affects secondary compound accumulation can be the stability of the compound either chemically or as affected by enzymes in the cells. Wink and Hartmann (1982) found that the concentration of quinolizidine alkaloids fluctuated diurnally in plant leaves and in photomixotrophic *Lupinus polyphyllus* and *Sarothamnus scoparius* suspension cultures grown under day–night light conditions. Cyclic degradation and synthesis apparently accounted for these fluctuations in the cultured cells since exogenous alkaloids were taken up and were degraded within 72 hr. A finding which could be similar was made by Robins *et al.* (1985) who found that *Humulus lupulus* (the hop plant) suspension culture cells contained no detectable α-acids, humulone and cohumulone. These compounds when added to the cultured cells were degraded by a peroxidase activity indicating that even if the cultures synthesized the compounds they would not accumulate because of degradation.

The revelation that degradation can be an important factor regulating secondary compound levels leads to the conclusion that in some cases this phenomenon needs to be investigated and schemes devised to combat the detrimental effect. Possible schemes for reducing compound degradation would include screening for mutants in the high producing strains which do accumulate the compound, attempting to block the degradation pathway using specific inhibitors or compounds similar in structure to the desired compound, and attempting to select mutants directly which lack the specific degradation system.

A most desirable characteristic for high producing strains would be the secretion of the compound from the cells. This would make the isolation of the compound easier and is probably a requirement when immobilized cell systems are employed. It should be possible to devise systems to select secreting cells from a large population of high yielding cells. One would think that blot tests like those of Catt (1981) and Knoop and Beiderbeck (1985), leakage into agar observed visually or with the help of a color reagent (Dorn, 1965), or simply picking cells in the population with low internal concentrations which might be due to secretion might be used to identify secreting variants.

Review of some of the previous examples would indicate that success may require multistep selection which could include selection first for precursor accumulation, then for synthetic enzyme alteration, then for decreased compound degradation, and finally for compound secretion. In most cases fewer steps will probably be needed depending on the results obtained.

The selection of mutants which overproduce compounds, as proposed in this chapter, will not eliminate the need to optimize culture conditions for the production of the compound. This is exemplified by the example of shikonin, the first commercially produced useful secondary compound from plant tissue cultures. Cell-aggregate cloning methods were used to produce strains with 10 times the shikonin derivative level of the original strain (reviewed by Tabata and Fujita, 1985). Then the media and conditions were optimized for shikonin production including a two-stage culture system. The strain was then recloned through protoplasts to obtain strains able to produce the highest levels of shikonin under the optimal conditions. It is of interest to note that the red shikonin derivatives are released from the cells as red granules which are embedded in the cell wall. Tabata and Fujita (1985) speculate that this secretion may increase the production by relieving any feedback control which might operate in the shikonin biosynthetic pathway.

So far there has only been limited success in selecting mutant strains which produce desirable compounds using methods described here. However, one would predict that as knowledge of pathways and of cell culture techniques increases and as more attempts are made these ideas may yield some dramatic successes.

ACKNOWLEDGMENT

The unpublished results were obtained with funds from the Illinois Agricultural Experiment Station and National Science Foundation Grant PCM-81-09808.

REFERENCES

Alston, R. E., and Hempel, K. (1964). Chemical documentation of interspecific hybridization. *J. Hered.* **55,** 267–269.

Berlin, J. (1980). *para*-Fluorophenylalanine resistant cell lines of tobacco. *Z. Pflanzenphysiol.* **97,** 317–324.

Berlin, J., and Sasse, F. (1985). Selection and screening techniques for plant cell cultures. *Adv. Biochem. Eng.* **31,** 99–131.

Berlin, J., and Widholm, J. M. (1977). Correlation between phenylalanine ammonia lyase activity and phenolic biosynthesis in *p*-fluorophenylalanine-sensitive and -resistant tobacco and carrot tissue cultures. *Plant Physiol.* **59,** 550–553.

Berlin, J., Witte, L., Hammer, J., Kuboschke, K. G., Zimmer, A., and Pape, D. (1982). Metabolism of *p*-fluorophenylalanine in *p*-fluorophenylalanine sensitive and resistant tobacco cell cultures. *Planta* **155,** 244–250.

Brown, S., Renaudin, J. P., Prévot, C., and Guern, J. (1984). Flow cytometry and sorting of plant protoplasts: Technical problems and physiological results from a study of pH and alkaloids in *Catharanthus roseus. Physiol. Veg.* **22,** 541–554.

Carlson, P. S. (1970). Induction and isolation of auxotrophic mutants in somatic cell cultures of *Nicotiana tabacum. Science* **168,** 487–489.

Catt, J. W. (1981). Cell-wall mutants from higher plants: A new method using cell-wall enzymes. *Phytochemistry* **20,** 2487–2488.

Deus, B., and Zenk, M. H. (1982). Exploitation of plant cells for the production of natural compounds. *Biotechnol. Bioeng* . **14,** 1965–1974.

Dorn, G. (1965). Genetic analysis of the phosphatases in *Aspergillus nidulans. Genet. Res.* **6,** 13–26.

Eichenberger, M. E. (1951). Sur une mutation survenue dans une culture de tissus de carotte. *C. R. Hebd. Seances Acad. Sci.* **145,** 239–240.

Ellis, B. E. (1985). Characterization of clonal cultures of *Anchuso officinalis* derived from single cells of known productivity. *J. Plant Physiol.* **119,** 149–158.

Gathercole, R. W. E., and Street, H. E. (1976). Isolation, stability and biochemistry of a *p*-fluorophenylalanine-resistant cell line of *Acer pseudoplatanus* L. *New Phytol.* **77,** 29–41.

Gebhardt, C., Schnebli, V., and King, P. J. (1981). Isolation of biochemical mutants using haploid mesophyl protoplasts of *Hyoscyamus muticus* II. Auxotrophic and temperature-sensitive clones. *Planta* **153,** 81–89.

Gonzales, R. A., and Widholm, J. M. (1985). Selection of plant cells for desirable characteristics: Inhibitor resistance. *In* "Plant Tissue Culture: A Practical Approach" (R. A. Dixon, ed.), pp. 67–78. Information Retrieval, London.

Gonzales, R. A., Das, P. K., and Widholm, J. M. (1984). Characterization of cultured tobacco cell lines selected for resistance to a methionine analog, ethionine. *Plant Physiol.* **74,** 640–644.

Horsch, R. B., King, J., and Jones, G. E. (1980). Measurement of cultured plant cell growth on filter paper discs. *Can. J. Bot.* **58,** 2402–2406.

Ikeda, T., Matsumoto, T., Obi, Y., Kisaki, T., and Noguchi, M. (1981). Characteristics of cultured tobacco cell strains producing high levels of ubiquinone-10 selected by a cell cloning technique. *Agric. Biol. Chem.* **45,** 2259–2263.

Knoop, B., and Beiderbeck, R. (1985). Adsorbent filter—A tool for the selection of plant suspension culture cells producing secondary substances. *Z. Naturforsch., C. Biosci.* **40C,** 297–300.

Matsumoto, T., Ikeda, T., Kanno, N., Kisaki, T., and Noguchi, M. (1980). Selection of high ubiquinone 10-producing strains of tobacco cultured cells by cell cloning technique. *Agric. Biol. Chem.* **44,** 967–969.

Müller, A. J., and Gräfe, R. (1978). Isolation and characterization of cell lines of *Nicotiana tabacum* lacking nitrate reductase. *Mol. Gen. Genet.* **161,** 67–76.

Naef, J., and Turian, G. (1963). Sur les carotinoides du tissu cambial de recine de carotte cultive in vitro. *Phytochemistry* **2,** 173–177.

Ogino, T., Hiraoka, N., and Tabata, M. (1978). Selection of high nicotine-producing cell lines of tobacco callus by single-cell cloning. *Phytochemistry* **17,** 1907–1910.

Palmer, J. E., and Widholm, J. M. (1975). Characterization of carrot and tobacco cells resistant to *p*-fluorophenylalanine. *Plant Physiol.* **56,** 233–238.

Polacco, J. C. (1979). Arsenate as a potential negative selection agent for deficiency variants in cultured plant cells. *Planta* **146,** 155–160.

Ranch, J. P., Rick, S., Brotherton, J. E., and Widholm, J. M. (1983). The expression of 5-methyltryptophan-resistance in plants regenerated from resistant cell lines of *Datura innoxia*. *Plant Physiol.* **71,** 136–140.

Robins, R. J., Furze, J. M., and Rhodes, M. J. C. (1985). α-Acid degradation by suspension culture cells of *Humulus lupulus*. *Phytochemistry* **24,** 709–714.

Sasse, F., Buchholz, M., and Berlin, J. (1983a). Site of action of growth inhibitory tryptophan analogues in *Catharanthus roseus* cell suspension cultures. *Z. Naturforsch., C: Biosci.* **38C,** 910–915.

Sasse, F., Buchholz, M., and Berlin, J. (1983b). Selection of cell lines of *Catharanthus roseus* with increased tryptophan decarboxylase activity. *Z. Naturforsch., C: Biosci.* **38C,** 916–922.

Schallenberg, J., and Berlin, J. (1979). 5-Methyltryptophan resistant cells of *Catharanthus roseus*. *Z. Naturforsch., C: Biosci.* **34C,** 541–545.

Schwartz, D., and Osterman, J. (1976). A pollen selection system for alcohol dehydrogenase-negative mutants in plants. *Genetics* **83,** 63–65.

Scott, A. I., Mizukami, H., and Lee, S. L. (1979). Characterization of a 5-methyltryptophan resistant strain of *Catharanthus roseus* cultured cells. *Phytochemistry* **18,** 795–798.

Sung, Z. R. (1979). Relationship of indole-3-acetic acid and tryptophan concentrations in normal and 5-methyltryptophan-resistant cell lines of wild carrots. *Planta* **145,** 339–345.

Tabata, M., and Fujita, Y. (1985). Production of shikonin by plant cell cultures. *In* "Biotechnology in Plant Science: Relevance to Agriculture in the Eighties" (M. Zaitlin, P. Day, and A. Hollaender, eds.), pp. 207–218. Academic Press, New York.

Widholm, J. M. (1972a). Cultured *Nicotiana tabacum* cells with an altered anthranilate synthetase which is less sensitive to feedback inhibition. *Biochim. Biophys. Acta* **261,** 52–58.

Widholm, J. M. (1972b). Anthranilate synthetase from 5-methyltryptophan-susceptible and -resistant cultured *Daucus carota* cells. *Biochim. Biophys. Acta* **279,** 48–57.

Widholm, J. M. (1977a). Selection and characterization of biochemical mutants. *In* "Plant Tissue Culture and Its Biotechnological Application" (W. Barz, E. Reinhard, and M. H. Zenk, eds.), pp. 112–122. Springer-Verlag, Berlin and New York.

Widholm, J. M. (1977b). Relation between auxin-autotrophy and tryptophan accumulation in cultured plant cells. *Planta* **134,** 103–108.

Widholm, J. M. (1980). Selection of plant cell lines which accumulate certain compounds. *In* "Plant Tissue Culture as a Source of Biochemicals" (E. J. Staba, ed.), pp. 99–113. CRC Press, Boca Raton, Florida.

Widholm, J. M. (1981). Utilization of indole analogs by carrot and tobacco cell tryptophan synthase *in vivo* and *in vitro*. *Plant Physiol.* **67,** 1101–1104.

Wink, M., and Hartmann, T. (1982). Diurnal fluctuation of quinolizidine alkaloid accumulation in legume plants and photomixotrophic cell suspension cultures. *Z. Naturforsch., C: Biosci.* **37C,** 369–375.

Zenk, M. H., El-Shagi, H., Arens, H., Stockigt, J., Weiler, E. W., and Deus, B. (1977). Formation of the indole alkaloids serpentine and ajmalicine in cell suspension cultures of *Catharanthus roseus*. *In* "Plant Tissue Culture and Its Biotechnological Application" (W. Barz, E. Reinhard, and M. H. Zenk, eds.), pp. 27–43. Springer-Verlag, Berlin and New York.

CHAPTER **8**

New Approaches to Genetic Manipulation of Plants

Denes Dudits

Institute of Genetics
Biological Research Center
Hungarian Academy of Sciences
Szeged, Hungary

I. INTRODUCTION

Recent progress in molecular and cell genetics has opened new avenues for manipulation of the plant genome. In attempts to alter plant cell metabolism, various methods such as DNA transformation or somatic cell fusion can be successfully used to extend genetic variability. These techniques share a common feature, namely, cultured somatic cells or tissues are the primary targets for introduction of new genes. Since cultured plant cells are directly applied to produce secondary metabolites or biologically active compounds, the use of genetically engineered cells has considerable significance in commercial production. In addition to the application in plant cell fermentation, the potential impact of novel genetic methods on improvement of medicinal plants has to be considered as an additional contribution to complement traditional breeding programs.

There are several major approaches to the introduction of genes into somatic cells. The use of advanced DNA transformation systems to integrate single genes into a host chromosome requires the successful cloning of the desired gene by recombinant DNA technology. If the gene of interest is not available as a cloned DNA sequence, the methods of cell fusion or organelle transfer offer an alternative solution. Frequently, there is a need to manipulate various traits simultaneously. For this purpose, techniques of cell genetics can also serve as tools. This chapter summarizes the characteristic features of various methods ranging from transfer of complete genomes by protoplast fusion to introduction of defined genes by DNA transformation. In many instances the potentials are demonstrated by test systems; further research is needed to extend the application to medicinal plants.

II. SOMATIC HYBRIDIZATION BY PROTOPLAST FUSION

Combination of cell organelles—nuclei, chromosomes, chloroplasts, mitochondria—with different genetic backgrounds can be accomplished parasexually by protoplast fusion. The resulting cell hybrids represent new genotypes combining characters from both parents. Among several requirements, the successful application depends on protoplast technology (isolation and culture of protoplasts) and on techniques for selection of fusion products. A regeneration system is needed also if the cell hybrids are produced for breeding purposes.

A. Protoplast Culture Systems for Medicinal Plants

Table I lists some achievements in culturing protoplasts from plant species with significance in secondary metabolite synthesis. Several species in the Solanaceae, such as *Nicotiana*, *Datura*, *Atropa*, and *Solanum*, have been extensively used in a wide variety of genetic manipulation experiments. The tissue culture systems are well developed for these plants. On the other hand, important plant species yielding natural products in plant cell cultures are missing from the list. Hopefully, in the future, establishment of new protoplast systems will contribute to the increase of cell productivity by genetic improvement.

In general, having a fine, fast growing cell suspension culture is sufficient to obtain dividing protoplast cultures. Incubation of a 2-day-old subculture with an equal volume of an enzyme solution—2% Onozuka R10 or RS (Yakult Biochemicals Co., Japan), 1% pectinase (Sigma), 0.5% Drisellase (Kyowa Hakko, Kogyo, Japan), 0.5% Rhozyme (Corning Glass, Corning, New York)—in 0.35 *M* sorbitol plus 0.35 *M* mannitol, 3 m*M* 2-(*N*-morpholino)ethanesulfonic acid, 6 m*M* $CaCl_2 \cdot H_2O$, 0.7 m*M* NaH_2PO_4 at pH 5.6—can result in reasonable yield of healthy protoplasts in a large number of species. Minor changes, e.g., in concentra-

TABLE I

Protoplast Culture of Phytochemical-Producing Species

Species	Source of protoplasts	Results	Reference
Atropa belladonna	Suspension	Plant	Gosch *et al.* (1975)
Datura innoxia	Mesophyll	Plant	Schieder (1975)
Hyoscyamus muticus	Mesophyll	Plant	Wernicke *et al.* (1979)
Nicotiana tabacum	Mesophyll	Plant	Nagata and Takebe (1970)
Nicotiana sp.	Mesophyll	Plant	Bourgin *et al.* (1979)
Nicotiana rustica	Mesophyll	Plant	Gill *et al.* (1979)
Solanum dulcamara	Mesophyll	Plant	Binding and Nehls (1977)
Solanum tuberosum	Mesophyll	Plant	Shepard and Totten (1977)
Digitalis lanata	Mesophyll	Plant	Li (1981)
Digitalis purpurea	Suspension	Callus	Diettrich *et al.* (1980)
Coffea arabica	Callus	Callus	Sondahl *et al.* (1980)
Rehmannia glutinosa	Mesophyll	Shoots	Xu and Davey (1983)
Trigonella foenumgraecum	Mesophyll	Shoot	Shekhawat and Galston (1983)

tion of osmotic stabilizer, might be needed in a few cases. Frequently, the nutritional requirements of protoplasts and cell suspensions are very similar. Glucose (0.4–0.6 *M*) is recommended as an osmoticum. Conditions for plant regeneration can differ and should be optimized for each species.

B. Protoplast Fusion and Hybrid Selection

Plant protoplasts surrounded by a negatively charged plasmalemma can be induced to fuse by chemical or physical stimuli. High frequency of membrane fusion and heterokaryon formation can be achieved by polyethylene glycol (PEG) treatment. Elution of PEG with solution of high pH and calcium ion concentration increases significantly the number of fusion products (Kao *et al.*, 1974). Also, in an ac electric field through dielectrophoresis the protoplasts form a pearl chain across the electrode gap. Application of a single dc square wave pulse induces electrofusion (Zimmermann and Scheurich, 1981).

Membrane fusion as a physicochemical process can occur between a variety of cell types. No incompatibility barrier is found at this stage. At present, protoplast fusion is not a limiting step. The available methods result in a sufficient number of fused cells for further culture and for production of cell hybrids.

After fusion treatment, the cultures consist of a mixed cell population. Identification and recovery of the fusion products can be carried out during the early development of the hybrids or later at the callus stage or after plant regeneration. Fusion of morphologically different protoplasts, isolated, e.g., from leaf mesophyll cells and cell suspension culture, or labeling the parental cells with viable stains makes it possible to identify and mechanically select the fused cells. Both manual isolation (Kao, 1977) and flow sorting of fluorescence-labeled protoplasts (Glimelius *et al.*, 1986) have been successfully used for selection. The selection scheme can be simplified by choosing a nondividing protoplast type as one of the parental cells. Use of irreversible biochemical inhibitors can help to eliminate certain fractions of the fused cell populations (Nehls, 1978). The fusion will activate mitosis in nuclei of nondividing parental cells or complement missing functions. The advantage of above-mentioned approaches is that they can be applied directly in a variety of fusion combinations and make unnecessary the introduction of selectable mutation which might alter the genotype of parental species. These methods are recommended for intra- and interspecific somatic hybridization. For wide hybridization where application of selective pressure may be essential to maintain foreign genetic material, these approaches show limitations.

Two major selection strategies can be designed for selection of somatic hybrids with biochemical mutants. The first one is based on recessive mutations such as chlorophyll deficiency or nutritional auxotrophy. Correction of these defects in fused cells results from genetic complementation between the parental genomes. Frequently, only one of the parental lines carries the mutation and the other partner is a wild-type protoplast. Albino or light-sensitive mutants have been successfully used in large number of somatic hybridization experiments (see Evans, 1983). Among auxotrophic mutants, nitrate reductase minus lines are favored fusion partners (Glimelius *et al.*, 1978; Somers *et al.*, 1986).

The second approach uses dominant selectable markers. Cell lines with resistance against metabolic inhibitors such as amino acid analogs (White and Vasil, 1979; Harms *et al.*, 1981), methotrexate (Dudits *et al.*, 1985), and picloram (Evola *et al.*, 1983) have been used as fusion partners. A powerful selection system with wide application can be developed by transformation of parental plants with Ti plasmid carrying kanamycin resistance (Deak *et al.*, unpublished). Tagging the cells with protein—neomycin phosphotransferase—and DNA markers offers significant advantages by reducing the time requirement for introduction of the marker into the parental genotype. A universal hybridizer was established by combination of azaguanine and α-amanitin resistance

(LoSchiavo *et al.*, 1983). Finally the fusion products can be identified at the plant stage by morphological (Dudits *et al.*, 1977) or biochemical (Wetter and Dyck, 1983) markers.

C. Gene Expression in Intra- and Intergenetic Somatic Hybrids

A majority of the somatic hybrids introduced thus far belong to the family Solanaceae (Evans, 1983). The few hybrids with *Daucus* and *Brassica* species extend the range of fusion combinations. In most of the cases, characterization of somatic hybrids is restricted to morphological traits, isozymes, and chromosome number. Data about phytochemical content of these hybrids were rarely published. Fusion between carrot genotypes with low and high carotenoid content resulted in white-rooted somatic hybrids (Dudits *et al.*, 1977). Pigmentation and anthocyanin formation as a genetic markers were also characterized in several somatic hybrid plants (Power *et al.*, 1978, 1980; Schieder, 1978; Evans *et al.*, 1981). Schieder (1984) reported important results about alkaloid production in *Datura* somatic hybrids. The content of scopolamine in *D. innoxia* + *D. stramonium* was 0.2% and in *D. innoxia* + *D. discolor*, 0.22%.

As far as the regulatory mechanisms are concerned, morphogenic differentiation and synthesis of secondary metabolites have similar features in genetic control of cell metabolism in suspension cultures. In both cases coordinated expression of several genes is required for special cell functions at the dedifferentiated callus stage. In higher plants, a wide range of somatic cell types have been shown to be totipotent. Uniquely among the eukaryotic cells, plants have a flexible genome able to reprogram gene expression patterns and to start a complete, new developmental program. In general, fully differentiated cells of mature plants serve as inocula for initiation of cell or tissue cultures. These cells with various differentiated functions like photosynthesis or accumulation of secondary metabolites are removed from their original environment into an artificial, *in vitro* condition, where they are exposed to various stress factors. The aphysiologically high concentration of plant hormones (auxins, cytokinins) induces a new, unscheduled series of cell divisions. Initiation of the cell cycle opens the possibility for reprogramming gene expression. It includes switching off several differentiated functions and activation of housekeeping genes required for the intensive proliferation in a fast growing callus culture. During these considerable changes in cell metabolism, various biosynthetic pathways can remain unaffected. Therefore certain differentiated functions can be active or the chromatin structure allows the return of genes to an active stage under defined conditions. Understanding the molecular basis of these complex regulatory processes is required to extend the use of cultured plant cells.

In plant cell culture systems genetic analysis of morphogenic and synthetic functions can be accomplished by the parasexual means of cell hybridization. There are several examples for restoration of plant regeneration capability through cell fusion (Maliga *et al.*, 1977; Melchers *et al.*, 1978; Gleba and Sytnik, 1982). More interestingly, fusion between two nonmorphogenic cell lines can complement the missing functions, and the fusion products form plants (Glimelius and Bonett, 1981; Dudits *et al.*. 1985). Behavior of a differentiation-dependent function was studied by fusion of cycloheximide (CH)-resistant and -sensitive cells (Lazar *et al.*, 1981). In embryogenic carrot cell suspension cultures expression of CH inactivation correlated with the differentiated stage of the culture. The dedifferentiated callus cells are sensitive to CH because of the lack of inactivation. Simultaneously with embryogenic differentiation the cells become CH resistant by decomposing the drug (Sung *et al.*, 1981). Combining the two genomes with different regulatory signals showed the dominance of the undifferentiated function. All the hybrids were found to be CH sensitive. The recessive nature of differentiation-related inactivation suggests a negative control for this trait. The CH inactivation can be a unique, special case. However, further studies are needed to outline a general conclusion. The question is still to be answered whether fusion between two types of cells, one with good production of secondary metabolite and the other one with a high growth rate, can result in hybrids which combine these two advantageous characteristics.

D. Intergeneric Somatic Hybridization

Methods of protoplast fusion have made it possible to produce hybrid cell lines from distantly related plant species (Dudits and Praznovszky, 1985). In intergeneric fusion products, chromosome composition and capability for plant regeneration strongly depend on the evolutionary distance between the parental species. In many experiments, the hybrid cell lines were not able to differentiate into plants, and nuclear hybrid formation was linked with an intensive and spontaneous chromosome elimination (Kao, 1977; Binding and Nehls, 1978; Chien *et al.*, 1982). A closer phylogenetic relation between the fusion partners can allow regeneration of hybrid plants with nearly complete parental genomes (Melchers *et al.*, 1978; Krumbiegel and Schieder, 1979; Gleba and Hoffmann, 1980; Gleba *et al.*, 1982). Partial elimination of chromosomes from one parent has also been observed. Biochemical analysis of intergeneric cell hybrids and plants revealed isozyme forms from both parents. These data showed the functional coexistence of two genomes in an artificially created hybrid genotype.

A special genomic construction is characteristic of a set of somatic hybrids after intergenic fusion. Spontaneous or induced chromosome elimination can result in asymmetric hybrids with reduced amount of gene material from one of the parental species. Up to now the following two approaches were found to be successful to control chromosome elimination.

1. Fusion between Dividing and Mitotically Inactive Protoplasts

Cytological studies proved that the mitotic partner can activate the nondividing cells (Kao and Michayluk, 1974), and in mitotic–interphase heterokaryons prematurely condensed chromosomes can be formed within few hours after fusion (Szabados and Dudits, 1980). In somatic hybrids of *Daucus carota* and *Aegopodium podagraria,* the hybrid plants retained *Aegopodium*-specific characters after elimination of *Aegopodium* chromosomes. In addition to restoring a nuclear albino mutation in carrot, the hybrids formed roots with low carotenoid content and neurosporene accumulation, characteristic of *Aegopodium* roots (Dudits *et al.*, 1979). Somatic hybridization of these species provides an example for transfer of phytochemical markers.

2. High Doses of Irradiation as a Tool to Induce Genome Instability in Intergeneric Somatic Hybrids

The successful use of irradiation to induce genome fragmentation and preferential chromosome loss was demonstrated in various experimental systems (Dudits *et al.*, 1980; Gupta *et al.*, 1982, 1984; Itoh and Futsuhara, 1983; Somers *et al.*, 1986). Maintenance of one or two chromosomes from the donor, irradiated partner may correlate with detection of specific biochemical and morphological traits in hybrids which are phenotypically similar to the recipient parent. Considering the practical requirements, this approach has an increasing significance in plant improvement at both the cell and the whole plant level.

III. ISOLATED PLANT CHROMOSOMES AS POTENTIAL VECTORS FOR GENE TRANSFER

Success in the use of mammalian chromosomes to transfer nuclear genes by integration of subchromosomal fragments into recipient genomes (Klobutcher and Ruddle, 1981) suggests a similar approach for developing a system for gene transfer in higher plants. Since the first attempts (Malmberg and Griesbach, 1980; Szabados *et al.*, 1981), significant progress has been made in isolating intact plant chromosomes from mitotic protoplasts (Hadlaczky *et al.*, 1983). Treatment of a mixture of chromosome suspension and plant protoplasts with PEG can cause an uptake of single chromosomes into cells (Szabados *et al.*, 1981). To follow the fate of foreign chromosomes, T. Praznovszky *et al.* (unpublished) have introduced chromosomes from a methotrexate-resistant carrot into protoplasts of *Triticum monococcum* cell suspensions. Cytological characterization of the recovered methotrexate-resistant *Triticum* clones showed that the majority of isolates carried no additional carrot chromosomes. However, in one of the lines, the presence of a single small carrot chromosome was detected after 3 months of culturing in methotrexate medium. These results encourage further studies to work out an efficient system which utilizes plant chromosomes as vectors in gene transfer.

IV. DNA TRANSFORMATION

Nowadays, well-developed vector systems are available to introduce single isolated genes into a plant genome. One of the most efficient methods of producing transgenic plants is based on the tumor-inducing (Ti) plasmid from *Agrobacterium tumefaciens* as a natural gene transfer system for dicotyledonous plants (Schell *et al.*, 1984). During interaction between wounded plant tissues or single cells and *Agrobacterium*, a specific segment (T-DNA) of the Ti plasmid is transferred into plant cells, where the T-DNA becomes integrated into the plant nuclear genome (Chilton *et al.*, 1980; Willmitzer *et al.*, 1980). The T-DNA is bounded by conserved 25-bp direct repeat sequences that are required for integration (Wang *et al.*, 1984). The transfer and integration of T-DNA is independent of expression of T-DNA coded genes, while the genes of the virulence (*vir*) region on the Ti plasmid encode trans-acting products involved in these processes (Joos *et al.*, 1983; Hille *et al.*, 1984). Inactivation or deletion of phytohormone genes on T-DNA results in "disarmed plasmids," allowing plant redifferentiation from transformed tissues (Lemmers *et al.*, 1980).

The use of non-oncogenic Ti plasmid derivatives requires an alternative selection method which is independent of tumor phenotype. Construction of chimeric genes conferring resistance to antibiotics have been used successfully for selection of transformants (Herrera-Estrella *et al.*, 1983; Fraley *et al.*, 1983). One of the sophisticated vectors known as the "binary vector" consists of two plasmids: one is a wide host-range cloning vector with functional T-DNA border sequences and the other contains the *vir* functions in trans (Hoekema *et al.*, 1983). In the vector systems developed by Koncz and Schell (1986), all the required elements—such as plant selectable and screenable genes flanked by the border sequences of T-DNA unique cloning sites, suitable bacterial markers, wide host-range replicon, and mobilization functions—were united in a single plant vector cassette.

For selection of transformants the vectors are generally constructed with plant–bacterial chimeric genes conferring resistance to various antibiotics. These constructs are equipped with a transcriptional control signal from the nopaline synthase gene (Fraley *et al.*, 1983; Herrera-Estrella *et al.*, 1983) or from other highly expressed, constitutive promoter sequences such as the 19S or 35S promoter regions of cauliflower mosaic virus (Rogers *et al.*, 1985). The introduced foreign genes can also be expressed by light-regulated (Lamppa *et al.*, 1985; Morelli *et al.*, 1985) or tissue-specific (Eckes *et al.*, 1986; Koncz and Schell, 1986) functions.

The rapidly expanding list of genes that have been introduced by transformation includes viral genes, encoding tobacco mosaic virus coat protein (Beachy *et al.*, 1985); plant genes such as phaseolin (Murai *et al.*, 1983); zein from corn (Matzke *et al.*, 1984); the small subunit of ribulose 1,5-bisphosphate carboxylase (Herrera-Estrella *et al.*, 1984); chlorophyll *a*/*b* binding protein (Simpson *et al.*, 1985); and 5-enolpyruvylshikimate-3-phosphate (Shah *et al.*, 1986). In addition to *Agrobacterium*-mediated gene transfer, efficient techniques have been developed to introduce plasmid DNA molecules directly into plant protoplasts (Paszkowski *et al.*, 1984).

V. CONCLUSION

The above discussion documents that a series of methods are available for plant genetic manipulation. Clearly, the new techniques can be applied to medicinal plants as well. Together with the improvements in culture conditions, genetic factors have a significant impact on product yield of cell cultures and cultivated plants. Alteration of the gene pool helps to analyze the molecular, biochemical, and physiological basis of biosynthetic functions and would allow improvements in productivity.

REFERENCES

Beachy, R. N., Abel, P., Oliver, M. J., Barun D. E., Fraley, R. T., Rogers, S. G., and Horsch, R. B. (1985). Potential for applying genetic transformation to studies of viral pathogenesis and cross-protection. *In* "Biotechnology in Plant Science: Relevance to Agriculture in the Eighties" (M. Zaitlin, P. Day, and A. Hollaender, eds.), pp. 265–275. Academic Press, Orlando, Florida.

Binding, H., and Nehls, R. (1977). Regeneration of isolated protoplasts to plants in *Solanum dulcamara* L. *Z. Pflanzenphysiol.* **85,** 279–280.

Binding, H., and Nehls, R. (1978). Somatic cell hybridization. *Vicia faba + Petunia hybrida. Mol. Gen. Genet.* **164,** 137–143.

Bourgin, J. P., Chupeau, U., and Missonier, C. (1979). Plant regeneration from mesophyll protoplasts of several *Nicotiana* species. *Physiol. Plant.* **45,** 288–292.

Chien, Y. C., Kao, K. N., and Wetter, L. R. (1982). Chromosomal and isozyme studies of *Nicotiana tabacum–Glycine max* hybrid cell lines. *Theor. Appl. Genet.* **62,** 301–304.

Chilton, M. D., Saiki, R. K., Yadav, N., Gordon, M. P., and Quetier, F. (1980). T-DNA from *Agrobacterium* Ti-plasmid is in the nuclear DNA. *Proc. Natl. Acad. Sci. U.S.A.* **77,** 4060–4064.

Diettrich, B., Neumann, D., and Luckner, M. (1980). Protoplast derived clones from cell cultures of *Digitalis purpurea. Planta Med.* **38,** 375–382.

Dudits, D., and Praznovszky, T. (1985). Intergeneric gene transfer by protoplast fusion and uptake of isolated chromosomes. *In* "Biotechnology in Plant Science: Relevance to Agriculture in the Eighties"(M. Zaitlin, P. Day, and A. Hollaender, eds.), pp. 115–127. Academic Press, Orlando, Florida.

Dudits, D., Hadlaczky, Gy., Levi, E., Fejer, O., Haydu, Zs., and Lazar, G. (1977). Somatic hybridization of *Daucus carota* and *D. capillifolius* by protoplast fusion. *Theor. Appl. Genet.* **51,** 127–132.

Dudits, D., Hadlaczky, Gy., Bajszar, Gy., Koncz, Cs., Lazar, G., and Horvath, G. (1979). Plant regeneration from intergeneric cell hybrids. *Plant Sci. Lett.* **15,** 101–112.

Dudits, D., Fejer, O., Hadlaczky, Gy., Koncz, Cs., Lazar, G. B., and Horvath, G. (1980). Intergeneric gene transfer mediated by protoplast fusion. *Mol. Gen. Genet.* **179,** 283–288.

Dudits, D., Maroy, E., Praznovszky, T., and Feher, A. (1985). Expression of methotrexate resistance in carrot intraspecific and intergeneric somatic hybrids. *In* "Proceedings of International Workshop on Somatic Embryogenesis" (M. Terzi *et al.*, eds.), pp. 108–112. IPRA, Italy.

Eckes, P., Rosahl, S., Schell, J., and Willmitzer, L. (1986). Isolation and characterization of a light-inducible, organ-specific gene from potato and analysis of its expression after tagging and transfer into tobacco and potato shoots. *Mol. Gen. Genet.* **205,** 14–22.

Evans, D. A. (1983). Protoplast fusion. *In* "Handbook of Plant Cell Culture" (D. A. Evans,

R. Sharp, D. Ammirato, and Y. Yamada, eds.), Vol. 1, pp. 291–321. Macmillan, New York.

Evans, D. A., Flick, C. E., and Jensen, R. A. (1981). Disease resistance: Incorporation into sexually incompatible somatic hybrids in the genus *Nicotiana*. *Science* **213,** 907–909.

Evola, S. V., Earle, E. D., and Chaleff, R. S. (1983). The use of genetic markers selected *in vitro* for the isolation and genetic verification of intraspecific somatic hybrids of *Nicotiana tabacum* L. *Mol. Gen. Genet.* **189,** 441–446.

Fraley, R. T., Rogers, S. G., Horsch, R. B., Sanders, P. R., Flick. J. S., Adams, S. P., Bittner, M. L., Brand, L. A., Fink, C. L., Fry, J. S., Galluppi, G. R., Goldberg, S. B., Hoffmann, N. L., and Woo, S. C. (1983). Expression of bacterial genes in plant cells. *Proc. Natl. Acad. Sci. U.S.A.* **80,** 4803–4807.

Gill, R., Rashid, A., and Maheshwari, S. C. (1979). Isolation of mesophyll protoplasts of *Nicotiana rustica* and their regeneration into plants flowering *in vitro*. *Physiol. Plant.* **47,** 7–10.

Gleba, Y. Y., and Hoffmann, F. (1980). "Arabidobrassica": A novel plant obtained by protoplast fusion. *Planta* **149,** 112–117.

Gleba, Y. Y., and Sytnik, K. M. (1982). "Protoplast Fusion and Genetic Engineering of Higher Plants." Springer-Verlag, Berlin and New York.

Gleba, Y. Y., Momot, V. P., Cherep, N. N., and Skarzynskaya, M. V. (1982). Intertribal hybrid cell lines of *Atropa belladonna* × *Nicotiana chinensis* obtained by cloning individual protoplast fusion products. *Theor. Appl. Genet.* **62,** 75–79.

Glimelius, K., and Bonett, H. T. (1981). Somatic hybridization in *Nicotiana:* Restoration of photoautotrophy to an albino mutant with defective plasmids. *Planta* **153,** 497–503.

Glimelius, K., Eriksson, T., Grafe, R., and Muller, A. J. (1978). Somatic hybridization of nitrate reductase-deficient mutants of *Nicotiana tabacum* by protoplast fusion. *Physiol. Plant.* **44,** 273–277.

Glimelius, K., Djupsjobacka, M., and Fellner-Feldegg, H. (1986). Selection and enrichment of plant protoplast heterokaryons of Brassicaceae by flow sorting. *Plant Sci.* **45,** 133–141.

Gosch, G., Bajaj, Y. P. S., and Reinert, J. (1975). Isolation, culture, and induction of embryogenesis in protoplasts from cell suspension of *Atropa belladonna*. *Protoplasma* **86,** 405–410.

Gupta, P. P., Gupta, M., and Schieder, O. (1982). Correction of nitrate reductase defect in auxotrophic plant cells through protoplast-mediated intergeneric gene transfer. *Mol. Gen. Genet.* **188,** 378–383.

Gupta, P. P., Schieder, O., and Gupta, M. (1984). Intergeneric nuclear gene transfer between somatically and sexually incompatible plants through asymmetric protoplast fusion. *Mol. Gen. Genet.* **197,** 30–35.

Hadlaczky, Gy., Bisztray, Gy., Praznovszky, T., and Dudits, D. (1983). Mass isolation of plant chromosomes and nuclei. *Planta* **157,** 278–285.

Harms, C. T., Potrykus, I., and Widholm, J. M. (1981). Complementation and dominant expression of amino acid analogue resistance markers in somatic hybrid clones from *Daucus carota* after protoplast fusion. *Z. Pflanzenphysiol.* **101,** 377–390.

Herrera-Estrella, L., De Block, M., Messens, E., Hernalsteens, J. P., Van Montagu, M., and Schell, J. (1983). Chimaeric genes as dominant selectable markers in plant cells. *EMBO J.* **2,** 987–995.

Herrera-Estrella, L., Van Den Broeck, G., Maenhaut, R., Van Montagu, M., Schell, J., Timko, M., and Cashmore, A. (1984). Light-inducible and chloroplast-associated expression of chimaeric gene introduced into *Nicotiana tabacum* using a Ti plasmid vector. *Nature (London)* **310,** 115–120.

Hille, J., Van Kan, J., and Schilperoort, R. (1984). Trans acting virulence functions of the octopine Ti plasmid from *Agrobacterium tumefaciens*. *J. Bacteriol.* **158,** 754–756.

Hoekema, A., Hirsch, P. R., Hooykaas, P. J., and Schilperoort, R. A. (1983). A binary plant vector strategy based on separation of Vir- and T-region of the *Agrobacterium tumefaciens* Ti-plasmid. *Nature (London)* **303,** 179–180.

Itoh, K., and Futsuhara, Y. (1983). Intergeneric transfer of only part of genome by fusion between non-irradiated protoplasts of *Nicotiana glauca* and X-ray irradiated protoplasts of *N. langsdorffii*. *Jpn. J. Genet.* **58,** 545–553.

Joos, H., Inzew, D., Caplan, A., Sormann, M., Van Montagu, M., and Schell, J. (1983). Genetic analysis of T-DNA transcripts in nopaline crown galls. *Cell (Cambridge, Mass.)* **32,** 1057–1067.

Kao, K. N. (1977). Chromosomal behaviour in somatic hybrids of soybean–*Nicotiana glauca*. *Mol. Gen. Genet.* **150,** 225–230.

Kao, K. N., and Michayluk, M. R. (1974). A method for high-frequency intergeneric fusion of plant protoplasts. *Planta* **115,** 335–367.

Kao, K. N., Constabel, F., Michayluk, M. R., and Gamborg, O. L. (1974). Plant protoplast fusion and growth of intergeneric hybrid cells. *Planta* **120,** 215–227.

Klobutcher, L. A., and Ruddle, F. H. (1981). Chromosome mediated gene transfer. *Annu. Rev. Biochem.* **50,** 533–554.

Koncz, Cs., and Schell, J. (1986). The promoter of Ti-DNA gene 5 controls the tissue-specific expression of chimaeric genes carried by a novel type of *Agrobacterium* binary vector. *Mol. Gen. Genet.* **204,** 383–396.

Krumbiegel, G., and Schieder, O. (1979). Selection of somatic hybrids after fusion of protoplasts from *Datura innoxia* and *Atropa belladonna*. *Planta* **145,** 371–375.

Lamppa, G., Nagy, F., and Chua, N. H. (1985). Light-regulated and organ specific expression of a wheat *Cab* gene in transgenic tobacco. *Nature (London)* **316,** 750–752.

Lazar, G. B., Dudits, D., and Sung, Z. R. (1981). Expression of cycloheximide resistance in carrot somatic hybrids and their segregants. *Genetics* **98,** 347–356.

Lemmers, M., De Beuckeleer, M., Holsters, M., Zambryski, P., Depicker, A., Hernalsteens, J. P., Van Montagu, M., and Schell, J. (1980). Internal organization, boundaries and integration of Ti-plasmid DNA in nopaline crown gall tumours. *J. Mol. Biol.* **144,** 353–376.

Li, X. H. (1981). Plantlet regeneration from mesophyll protoplasts of *Digitalis lanata* Ehrh. *Appl. Genet.* **60,** 345–347.

LoSchiavo, F., Giovinazzo, G., and Terzi, M. (1983). 8-Azaguanine resistant carrot cell mutants and their use as universal hybridizers. *Mol. Gen. Genet.* **192,** 326–329.

Maliga, P., Lazar, G., Joo, F., Nagy, A. H., and Menczel, L. (1977). Restoration of morphogenic potential in *Nicotiana* by somatic hybridization. *Mol. Gen. Genet.* **157,** 291–296.

Malmberg, R., and Griesbach, R. I. (1980). The isolation of mitotic and meiotic chromosomes from plant protoplasts. *Plant Sci. Lett.* **17,** 141–147.

Matzke, A. M., Susani, M., Binns, A. N., Lewis, E. D., Rubenstein, I., and Matzke, A. J. M. (1984). Transcription of a zein gene introduced into sunflower using a Ti plasmid vector. *EMBO J.* **3,** 1525–1531.

Melchers, G., Sacristan, M. D., and Holder, A. A. (1978). Somatic hybrid plants of potato and tomato regenerated from fused protoplasts. *Carlsberg Res. Commun.* **43,** 203–218.

Morelli, G., Nagy, F., Fraley, R. T., Rogers, R. G., and Chua, N. H. (1985). A short conserved sequence is involved in the light-inducibility of a gene encoding ribulose-1,5-bisphosphate carboxylase small subunits of pea. *Nature (London)* **315,** 200–204.

Murai, N., Sutton, D. W., Murray, M. G., Slightom, J. L., Merlo, D. J., Reichert, N. A., Sengupta-Gopalan, C., Stock, C. A., Barker, R. F., Kemp, J. D., and Hall, T. C. (1983). Phaseolin gene from bean is expressed after transfer to sunflower via tumor-inducing plasmid vectors. *Science* **222,** 476–482.

Nagata, T., and Takebe, I. (1970). Cell wall regeneration and cell division in isolated tobacco mesophyll protoplasts. *Planta* **92,** 301–308.

Nehls, R. (1978). The use of metabolic inhibitors for selection of fusion products of higher plant protoplasts. *Mol. Gen. Genet.* **166,** 117–118.

Paszkowski, J., Shillito, R. D., Saul, M., Mandak, V., Hohn, T., Hohn, B., and Potrykus, I. (1984). Direct gene transfer to plants. *EMBO J.* **3,** 2717–2722.

Power, J. B., Sink, K. C., Berry, S. F., Burns, S. F., and Cocking, E. C. (1978). Somatic and sexual hybrids of *Petunia hybrida* and *Petunia parodii. J. Hered.* **69,** 373–376.

Power, J. B., Berry, S. F., Chapman, J. V., and Cocking, E. C. (1980). Somatic hybridization of sexually incompatible petunias: *Petunia parodii, Petunia parviflora, Theor. Appl. Genet.* **57,** 1–4.

Rogers, S. G., O'Connell, K. O., Horsch, R. B., and Fraley, R. T. (1985). Investigation of factors involved in foreign protein expression in transformed plants. *In* "Biotechnology in Plant Science: Relevance to Agriculture in the Eighties" (M. Zaitlin, P. Day, and A. Hollaender, eds.), pp. 219–235. Academic Press, New York.

Schell, J., Van Montagu, M., Willmitzer, L., Leemans, J., Deblaere, R., Joos, H., Inzew, D., Wostemeyer, A., Otten, L., and Zambryski, P. (1984). Transfer of foreign genes to plants and its use to study developmental processes. *In* "Cell Fusion: Gene Transfer and Transformation" (R. F. Beers and E. G. Bassett, eds.). Raven, New York.

Schieder, O. (1975). Regeneration of haploid and diploid *Datura innoxia* Mill. mesophyll protoplasts to plants. *Z. Pflanzenphysiol.* **76,** 462–466.

Schieder, O. (1978). Somatic hybrids of *Datura innoxia* Mill. + *Datura discolor* Bernh. and of *Datura innoxia* Mill. + *Datura stramonium* L. var. *tatula* L. I. Selection and characterization. *Mol. Gen. Genet.* **162,** 113–119.

Schieder, O. (1984). Aktuelle Zuchtungsforschung mit Arzneipflanzen: Ergebnisse und Perspectiven. *In* "Biogene Arzneistoffe" (F. C. Cygan, ed.), pp. 177–200. Vieweg Verlag Braunschweig, Wiesbaden.

Shah, D. M., Horsch, R. B., Klee, H. J., Kishore, G. M., Winter, J. A., Tumer, N. E., Hironaka, C. M., Sanders, P. R., Gasser, C. S., Aykent, S., Siegel, N. R., Rogers, S. G., and Fraley, R. T. (1986). Engineering herbicide tolerance in transgenic plants. *Science* **233,** 478–481.

Shekhawat, N. S., and Galston, A. W. (1983). Mesophyll protoplasts of Fenugreek (*Trigonella foenumgraecum*): Isolation, culture and shoot regeneration *Plant Cell Rep.* **2,** 119–121.

Shepard, J. F., and Totten, R. E. (1977). Mesophyll cell protoplasts of potato: Isolation, proliferation, and plant regeneration. *Plant Physiol.* **60,** 313–316.

Simpson, J., Timko, M., Cashmore, A. R., Schell, J., Van Montagu, M., and Herrera-Estrella, L. (1985). Light-inducible and tissue specific expression of a chimaeric gene under control of a pea chlorophyll *a/b*-binding protein gene. *EMBO J* **4,** 2723–2729.

Somers, D. A., Narayanan, K. R., Kleinhofs, A., Cooper-Bland, S., and Cocking, E. C. (1986). Immunological evidence for transfer of barley nitrate reductase structural gene to *Nicotiana tabacum* by protoplast fusion. *Mol. Gen. Genet.* **204,** 296–301.

Sondahl, M. R., Chapman, M. S., and Sharp, W. R. (1980). Protoplast liberation, cell wall reconstruction, and callus proliferation in *Coffea arabica* L. callus tissues. *Turrialba* **30,** 161–165.

Sung, Z. R., Lazar, G. B., and Dudits, D. (1981). Cycloheximide resistant trait in carrot culture: A differentiated function. *Plant Physiol.* **68,** 261–264.

Szabados, L., and Dudits, D. (1980). Fusion between interphase and mitotic plant protoplasts. *Exp. Cell Res.* **127,** 442–446.

Szabados, L., Hadlaczky, Gy., and Dudits, D. (1981). Uptake of isolated plant chromosomes by plant protoplasts. *Planta* **151,** 141–145.

Wang, K., Herrera-Estrella, L., Van Montagu, M., and Zambryski, P. (1984). Right 25 bp terminus sequence of the nopaline T-DNA is essential for and determines direction of DNA transfer from *Agrobacterium* to the plant genome. *Cell (Cambridge, Mass.)* **38,** 455–462.

Wernicke, W., Lorz, H., and Thomas, E. (1979). Plant regeneration from leaf protoplasts of haploid *Hyoscyamus muticus* L. produced via anther culture. *Plant Sci. Lett.* **15,** 239–249.

Wetter, L., and Dyck, J. (1983). Isoenzyme analysis of cultured cells and somatic hybrids. *In* "Handbook of Plant Cell Culture" (D. A. Evans, R. Sharp, D. Ammirato, and Y. Yamada, eds.), Vol. 1, pp. 607–628. Macmillan, New York.

White, D. W. R., and Vasil, I. K. (1979). Use of amino acid analogue-resistant cell lines of selection of *Nicotiana sylvestris* somatic cell hybrids. *Theor. Appl. Genet.* **55,** 107–112.

Willmitzer, L., De Beuckeleer, M., Lemmers, M., Van Montagu, M., and Schell, J. (1980). DNA from Ti plasmid present in the nucleus and absent from plastids of grown gall plant cells. *Nature (London)* **287,** 359–361.

Xu, X. H., and Davey, M. R. (1983). Shoot regeneration from mesophyll protoplasts and leaf explants of *Rehmannia glutinosa. Plant Cell Rep.* **2,** 55–57.

Zimmermann, U., and Scheurich, P. (1981). High frequency fusion of plant protoplasts by electric fields. *Planta* **151,** 26–32.

CHAPTER 9

Elicitation: Methodology and Aspects of Application

Udo Eilert

Institut für Pharmazeutische Biologie
Technische Universität Braunschweig
D-3300 Braunschweig, Federal Republic of Germany

I. INTRODUCTION

Elicitor-induced accumulation of secondary metabolites is receiving increasing attention, and several recent reviews have focused on its various aspects, e.g., host–pathogen interaction (Dixon, 1980; Helgeson, 1983; Miller and Maxwell, 1983; Darvill and Albersheim, 1984; Dixon, 1986; Keen, 1986), expression and regulation of secondary metabolism (Dixon *et al.*, 1983a; Ebel, 1986), and synthesis and accumulation of secondary metabolites by cell cultures (Wolters and Eilert, 1983; DiCosmo and Misawa, 1985). The objective of this chapter is to summarize and discuss experimentation performed to date, to present the procedures and techniques employed, and to highlight the various areas of application.

II. TERMINOLOGY

Antibiotically active compounds, secondary metabolites, which are accumulated by a plant in response to microbial attack are referred to as phytoalexins (Müller, 1956; Kuć, 1972). Accumulation of phytoalexins which results in chemical resistance is an important factor in plant defense (Cruickshank, 1980; Darvill and Albersheim, 1984) and has been demonstrated for a wide variety of species (Bailey and Mansfield, 1982).

When studying the induction of phytoalexin accumulation, compounds of pathogen origin were found to cause the same response in the plant as the pathogen itself. These compounds are termed elicitors (Keen *et al.*, 1972). Plant-derived, so-called endogenous elicitors, have been added (see Darvill and Albersheim, 1984), and together with the microorganism-derived elicitors they comprise the group of biotic elicitors. Physical and chemical stresses like UV irradiation, exposure to cold or heat, ethylene, fungicides, and antibiotics, salts of heavy metals, or high salt concentrations, which can also induce product accumulation, are stress agents, sometimes referred to as abiotic elicitors (Yoshikawa, 1978; Moesta and Grisebach, 1981; Tietjen and Matern, 1983; Tietjen *et al.*, 1983; Davis *et al.*, 1986). A more precise use of the term elicitor was demanded at the Sixth International Congress for Plant Tissue and Cell Culture, 1986. The proposal was to apply the term elicitor to compounds of biological origin only, while the other treatments could be designated as abiotic stress. This chapter will focus on the effect and use of biotic elicitors and coculture systems with living organisms.

III. ELICITORS AND THEIR MODE OF ACTION

Studies of plant–microbe recognition led to the identification of four different types of interaction:

1. Direct release of the elicitor by the microorganism and recognition by the plant cell (West, 1981; Brindle and Threlfall, 1983). A receptor site for such an elicitor seems to be located on plasma membranes of soybean cells (Yoshikawa *et al.*, 1983).
2. Microbial enzymes release plant cell wall components, which then act as elicitors. Endopolygalacturonic acid lyase from *Erwinia carotovora* (Davis *et al.*, 1984) and from *Rhizopus stolonifer* (Lee and West, 1981) as well as pectinolytic enzymes from various fungi (Amin *et al.*, 1986) were shown to induce phytoalexin formation via this mechanism.
3. Plant enzymes release cell wall components from the microorganism, which in turn induce phytoalexin formation in the plant cells. Examples of this type of interaction were presented by Hadwiger and Beckman (1980), who showed enzymatic release of chitosan from *Fusarium* cell walls, and by Keen and Yoshikawa (1983), who isolated a β-1,3-endoglucanase from soybean which released elicitor-active components from *Phytophthora* cell walls.

4. Elicitor compounds, endogenous and constitutive in nature, are formed or released by the plant cell in response to various stimuli or can even be extracted artificially. In part they may be identical to plant cell wall-derived elicitors, released by fungal enzymes (group 2, above). Hahn *et al.* (1981) obtained oligosaccharides with elicitor activity from cell walls of plant cells by partial acid hydrolysis or simply by extraction with hot water. Hargreaves and Bailey (1978) reported the release of elicitor-active metabolites after damage of cells. Kurosaki *et al.* (1985b) were able to induce formation of 6-methoxymellein in carrot cultures with partial hydrolysates obtained by endopolygalacturonidase or endopectinolyase digests of cultured carrot cells. Acid hydrolysates of cell walls of parsley leaves or *in vitro*-cultured cells induced furanocoumarin formation in parsley cultures (Kombrink and Hahlbrock, 1985). Proteinase inhibitor formation in tomato cultures was induced by addition of a uronic acid-rich polysaccharide, which was extracted and purified from spent culture medium (Walker-Simmons and Ryan, 1986).

Some abiotic stress agents (elicitors), e.g., mercuric chloride, also were shown to act via release of an endogenous elicitor compound (Moesta and Grisebach, 1980, 1981). Yoshikawa (1978), however, postulated a different mode of action for biotic and abiotic elicitors. His experimental results suggested an inhibition of the degradative turnover of secondary metabolites by salts of heavy metals.

The release of an endogenous elicitor after treatment with an abiotic elicitor was demonstrated in *Phaseolus vulgaris* cell cultures (Dixon *et al.*, 1983b). Denatured ribonuclease A induced enzymes of the isoflavonoid pathway [phenylalanine ammonia-lyase (PAL), chalcone isomerase, and chalcone synthase] in suspension-cultural bean cells. Cells separated from the RNAase-treated ones by a dialysis membrane also responded with an increase of enzyme activity and isoflavon accumulation. Dixon *et al.* (1983c) concluded that a low molecular weight compound of plant cell origin is responsible for the intercellular transmission.

DiCosmo and Misawa (1985) extensively discussed a possible involvement of a second messenger in the process of elicitation. This compound would transmit signals from the plasma membrane and thus trigger transcription and translation of enzymes. This system of intracellular communication functions in animal cells where cyclic AMP serves as second messenger. No experimental support of this hypothesis exists, however, and results by Hahn and Grisebach (1983) contradict such a function of cAMP in plant cells. Köhle *et al.* (1985) observed that chitosan-elicited callose synthesis in soybean cells is a Ca^{2+}-dependent process. The polycation chitosan inflicts membrane damage, and Ca^{2+} influx into cells becomes possible. Here, calcium ions would represent a type of second messenger. Ethylene released from wounded and infected tissues has also been discussed as a second messenger (e.g., Boller, 1983). Esquerre-Tugaye *et al.* (1985) reported the importance of ethylene in the elicitation of defense responses including phytoalexin accumulation in melon. In the soybean system it seems, however, that ethylene is an indicator rather than an inducer of phytoalexin synthesis (Paradies *et al.*, 1980). An observation by Strasser *et al.* (1983) on cellular phosphate levels should be mentioned in this context. On elicitation the cytoplasmic phosphate level in cultured parsley cells decreased temporarily, while the vacuolar level increased. This change in distribution was not affected by the phosphate level in the medium and may be linked to the redirection of the phenylpropanoid pathway. The various types of interaction described above must be considered when selecting eliciting agents.

IV. METHODOLOGY OF ELICITATION

A. Selection of Microorganisms

In studies of host–pathogen interaction the problem itself dictates which microorganism to use. Where the interest is focused on simply triggering secondary metabolism, virus preparations, bacteria, blue–green algae, and most frequently fungi can successfully be used (Table I). The problem of quantitation of the inoculum of fungi can be overcome using conidiospores, zoospores, or sporangia; their elicitor properties, however, may be different from those of mycelium. The response of a cell culture to a variety of microorganisms should be investigated, as there is no way of predicting which organism will result in elicitation. The final selection should include various types of fungi, which need not be plant pathogens but may be saprophytes. Compatible strains of plant-pathogenic fungi may be avoided; they may be successful pathogens because the plant's inducible defense mechanisms fail in their presence.

B. Coculture

Tissue cultures are generally kept under sterile conditions. The deliberate infection with a living microorganism, however, can be of interest

TABLE I

Elicitor-Stimulated Accumulation of Secondary Metabolites in Cells Cultured *in Vitro*

Compound	Amount (μg/gFW)	Plant species	Type of culture[a]	Medium[b]	Hormones[c] (μg/liter)					Elicitor	Elicitor concentration (per ml)	References
					b	k	D	I	N			
Isoflavonoids												
Daidzein 7-*O*-β-D-glucoside;	14	*Vigna angularis*	S	MS			1			Nigeran	50 μg	Hattori and Ohta (1985)
										Phytophthora megasperma glucan	10 μg	
Daidzein 7,4′-di-*O*-β-D-glucoside;	18											
2-Hydroxy daidzein-7,4′-di-*O*-β-D-glucoside	20–60									RNAase	0.5 mg	
										Na_3VO_4	50 μ*M*	
Glyceollin	430–980	*Glycine max*	C	LS		1	1			*Phytophthora megasperma* spores	10^5	Keen and Horsch (1972)
	67		S	B5		2				*Phytophthora megasperma* cell wall released	1 μg	Ebel *et al.* (1976)
	67		S	B5		2				Nigeran	40 μg	Ebel *et al.* (1976)
	20–60		S	B5						Chitosan	1 mg	Köhle *et al.* (1984)

(continued)

TABLE I

(Continued)

Compound	Amount (μg/gFW)	Plant species	Type of culture[a]	Medium[b]	Hormones[c] (μg/liter)					Elicitor	Elicitor concentration (per ml)	References
					b	k	D	I	N			
Kievitone	151	*Phaseolus vulgaris*	S	MS		0.1	0.5			*Phaseolus vulgaris* hypocotyl extract	50 μl	Hargreaves and Selby (1978)
	12		S	SH						*Colletotrichum lindemuthianum* cell wall released	ns[d]	Robbins *et al.* (1985)
Medicarpin	60–100	*Trifolium repens*	C	B5		0.5	1.25		0.5	$HgCl_2$	1.85 mg	Gustine (1981)
	60	*Canavallia ensiformis*	C	M		0.25			0.4	*Pithomyces chartarum* spores	10^7	Gustine *et al.* (1978)
	25–40	*Medicago sativa*	C	UM		0.25	2			*Verticillium albo-atrum* spores	ns	Latunde-Dada and Luca (1985)
Phaseollin	34	*Phaseolus vulgaris*	S	SH		1.08			0.93	*Botrytis cinerea* filtrate	100 μl	Dixon and Bendall (1978)
	20									Nigeran	0.5 mg	
	10									Cupric sulfate	125 μg	
	10									Poly-L-lysine (av. MW 30,000)	1 mg	
	58–80									Denatured RNAase A	0.5 mg	
	160									*Colletotrichum lindemuthianum*	ns	Robbins *et al.* (1985)

Phaseollidin	31	*Phaseolus vulgaris*	S	MS	0.1	0.5	*Phaseolus vulgaris* hypocotyl extract	50 μl	Hargreaves and Selby (1978)
Phaseollin isoflavon	207	*Phaseolus vulgaris*	S	MS	0.1	0.5	*Phaseolus vulgaris* hypocotyl extract	50 μl	Hargreaves and Selby (1978)
Furanocoumarins									
Psoralen, xanthotoxin, graveolone	0.75–2.5 (/ml)	*Petroselinum hortense*	S	B5		1	*Phytophthora megasperma* cell wall released	100	Tietjen *et al.* (1983)
Bergaptene, isopimpinelline, graveolone	0.75–2.5 (ml)	*Petroselinum hortense*	S	B5		1	*Alternaria carthami* cell wall released	125	Tietjen *et al.* (1983)
Anthraquinones									
Not specified	500	*Cinchona ledgeriana*	S	B5	0.2	1	*Aspergillus niger*	0.5 mg	Wijnsma *et al.* (1985)
	250		S	B5	0.2	1	*Phythophthora cinnamomi*	1.1 mg	Wijnsma *et al.* (1985)
Benzophenanthridine-alkaloids									
Sanguinarine	400–1200	*Papaver somniferum*	S	B5C		1	*Botrytis* sp. homogenate	10 μl	Eilert *et al.* (1985a)
			S	B5C		1	*Rhodotorula rubra* homogenate	5 μl	Eilert *et al.* (1985a)
Dihydrosanguinarine	100–200	*Papaver somniferum*	S	B5C		1	*Botrytis* sp. homogenate	10 μl	Eilert *et al.* (1986b)

(*continued*)

TABLE I

(Continued)

Compound	Amount (μg/gFW)	Plant species	Type of culture[a]	Medium[b]	Hormones[c] (μg/liter)					Elicitor	Elicitor concentration (per ml)	References
					b	k	D	I	N			
Acridone alkaloids												
Rutacridone epoxide	150	*Ruta graveolens*	C	E		0.02			1	*Botrytis allii* filtrate	2.5 ml/cult	Wolters and Eilert (1982)
	120		C	E		0.02			1	*Rhodotorula rubra*	coculture	Wolters and Eilert (1982)
	1.5		S	MS		0.25		0.1		*Rhodotorula rubra*	coculture 6 μl	Eilert *et al.* (1984)
Hydroxyrut-acridone-epoxide	135	*Ruta graveolens*	C	E		0.02			1	*Botrytis allii* filtrate	2.5 ml/culture	Wolters and Eilert (1982)
	130		C	E		0.02			1	*Rhodotorula rubra*	coculture	Wolters and Eilert (1982)
	38		S	MS		0.25		0.1		*Rhodotorula rubra*	coculture 6 μl	Eilert *et al.* (1984)
Benzoxazinone												
Diantha-lexin	50–100	*Dianthus caryophyllus*	S	SH		0.1	0.1			*Phytophthora parasitica*	80 μl	Gay (1985)
Morphinan alkaloids												
Codeine	56	*Papaver somniferum*	S	ns						*Verticillium dahliae*	Conidia	Heinstein (1985a)
	70		S	ns						*Fusarium moniliformae*	Conidia	Heinstein (1985a)
Morphine	60	*Papaver somniferum*	S	ns						*Verticillium dahliae*	Conidia	Heinstein (1985a)
	70		S	ns						*Fusarium moniliformae*	Conidia	Heinstein (1985a)

Indole alkaloids									
N-Acetyl-tryptamine	9	*Catharanthus roseus*	S	B5		1	*Pythium aphanidermatum* homogenate or *Rhodotorula rubra*	50 5	Eilert *et al.* (1986a)
Strictosidine lactam	2–10	*Catharanthus roseus*	S	B5		1			
Ajmalicine	4–7	*Catharanthus roseus*	S	B5		1			
Catharanthine	10–20	*Catharanthus roseus*	S	B5		1			
Tabersonine	trace	*Catharanthus roseus*	S	B5		1			
Loch nericine	trace	*Catharanthus roseus*	S	B5		1			
Cephalotaxus alkaloids	150–250	*Cephalotaxus harringtonia*	S	ns			*Verticillium dahliae* spore susp.	12 μl	Heinstein (1985a)
Sesquiterpenes									
Capsidiol	nd[d]	*Nicotiana* tabacum	C	LS	0.2	0.1	*Phytophthora parasitica*	ns	Helgeson *et al.* (1978)
	nd		C	ns			spores and cell wall released	ns	Budde and Helgeson (1981a)
Sesquiterpenoids	200	*Nicotiana tabacum*	C	ns			*Phytophthora parasitica* spores and cell wall released	ns	Budde and Helgeson (1981b)

(*continued*)

TABLE I

(Continued)

Compound	Amount (μg/gFW)	Plant species	Type of culture[a]	Medium[b]	Hormones[c] (μg/liter)					Elicitor	Elicitor concentration (per ml)	References
					b	k	D	I	N			
Capsidiol	2900	*Capsicum annum*	S	MS		0.1	1			Cellulase; *Gliocladium deliquescens* cellulase	5–20 μg 10 μg	Brooks *et al.* (1986)
Debneyol	nd	*Nicotiana tabacum*	S	MS		0.2	1				10 μg	Watson *et al.* (1985)
Hemigossypol; hemigossypolone; gossypol	720–900 675–765	*Gossypium arboreum*	S	LS		0.22			0.18	*Verticillium dahliae* spores	10^3	Heinstein (1982)
			S	LS		0.22			0.18	*V. dahliae* culture filtrate	ns	Heinstein (1982)
	675–810		S	LS		0.22			0.18	*Saccharomyces cerevisiae* spores	10^5	Heinstein (1982)
	1600		S	LS		0.22			0.18	*Verticillium dahliae* spores	10^3	Heinstein (1985b)
	640		S	LS		0.22			0.18	*V. dahliae* culture filtrate	60 μl	Heinstein (1985b)
Lubimin	35	*Solanum tuberosum*	S	RM						*Phytophthora infestans* spores	10 μl	Brindle *et al.* (1983)
Phytuberin	2–4	*Nicotiana tabacum*	C	LS		0.1		3	3	*Pseudomonas solanacearum* cells	10^6	Fujimori *et al.* (1983)

	2–10	*Nicotiana tabacum*	C	LS	0.1		3	3	*Pseudomonas solanacearum cells*	10^7	Tanaka and Fujimori (1985)
	5–10	*Solanum tuberosum*	C	W	1		2		*Phytophthora infestans* homogenate	4 ml/culture	Érsek and Sziraki (1980)
	nd	*Nicotiana tabacum*	S	MS	0.2		1		Cellulase	10 μg	Watson *et al.* (1985)
Phytuberol	nd	*Nicotiana tabacum*	S	MS	0.2		1		Cellulase	10 μg	Watson *et al.* (1985)
	0.2–0.4		C	LS	0.1		3	3	*Pseudomonas solanacearum* cells	10^6	Fujimori *et al.* (1983)
	0.5		C	LS	0.1		3	3	*Pseudomonas solanacearum* cells	10^7	Tanaka and Fujimori (1985)
Rishitin	9	*Solanum tuberosum*	S	ns					*Phytophthora infestans* spores	10 μl	Brindle *et al.* (1983)
	15–30		C	W	1	2			*Phytophthora infestans* homogenate	4 ml/culture	Ersek and Sziraki (1980)
	nd	*Nicotiana tabacum*	C	nd					*Phytophthora parasitica*	ns	Budde and Helgeson (1981a)
Solavetivone	11	*Solanum tuberosum*	S	ns					*Phytophthora infestans* spores	10 μl	Brindle *et al.* (1983)

(*continued*)

TABLE I

(Continued)

Compound	Amount (μg/gFW)	Plant species	Type of culture[a]	Medium[b]	Hormones[c] (μg/liter)					Elicitor	Elicitor concentration (per ml)	References
					b	k	D	I	N			
Steroids												
Diosgenin	4000 (70% increase)	*Dioscorea deltoidea*	S	MS			0.1			*Rhizopus arrhizus* homogenate	5 μl	Rokem *et al.* (1984)
Steroidal alkaloids	increased	*Solanum laciniatum*	S							*Chlorogloeopsis fritschii*	coculture	Gorelova *et al.* (1984)
Polyacetylenes												
1-Phenyl-hepta-1,3,5-triene	3.2	*Bidens pilosa*	C	SH					4	*Pythium aphanidermatum* culture filtrate	100 μl	DiCosmo *et al.* (1982)
Safynol	0.04	*Carthamus tinctorius*	S	LS		0.2		2		*Phytophthora megasperma* cell wall released	250 μg	Tietjen and Matern (1984)
Dehydro-safynol	0.15	*Carthamus tinctorius*	S	LS		0.2		2		*Phytophthora megasperma* cell wall released	250 μg	Tietjen and Matern (1984)
Safynol	0.09 (/ml)	*Carthamus tinctorius*	S	LS		0.2		2		*Phytophthora megasperma* cell wall released	30 μg	Tietjen and Matern (1984)

Dehydrosafynol	0.25	*Carthamus tinctorius*	S	LS	0.2		2	*Phytophthora megasperma* cell wall released	30 μg	Tietjen and Matern (1984)
Isocoumarins										
6-Methoxymellein	0.4 (/ml)	*Daucus carota*	S	MS				*Chaetomium globosum*	9×10^5 spores	Kurosaki and Nishi (1983)
	2		S	MS	1			*Daucus carota* hydrolysate	—	Kurosaki *et al.* (1985a)
	2		S	MS	1			Pectinase	10 mU	Kurosaki *et al.* (1985a)
	2		S	MS	1			Trypsin	10 mU	Kurosaki *et al.* (1985a)
	—		S	MS		1		*Daucus carota* hydrolysate	—	Kurosaki *et al.* (1985a)
	—		S	MS		—		*Daucus carota* hydrolysate	—	Kurosaki *et al.* (1985a)
Terpenoid esters of caffeic acid										
(*Z,E*)-2(3,4-Dihydroxyphenyl)ester of 3(3,4-dihydroxyphenyl)-2-propenoic acid and of caffeic acid	increase of production	*Lavandula angustifolia*	S	MS			2	Accidental infection with fungi or bacteria		Banthorpe *et al.* (1985)

(*continued*)

TABLE I

(Continued)

Compound	Amount (μg/gFW)	Plant species	Type of culture[a]	Medium[b]	Hormones[c] (μg/liter)					Elicitor	Elicitor concentration (per ml)	References
					b	k	D	I	N			
Aromatic acids												
Salicylic acid, vanilic acid, 4-hydroxybenzoic acid	ns	*Solanum tuberosum*								*Phytophthora infestans*		Robertson *et al*. (1968)
Naphthoquinones												
Shikonin	2000	*Lithospermum erythrorhizon*	S	LS			3	0.2		Agar (agaropectin)	2% (0.5%)	Fukui *et al*. (1983)

[a] C, Callus; S, suspension culture.

[b] Media: B5, Gamborg *et al*. (1968); B5C, B5 medium sublemented with casein; E, Eriksson (1965); LS, Linsmeyer and Skoog (1965); M, modified Miller's medium, Krasnuk *et al*. (1971); MS, Murashige and Skoog (1962); SH, Schenk and Hildebrandt (1972); W, modified White's medium, Okazawa *et al*. (1967).

[c] Growth factors: b, benzyladenine; k, kinetin (6-furfurylaminopurin); D (2,4-D), 2,4-dichlorophenoxyacetic acid; I (IAA), indolacetic acid; N (NAA), naphthaleneacetic acid.

[d] ns, Not shown; nd, not determined.

for the following reasons: (a) culture of obligate parasites, (b) study of various processes of infestation, (c) general tests of susceptibility or resistance, (d) tests for toxin resistance, and (e) stimulation of secondary metabolism in cultured cells (phytoalexin induction). Handbooks edited by Ingram and Helgeson (1980), Helgeson (1983), and Dixon (1985) provide extensive information on techniques concerning aspects (a)–(d); here, aspect (e) is the focal point.

For stimulation of secondary metabolism, callus cultures are commonly transferred to fresh medium 1–4 weeks and suspension cultures 3–14 days prior to inoculation with a microorganism. Though trivial, it is important to assure that in case of coculture the plant cell culture medium allows for survival of the microorganism. With callus cultures either the tissue is directly infected or the microorganism is placed next to the callus but direct contact is avoided. In the first instance the responses triggered require direct contact of plant cell and inoculum. Depending on the incubation period this direct contact can lead to extensive cell damage. A separation of plant tissue and inoculum usually becomes impossible, thus posing difficulties for the determination of metabolic changes. This problem is overcome by a coculture method where the organisms are kept separate. For coculturing without direct contact, only nonmobile bacteria, yeasts, or slow-growing filamentous fungi should be selected. Interaction is possible by diffusion of compounds in the medium and via the gaseous phase. This method is especially suited if the microbial inducing factors are heat labile, e.g., enzymes which will release the actual elicitor compound. The microorganism may continuously provide elicitor, and cell damage will occur to a much lesser extent. Separate coculture is a method for first tests of interactions in any new plant cell culture/microorganism system.

Suspension culture cells can be elicited simply by addition of microbial inoculum. Contact of the two organisms and exposure to their excreted products is very intense; the constant mechanical perturbation, however, may slow down the invasion of cells as compared to inoculated callus. This type of coculture interaction is the one used most frequently (Table I), as suspension cultures in general present a more homogeneous cell material than callus, and the concentration of inoculum can be controlled more easily.

Direct contact of plant cells and microorganisms can be prevented by entrapment of either of the partners in gels, as used for immobilization. Usually the microorganism will be immobilized to avoid any disturbance of cell culture growth. The gels also may interfere with product isolation if the cells are entrapped. For the purpose of secondary metabolite production, however, the entrapment of the plant cells may be ideal, if they excrete the desired compounds into the medium. Such an immobilized system might result in semicontinuous or continuous production. Eilert *et al.* (1984) observed that the entrapment gel, calcium alginate, itself caused a weak elicitation. Elicitor activity has been shown for other polysaccharides used for immobilization like agaropectin (Fukui *et al.*, 1983) and especially chitosan (Hadwiger and Beckman, 1980). The possibility of such an interaction and its implications have to be considered when working with immobilized systems.

C. Chemically Defined Elicitors and Complex Preparations

In coculture systems with living microbes synergistic and inhibiting effects are possible (see Section III). The advantage of a wide range of interactions covered in coculture systems is accompanied by the disadvantage that coculture systems, especially with filamentous fungi, easily lose balance and get out of control, resulting in severe damage to the plant cells. Use of complex sterile culture homogenates avoids this problem while maintaining a wide range of possible interactions. Complex elicitor preparations are easily obtained by culturing the selected microorganisms, preferably on the tissue culture medium, then homogenization and autoclaving of the entire culture. The autoclaving process can lead to the release or formation of elicitor compounds (Hahn *et al.*, 1981). Heat-labile compounds, like enzymes, however, are destroyed. Sterile filtered culture homogenates will maintain this activity unless the activity resides in insoluble cell wall components (Darvill *et al.*, 1985).

For optimum elicitor activity of complex homogenates the culture period of the elicitor-yielding organism can be of importance, as metabolism changes during a growth cycle and different stages (mycelia, spores) may possess different elicitor activity. Autoclaved conidiospores or zoospores are also common as elicitors (Table I; for preparation, see Ayers *et al.*, 1976; Keen *et al.*, 1983).

Complex elicitor preparations are the material of choice in screening programs. Attempts to characterize the elicitor compound of complex preparations frequently revealed the presence of more than one active component. Additive and even synergistic action of elicitors has been demonstrated (Bostock *et al.*, 1982; Davis *et al.*, 1986), resulting in higher activity of the complex homogenate than of the isolated compounds. On the other hand, Ziegler and Pontzen (1982) showed that a mannan glycoprotein of fungal origin inhibited the elicitor activity of a cell wall-derived glucan elicitor of the same fungus. In this case a complex homogenate will show low or no activity at all. Addition of elicitors will not always induce the expression of the biosynthetic potential in cultured cells but may also cause its suppression. Oba and Uritani (1979) observed this effect in sweet potato cultures which produced furanoterpenoid phytoalexins under normal culture conditions; elicitor addition suppressed this production.

The complex composition of culture homogenates can hamper reproducibility of results. Experiments which require high reproducibility should preferably be performed with chemically defined elicitors. Variation, however, will not be eliminated, when generated by the cultured plant cells. Hahlbrock *et al.* (1981), for instance, had difficulties in reproducing absolute values of elicitor-induced enzyme activities.

The list of chemically defined biotic and abiotic elicitors is growing steadily. Table I lists those compounds used successfully to elicit cultured plant cells. For further information the reader is referred to reviews by West (1981), Wolters and Eilert (1983), Darvill and Albersheim (1984), and Darvill *et al.* (1985).

V. FACTORS OF ELICITATOR-INDUCED ACCUMULATION

Stimulation of secondary metabolism by elicitation is the result of a complex interaction between the elicitor and the plant cell. The response of cultured plant cells is affected by a number of factors, some of which are linked to the elicitor, others to the cells cultured. Detailed predictions for a system cannot be made. Optimization of the elicitor response can only be achieved empirically, but conditions which have been found favorable in previous experimentation may provide guidelines when a new system is developed.

A. Elicitor Specificity

When elicitation is envisaged as a tool to induce secondary metabolism, the question of elicitor specificity has to be considered. The data in Table I show that coculturing with *Rhodotorula rubra*, for example, results in acridone epoxide accumulation in *Ruta graveolens*, indole alkaloid formation in *Catharanthus roseus*, and sanguinarine formation in *Papaver somniferum* cultures. Culture homogenates of *Botrytis* species induced isoflavon accumulation in *Phaseolus vulgaris*, acridone epoxide formation in *Ruta*, and sanguinarine production by *Papaver* cultures, but induction of indole alkaloid accumulation in *Catharanthus roseus* was not observed. *Phytophthora megasperma* glucan, as a third example, induced isoflavon phytoalexin formation in cell cultures of several different Fabaceae. The particular compound which was accumulated was specific for each culture. These results show that the same elicitor preparation will often stimulate secondary metabolism in different cell cultures. The products formed in response are specific for the plant cell culture and, with some exceptions, are not affected by the elicitor. In other words, treatment of a particular cell culture with different elicitors will result in the accumulation of the same compounds, if the culture will respond at all. Most elicitors have shown activity in plant cell cultures from various plant families. The unsaturated fatty acids arachidonic and eicosapentenoic acid displayed activity only in cultures of solanaceous plants, inducing sesquiterpenoid formation.

Differential response to different stress agents was noted in cultured cells of *Petroselinum hortense*. Irradiation with UV light induced formation of flavonoids (Hahlbrock, 1981). Addition of fungal elicitors, in contrast, caused formation and accumulation of furanocoumarins (Tietjen and Matern, 1983; Tietjen *et al.*, 1983). Although a differential response was noted here with both the qualitative and quantitative furanocoumarin pattern being affected by the elicitor preparation, all compounds belong to the same class. Only different kinds of stress may result in an induction of different pathways of secondary metabolism. A combined UV light and elicitor treatment of parsley cells did not yield any significant product accumulation at all (Hahlbrock *et al.*, 1981).

B. Elicitor Concentration

The elicitor concentration is a factor which strongly affects the intensity of the response. Table I lists the elicitor concentrations used to obtain optimum product accumulation. Purified elicitor compounds may be active in trace amounts, while concentrations up to 5% are required for optimum activity of complex fungal homogenates. In the literature (Table II) two types of dose–response curves have been described. One type shows the typical form of a saturation curve. Overdosage of elicitor will have no adverse effects. More frequently, dose–response curves with a sharp optimum were found. A complex elicitor homogenate even showed two optima, each one probably belonging to a different elicitor compound (Dixon *et al.*, 1981). A similar effect was observed for endopolygalacturonidase-induced 6-methoxymellein formation in carrot cells (Kurosaki *et al.*, 1985b). With several different elicitors Kombrink and Hahlbrock (1986) performed an extensive study of the dose dependency of the level of coumarin accumulation and induction of several enzymes in parsley cultures. Optimum elicitation was achieved at different concentrations of each elicitor preparation, and the optimum elicitor concentration was different for each group of biosynthetic pathway enzymes. Even the quantitative ratio of furanocoumarins varied considerably with elicitor concentration. These data explain the importance of determining the optimum elicitor concentration. The optimum amount of one elicitor may be similar for elicitation of different cell cultures (see Table I). Screening programs require testing of each elicitor in a range of concentrations.

C. Period of Elicitor Contact

Generally the elicitor will be in contact with the cultured plant cells until harvest, which in turn is dependent on the kind of cell culture. The time course of accumulation is an attribute of the plant cell. The minimal time required, however, is of interest for the general understanding of the elicitation process. Very few data have been communicated on this topic. Strasser and Matern (1986) investigated this question in the parsley—*Alternaria carthami* elicitor system. They determined that a minimal exposure time of 20 min was required for induction of a lasting effect. In this context results by C. L. Cramer (personal communication, 1986) should be mentioned; formation of mRNA coding for PAL occurred as quickly as 2 min after addition of *Colletotrichum* cell wall elicitor to *Phaseolus vulgaris* cultures.

D. Importance of the Cell Culture Line

The phenomenon of variation among cell culture lines derived from the same plant species or even the same plant with respect to expression of biochemical capabilities is well documented (Constabel *et al.*, 1982; Evans, 1986). Almost no data on the response of different cell lines are available. The importance of this question was demonstrated by Eilert *et al.* (1986a). When subjecting a number of variant cell lines of *Catharanthus roseus* to elicitor treatment, only one line was detected which accumulated indole alkaloids in response. Schmelzer *et al.* (1985) observed differences in rates of transcription on elicitor treatment of differ-

TABLE II

Dose–Response Relationship of Elicitor-Induced Phytoalexin Accumulation in Suspension Culture Plant Cells

Type of dose–response curve	Elicitor	Induced compound	Cell culture	Reference
Saturation curve	Cellulase	Capsidiol	*Capsicum annuum*	Brooks *et al*. (1986)
	Phytophthora parasitica (extract)	Dianthalexins	*Dianthus caryophyllus*	Gay (1985)
	Alternaria carthami glucan	Polyacetylenes	*Carthamus tinctorius*	Tietjen and Matern (1984)
	Pythium aphanidermatum homogenate	Indole alkaloids	*Catharanthus roseus*	Eilert *et al*. (1986a)
Optimum curve	*Botrytis* sp. homogenate	Sanguinarine	*Papaver somniferum*	Eilert *et al*. (1985)
	Rhodotorula rubra	Sanguinarine	*Papaver somniferum*	Eilert *et al*. (1985)
	Verticillium dahliae spores	Sesquiterpenes	*Gossypium arboreum*	Heinstein (1985b)

Phytophthora megasperma glucan	Furanocoumarins	*Petroselinum hortense*	Kombrink and Hahlbrock (1986)
	PAL, 4-coumarate: CoA ligase	*Petroselinum hortense*	Hahlbrock *et al.* (1981)
	PAL	*Glycine max*	Ebel *et al.* (1976)
	PAL, chalcone synthase	*Glycine max*	Hille *et al.* (1982)
Colletotrichum lindemuthianum cell wall	Phaseollin	*Phaseolus vulgaris*	Dixon and Lamb (1979)
	PAL, flavone synthase	*Phaseolus vulgaris*	Dixon *et al.* (1981)
	PAL mRNA, chalcone synthase mRNA	*Phaseolus vulgaris*	Lawton *et al.* (1983b)
Endopolygalacturonidase, endopectinlyase	6-Methoxymellein	*Daucus carota*	Kurosaki *et al.* (1985b)

ent strains of cultured parsley cells. Thus inducibility of secondary metabolism is a variant quality of cell cultures. This result certainly is not surprising if one considers the complexity of the interaction. It is rather amazing that so many dedifferentiated cell cultures respond to elicitors in the way the differentiated plant does. As already suggested by Brindle *et al.* (1983), contradictory results obtained with elicitor treatment of different cell cultures of the same plant species may be explained by variance at the cell culture level. The variance may be exploitable to optimize yield when secondary metabolite production is the target.

E. Time Course of Elicitation

In most systems the time course of elicitation has been investigated in detail. Accumulation of secondary compounds starts within hours after elicitor treatment and is preceded by an increase of biosynthetic enzyme activity. Maximum levels will be reached after 12 hr to 5 days, generally followed by a rapid decline. Thus, knowledge of the time course is a valuable factor when optimizing yields. In Table I the period of maximum product content of elicited cells is listed for each system. In a cell line time courses of accumulation may differ for different compounds, e.g., accumulation of kievitone and phaseollin in *Phaseolus vulgaris* (Robbins *et al.*, 1985), phytuberin and phytuberol accumulation in *Nicotiana* callus cultures (Tanaka and Fujimori, 1985), or furanocoumarins in parsley cultures (Kombrink and Hahlbrock, 1986). Precursors may accumulate first in substrate-regulated systems and are then metabolized to a product further down the biosynthetic pathway. On the other hand, feedback inhibition will lead to the accumulation of the end product first, followed by an increase of the level of precursors.

The time required for activation of the biosynthetic pathway is plant specific, but there is some evidence that the amount and type of elicitor may affect or modulate the time course (Kombrink and Hahlbrock, 1986). In related plant species which form structurally related compounds in response to elicitation (e.g., isoflavonoids in Fabaceae, sesquiterpenoids in Solanaceae) the time courses of accumulation are similar. When investigating a new system, these findings should be considered.

F. Growth Stage of Cell Culture

The relationship between growth phase and formation of secondary metabolites in cultured plant cells is a topic of numerous publications

TABLE III

Dependence of Elicitor-Induced Accumulation of Secondary Metabolites on the Growth Stage of the Cell Culture

Cell culture	Product	Days after subculturing	Growth stage	Reference
Glycine max	Isoflavonoids	8–12	End of the growth phase	Ebel *et al.* (1976)
Phaseolus vulgaris	Isoflavonoids	5–10	End of the growth phase	Hargreaves and Selby (1978)
Ruta graveolens	Acridone epoxides	No effect		Eilert *et al.* (1984)
Dianthus caryophyllus	Dianthalexins	8–12	End of the growth phase	Gay (1985)
Petroselinum crispum	Furanocoumarins	4–9	Optimum at the end of the growth phase	Kombrink and Hahlbrock (1985)
Daucus carota	6-Methoxymellein	5–10	End of the growth phase–early stationary phase	Kurosaki *et al.* (1985a)
Catharanthus roseus	Indole alkaloids	5–10	End of the growth phase	Eilert *et al.* (1986a)
Thalictrum rugosum	Berberin	7	End of the growth phase	Brodelius *et al.* (1986a)
Glycine max	Enzymes of the isoflavonoid pathway	Two peaks		Hille *et al.* (1982)
		0–2	Lag phase	
		5–10	End of the growth phase	

and has been summarized repeatedly in reviews. The majority of compounds is formed in the stationary phase. The elicitor response also proved to be dependent on the growth stage in most of the systems (Table III). With the exception of *Ruta graveolens,* all cultures showed response to elicitation only during the growth phase. The growth stage of a culture may affect not only the quantitative response to elicitor treatment but also the product pattern. *Pythium* culture homogenate stimulated *N*-acetyltryptamine formation in 5-day-old *Catharanthus roseus* cultures; 10-day-old cells, in contrast, accumulated a whole spectrum of monoterpene indole alkaloids (Eilert *et al.*, 1986a). Elicitor treatment at a time when a culture has already started to accumulate the inducible compounds—e.g., furanocoumarins in parsley (Kombrink and Hahlbrock, 1985) and indole alkaloids in *Catharanthus roseus* (Eilert *et al.*, 1986a)—does not enhance or accelerate accumulation and can even suppress already activated biosynthesis (Oba and Uritani, 1979). Kombrink and Hahlbrock (1985) speculate that the formation of furanocoumarins in parsley cultures in the stationary phase might be a result of autoelicitation; on lysis of cells endogenous elicitor might be released. Kurosaki *et al.* (1985b) found that the activity of elicitor which can be released from cultured carrot cells by pectinase or trypsin treatment proved to be dependent on the growth stage.

Elicitor treatment in the growth phase led to an instantaneous temporary or permanent arrest of cell growth in several cultures (Ebel *et al.*, 1976; Hargreaves and Selby, 1978; Kombrink and Hahlbrock, 1985). This rapid change of the physiological stage could trigger the expression of secondary metabolism, which otherwise is activated at a later time. The influence of the growth stage of cultured cells is obvious and of importance not only for purposes of production, but also for experimentation aimed at the investigation of physiological aspects of elicitation.

G. Influence of Growth Regulators

The concentration of growth regulators in the nutrient medium can affect expression of secondary metabolism in cultured cells quite dramatically. Only a few studies have dealt with the influence of this factor on cell response to elicitation. Haberlach *et al.* (1978) examined the effect of cytokinin and auxin concentration on resistance of *Nicotiana tabacum* callus cultures against *Phytophthora parasitica.* Susceptibility or resistance was a function of the phytohormone balance. A high cytokinin ratio favored infestation and suppressed the hypersensitive response. Holliday and Klarmann (1979) demonstrated that resistance of soybean callus against colonization by *Phytophthora megasperma* was related to the source plant, but could be modified by the growth factor regime. Working with cotyledons of *Phaseolus vulgaris,* Gossens and Vendrig (1982) observed induction of phaseollin and kievitone after incubation with abscisic acid and/or benzylaminopurine solutions (10^{-4} *M*). After treatment with $HgCl_2$ solution, absciscic acid produced a stimulating effect. In suspension cultures of bean the ratio of kinetin and auxin as well as the type of auxin markedly affected *Botrytis cinerea* homogenate-induced phaseollin accumulation (Dixon and Fuller, 1978). Cultures grown in the presence of 2,4-dichlorophenoxyacetic acid (2,4-D) accumulated lower levels than those grown with naphthaleneacetic acid (NAA) as auxin. Yields also decreased with increasing auxin concentration.

The production of 6-methoxymellein by *Daucus carota* cultures was promoted by the auxin 2,4-D as compared to indoleacetic acid (IAA).

Carrot cells cultured without auxin did not accumulate detectable amounts of phytoalexins on elicitation (Kurosaki *et al.*, 1985a). The induction of indole alkaloid accumulation by *Catharanthus roseus* cultures also offers interesting insights into the influence of the growth hormone regime of the nutrient medium (Eilert *et al.*, 1986a). Cells cultured on growth medium (B5; 1 mg/liter 2,4-D) normally do not accumulate indole alkaloids. Elicitor treatment between days 7 and 10 resulted in a similar alkaloid accumulation pattern as formed after transfer to alkaloid production medium (after Zenk *et al.*, 1977). In cells grown in production medium, elicitor treatment induced alkaloid accumulation only prior to the onset of media-related induction. Treatment at a later stage did not affect alkaloid accumulation at all. Elicitor induction and medium induction yielded similar amounts. Culture of the cell line under hormone autotrophic conditions resulted in a constantly high alkaloid level over the entire culture period, and at no time did elicitor treatment induce any marked response (U. Eilert, unpublished results). Obviously, once the process of indole alkaloid accumulation has been induced, additional stimuli will not increase yield. The storage capacity of the cells may become the limiting factor.

Dixon and Fuller (1978) emphasize that one has to differentiate carefully between level of accumulation (yield) and induction rate. Under conditions allowing for constitutive accumulation, i.e., without elicitation, the rate of induction may be lower than under conditions which suppress constitutive product formation. The ideal tissue culture system to study the physiology and biochemistry of elicitation should possess low background rates of synthesis and produce a maximum amount of product on elicitor treatment. Modification of the culture conditions can help reach this target.

H. Culture Conditions and Nutrient Composition

Light is a crucial factor in plant culture, as seen with the regulation of UV-induced flavonoid formation while elicitors modulate the metabolism of parsley cells to form furanocoumarins (Hahlbrock *et al.*, 1981; Tietjen and Matern, 1983; Tietjen *et al.*, 1983). Yields of acridone epoxides, which in plants accumulate primarily in roots (see Eilert, Chapter 24, Volume 5, this treatise), are considerably higher in dark-grown cultures. Mere elevation of the pH or addition of vanadate (50 mM) was sufficient to induce PAL in suspension cultures of *Vigna angularis* (Hattori and Ohta, 1985). Whether isoflavonoid accumulation equalled elicitor-induced formation qualitatively and quantitatively was not reported. The influence of the composition of the culture medium on elicitor-induced accumulation of secondary metabolites has generally been neglected.

VI. ELICITATION OF PROTOPLASTS AND ISOLATED CELLS

The role of the cell wall and the plasma membrane in the process of elicitation has still not been completely elucidated (Darvill and Albersheim, 1984). Protoplasts could offer an ideal system to examine reactions between plasma membrane and elicitor. Very few studies, however, have focused on the elicitation of protoplasts. Peters *et al.* (1978) exposed protoplasts obtained from potato leaves to cell wall glucan elicitor from *Phytophthora megasperma.* The result was rapid agglutination of protoplasts and subsequent death. A variety of other polysaccharides, e.g., laminarin, did not induce agglutination. The authors concluded from their experiments that specific receptor sites for elicitor molecules are located at the outer surface of the plasma membrane. Doke and Tomiyama (1980a,b) investigated whether host–pathogen specificity was apparent in the response of protoplasts. They treated protoplasts from potato tubers with cell wall preparations of different races of the parasite *Phytophthora infestans.* Wall components of all races, independent of their degree of pathogenicity, induced agglutination of the protoplasts followed by lysis. Intensity of the response of the protoplasts increased with the degree of field resistance of the source plant. This hypersensitive response was suppressed in protoplasts which were preincubated with soluble glucans obtained from the *Phytophthora* cultures. These soluble glucans by themselves did not induce any defense response in the protoplasts, but their activity to suppress elicitor activity varied with their origin from different fungal races. Furthermore, Doke and Furuichi (1982) observed a close correlation between the number of killed protoplasts on elicitor treatment and the degree of hypersensitive response of a cultivar on infection with an incompatible race.

Recently, Mieth *et al.* (1986) studied the possibility of employing protoplasts for elucidation of the process of phytoalexin accumulation in soybean. They observed that the process of protoplasting itself induced PAL and chalcone synthase activity. This result is not surprising considering that enzymatically released plant cell components are known to act as potential phytoalexin elicitors (see Section III). The data prove that the enzyme treatment involuntarily triggered stress responses. These may have caused the well-known problems of maintaining the viability of protoplasts. Addition of *Phytophthora* glucan elicitor did not further affect glyceollin formation by the protoplasts. The somehow disappointing results of this important study demonstrate that protoplasts prepared by enzymatic removal of the cell wall are not a suitable system to investigate the early processes of elicitation.

In this context, an interesting system was developed by Callow and Dow (1980). Mesophyll cells from tomato leaves were prepared by a rapid enzymatic maceration technique and successfully employed to study phytoalexin induction. The preparation technique also provided a more representative cell population than available with cells cultured *in vitro*.

VII. ELICITATION AND CELL ULTRASTRUCTURE

Elicitation results in considerable changes of cellular metabolism. The question whether the alterations in metabolism are accompanied by ultrastructural changes was investigated in three different systems. *Papaver somniferum* cells showed stacking of endoplasmic reticulum and electron-dense precipitate lining the tonoplast within 24 hr after elicitation (Eilert and Constabel, 1985). Elicitor treatment of *Ruta graveolens* cells also resulted in deposition of electron-dense material at the tonoplast, the only change observed; acridone idioblasts remained unaltered (Eilert *et al.*, 1986d). The third system examined was elicitor-treated cell cultures of *Catharanthus roseus*. Here, within 6 hr after elicitation, large lipid droplets appeared in the cytoplasm (U. Eilert *et al.*, unpublished results). Overall the dramatic changes of cellular metabolism induced by elicitation were not paralleled by equally dramatic changes of cellular ultrastructure.

VIII. INDUCTION OF PROTEINASE INHIBITORS

Although inducible, formation of proteinase inhibitors differs from phytoalexin formation. The defense mechanism is directed against a different class of enemies, namely, insects and mammals, rather than microbes. The product is a protein but can be considered a secondary metabolite by its physiological function. Walker-Simmons and Ryan (1986) have reported on elicitor-induced formation of proteinase inhibitors in tomato suspension cultures. During the growth cycle untreated cultures accumulate proteinase inhibitor with depletion of sucrose, i.e., at the beginning of the stationary phase. Treatment with either trigalacturonic acid, ethylene glycol, chitin, chitosan, or an uronic acid-rich extracellular polysaccharide of cell culture origin induced the accumulation at earlier stages of the growth cycle. Maximum accumulation was reached after 48 hr of elicitor contact and exceeded the age-induced accumulation by 50–100%. The extracellular polysaccharide, which appears with age in the culture medium, may self-induce the culture, thus causing the "age-induced" accumulation.

IX. BIOCHEMICAL AND MOLECULAR-GENETIC ASPECTS

Our understanding of the biochemical and molecular events triggered by elicitors has been greatly enlarged by studies conducted with either plant tissue systems (i.e., cotyledons, seedlings) or cells cultured *in vitro* (cf. Dixon *et al.*, 1983a,b). The induction of biosynthetic pathways is assumed to be the general mechanism but, so far, has been demonstrated in few tissue culture systems only: elicitor-induced formation of isoflavonoids in suspension cultures of *Glycine max* and *Phaseolus vulgaris,* and furanocoumarin formation in suspension cultures of *Petroselinum.* Coordinated induction of several general and specific pathway enzymes on the transcriptional and translational level has been shown. Table IV presents a comprehensive summary of the data obtained with these three systems.

Elicitor treatment induced rapid transient transcription and translation of enzymes of the general phenylpropanoid pathway, i.e., PAL and 4-coumarate:CoA ligase, in all three systems. In *Petroselinum* as well as in *Glycine,* optimum induction of PAL was achieved with similar amounts of the same elicitor (Hahlbrock *et al.*, 1981). Increase of PAL activity, however, does not necessarily result in phytoalexin accumulation. When *Glycine* cultures were treated with *Phytophthora megasperma* cell wall glucan, extracellular polysaccharide of *Xanthomonas,* or endopolygalacturonidase, all three agents induced PAL mRNA formation

TABLE IV

Elicitor Induction of Transcription and Translation of Biosynthetic Enzymes in Three Cell Culture Systems

Induced enzyme activity	*Phaseolus vulgaris*	*Glycine max*	*Petroselinum hortense*
General phenyl propanoid pathway			
PAL	Lamb and Dixon (1978), Dixon and Lamb (1979), Lamb *et al.* (1980), Lawton *et al.* (1980, 1983a), Dixon *et al.* (1981), Bolwell *et al.* (1985), Robbins *et al.* (1985), Edwards *et al.* (1985)	Ebel *et al.* (1976, 1984), Hille *et al.* (1982)	Hahlbrock *et al.* (1981), Tietjen and Matern (1983), Tietjen *et al.* (1983), Schmelzer *et al.* (1985), Kombrink and Hahlbrock (1986)
4-Coumarate : CoA ligase	—	Hille *et al.* (1982)	Hahlbrock *et al.* (1981), Tietjen and Matern (1983), Schmelzer *et al.* (1985), Kombrink and Hahlbrock (1986)
Flavonoid pathway			
Chalcone synthase	Lawton *et al.* (1983a), Cramer *et al.* (1985b), Robbins *et al.* (1985)	Hille *et al.* (1982) Ebel *et al.* (1984)	No induction, Hahlbrock *et al.* (1981), Tietjen and Matern (1983)
Chalcone isomerase	Dixon and Bendall (1978), Dixon *et al.* (1983a), Robbins and Dixon (1984), Cramer *et al.* (1985b)	—	No induction Hagmann *et al.* (1983)
Acetyl-CoA carboxylase	Relatively unaffected, Robbins *et al.* (1985)	Ebel *et al.* (1984)	—
Isoflavone synthase	—	Hagmann and Grisebach (1984)	—

(*continued*)

TABLE IV

(Continued)

Induced enzyme activity	*Phaseolus vulgaris*	*Glycine max*	*Petroselinum hortense*
Flavanone synthase	Dixon *et al.* (1981)	—	—
3,9-Dihydroxy-pterocarpan 6α-hydroxylase	—	Leube and Grisebach (1983), Hagmann *et al.* (1984)	—
Furanocoumarin pathway			
Dimethylallyl-di-phosphate : umbelliferon-dimethyl-trans-ferase	—	—	Tietjen and Matern (1983)
S-Adenosyl-L-methionine : bergaptol *O*-methyltrans-ferase	—	—	Hauffe *et al.* (1986)
S-Adenosyl-L-methionine : xanthotoxol *O*-methyltrans-ferase	—	—	Hauffe *et al.* (1986)
5-*O*-(4-Couma-royl)shikimate 3′-hydroxylase	—	—	Heller and Kühnl (1985)
Phenylpropanoid pathway			
Cinnamic acid 4-hydroxylase	Bolwell *et al.* (1985)	—	—
4-Hydroxycin-namic acid 3-hydroxylase	Bolwell *et al.* (1985)	—	—
Induction of mRNA coding for			
PAL	Lawton *et al.* (1983b), Cramer *et al.* (1985b), Edwards *et al.* (1985)	Ebel *et al.* (1984), Schmelzer *et al.* (1984)	Hahlbrock *et al.* (1981), Schmelzer *et al.* (1985), Kuhn *et al.* (1984), Chappel and Hahlbrock (1984)

(continued)

TABLE IV

(Continued)

Induced enzyme activity	*Phaseolus vulgaris*	*Glycine max*	*Petroselinum hortense*
4-Coumarate : CoA ligase	—	—	Kuhn *et al.* (1984), Chappell and Hahlbrock (1984), Schmelzer *et al.* (1985)
Chalcone synthase	Lawton *et al.* (1983b), Ryder *et al.* (1984), Cramer *et al.* (1985a)	Ebel *et al.* (1984), Schmelzer *et al.* (1984); Grab *et al.* (1985)	Chappell and Hahlbrock (1984)
Chalcone isomerase	Cramer *et al.* (1985a)	—	—
Total translatable mRNA	—	Grab *et al.* (1985)	Somssich *et al.* (1986)

and an increase of extractable PAL activity. Only treatment with *Phytophthora* glucan, however, resulted in isoflavonoid formation (Ebel *et al.*, 1984). Similar results were obtained with parsley suspension cultures (Kombrink and Hahlbrock, 1986). Agents like chitosan, laminarin, poly-L-lysine, or gum xanthan did induce PAL and 4-coumarate–CoA ligase, but no subsequent coumarin formation was observed. An explanation for this result may be the formation of different isoenzymes. Dixon *et al.* (1983b) pointed to the possible role of the different forms of PAL in plant tissue. Hille *et al.* (1982) observed the induction of only one of two isoenzymes of 4-coumarate–CoA ligase. The particular isoenzyme is presumably specifically involved in flavonoid biosynthesis.

In parsley cells elicitation caused specific expression of the enzymes of the furanocoumarin pathway (see Table IV), while UV irradiation, in contrast, resulted in induction of the flavonoid pathway (Chappell and Hahlbrock, 1984). PAL activity increased in response to either type of stress and reached the same level of activity, and no structural differences of the enzymes were detected (Hahlbrock *et al.*, 1981). The time course of induction, however, varied significantly, thus showing the specificity of the response in dependence of the stress factor. Here, a regulatory role of PAL is not evident. The problem of the induction of PAL, its role in secondary metabolism, and its regulation is complex. A review by Jones (1984) discusses the matter in detail. In addition, a number of other enzyme activities changed in response to elicitation. Total translatable mRNA was found increased, coding for "pathogenesis-related" proteins (Somssich *et al.*, 1986). With enzymes of primary metabolism a rapid increase of glucose-6-phosphate dehydrogenase and a rapid decrease of pyrophosphate : fructose-6-phosphate phosphotransferase were observed, while phosphofructokinase activity remained unaffected (Kombrink and Hahlbrock, 1986). As a result of activated 1-aminocyclopropane-1-carboxylate synthase, ethylene biosynthesis was increased (Chappell *et al.*, 1984). Two hydrolytic enzymes, chitinase and 1,3-β-glucanase, proved to be inducible too (Chappell *et al.*, 1984; Kombrink and Hahlbrock, 1986).

Induction of flavonoid (biosynthetic) pathway enzyme activities concomitant with isoflavonoid accumulation by elicitation was found in tissue cultures of fabaceous plants. Gustine *et al.* (1978) detected an increase of PAL, isoliquiritigenin, daidzein, and genistein *O*-methyltransferase activities in $HgCl_2$-treated callus of *Canavalia ensiformis*. Elicitation with spores of *Phytophthora megasperma* did not induce *O*-methyltransferase activities. Phytoalexin accumulation, however, did take place. Hattori and Ohta (1985) also reported that a rapid transient increase of PAL preceded isoflavone accumulation in *Vigna angularis* cultures.

Eilert and Constabel (1986) were able to demonstrate the induction of sanguinarine biosynthesis in elicitor-treated *Papaver somniferum* cultures using inhibitors of protein synthesis and radiolabeled precursors. Brodelius *et al.* (1986a) were able to enhance production of berberine, another benzylisoquinoline alkaloid, by *Thalictrum rugosum* cell suspensions with a carbohydrate fraction of yeast extract as elicitor. Induction of tyrosine decarboxylase, the first step on the biosynthetic route to berberine, was found. Its activity and alkaloid production appeared correlated.

Detailed work was conducted by Eilert *et al.* (1986e) to demonstrate the induction of monoterpene indole alkaloid biosynthesis in elicitor-treated *Catharanthus roseus* cultures. In contrast to previous reports on the regulation of indole alkaloid formation in *C. roseus* cultures, tryptophan decarboxylase (TDC) and strictosidine synthase proved to be inducible in this particular cell line. While in the isoflavonoid-accumulating systems an increase of PAL activity precedes the increase of chalcon synthase, here strictosidine synthase activity increases first, followed by TDC activity.

Robbins *et al.* (1985) and Bolwell *et al.* (1985) investigated the effect of elicitor treatment on the general metabolism of suspension cultures of *Phaseolus vulgaris*. Time courses of the activities of about 15 enzymes were monitored. No induction of the shikimic acid pathway nor of acetyl-CoA carboxylase, both involved in the biosynthesis of distant precursors of isoflavonoids, was observed. Enzymes of ammonia assimilation with the exception of glutamate dehydrogenase remained unaffected. A glutamate dehydrogenase increase, occurring in the controls, was suppressed. Activities of hydrolases, which may act on fungal cell walls, also did not increase.

Although Farmer (1985) recorded drastic changes of the cell wall composition, induction of enzymes involved in lignin biosynthesis did not occur in cultured cells of either *Phaseolus* of *Glycine*. Wall-associated hydroxyproline content as well as phenolic material bound to cellulosic and hemicellulosic fractions increased, cinnamic acid 4-hydroxylase activity was induced, and the glucose content of the hemicellulosic fraction was decreased.

The induction of callose synthesis by chitosan treatment of soybean cells was not paralleled by any enzyme induction (Köhle *et al.*, 1985). Considering the rapidity of this induction, the involvement of a transcription step is excluded by the authors. A process of covalent modification of 1,3-β-glucan synthase activity is postulated, caused by either Ca^{2+} influx or proteolytic activity.

Results indicate that elicitor-induced gene expression is specific but complex. As demonstrated, induction of enzymes does not necessarily lead to product accumulation. The events observed in tissue cultures are not of mere biochemical interest, and their relevance for phytopathological considerations was demonstrated when the sequence of responses triggered by elicitation in cell cultures and plant tissues were compared (Schmelzer *et al.*, 1984; Grab *et al.*, 1985; Robbins *et al.*, 1985).

X. CONCLUSION

Elicitor induction offers a novel approach to achieve rapid accumulation of certain secondary metabolites. Compounds found in cultured cells in response to elicitation are listed in Table I and are grouped according to their biosynthetic relationships. The wide range of compounds illustrates the potential for a broad application of elicitor induction. Most of the compounds can be referred to as phytoalexins. Their induction was previously shown in natural plant–pathogen systems. Elicitor-induced accumulation of compounds like diosgenin, indole alkaloids, berberine, morphine, codeine, and sanguinarine, however, was first observed in artificial *in vitro* systems and may not occur in the differentiated plant, where the compounds may be present constitutively. Designation of these compounds as phytoalexins thus has its problems. Anthraquinone formation in elicitor-treated *Cinchona* cultures, another example, was interpreted as phytoalexin accumulation (Wijnsma *et al.*, 1985a). As there is a high level of anthraquinones present in untreated cultures and as the time required for induction is unusually long (8–15 days after treatment), the interpretation may not be correct, even though anthraquinones could be isolated only from diseased bark of *Cinchona* trees (Wijnsma *et al.*, 1985b). For classification of all secondary metabolites which accumulate in plant cell cultures on elicitor treatment, the term "elicitation metabolite" or "elicitation product" is suggested.

Elicitation can improve the efficiency of product accumulation in cell cultures in several ways:

1. The time required to obtain a product can be reduced
2. An exchange of medium may become superfluous
3. Metabolites may be accumulated not only within the cell but considerable amounts may also be excreted into the medium
4. Reelicitation may lead to a process for semicontinuous production

Furthermore, elicited cells may be used for biotransformations. Elicitation results in the induction of various enzymes. High activities could lead to a more rapid and efficient transformation than observed in untreated cultures. Additional pathways of metabolism may also be induced. Tabersonine, for example, is transformed into a number of different products by suspension cultures of *Catharanthus roseus*. After elicitor treatment tabersonine was transformed to hydroxytabersonine with an efficiency of 20–30% (U. Eilert, unpublished). Otherwise this compound is rarely detected in tissue cultures and cannot be obtained from tabersonine by chemical synthesis. In summary, predictions are that elicitation will find future application in the commercial production of secondary metabolites by plant cell cultures.

ACKNOWLEDGMENT

Work was supported by a Deutsche Forschungsgemeinschaft (DFG) grant.

APPENDIX

Elicitor-Induced Product Accumulation as Presented at the Sixth IAPTC Congress, Minneapolis, 1986

Elicitor	Preparation	Cell culture	Products	Reference
Pythium aphanidermatum, Eurotium rubrum, Micromucor isabellina, or *Chrysosporium palmorum*	Filtrates and extracts	*Catharanthus roseus*	Tryptamine, ajmalicine, and catharanthine	Tallevi *et al.* (1986)
Pythium aphanidermatum	Homogenate	*Catharanthus roseus*	Strictosidine, ajmalicine, tabersonine, lochnericine, and catharanthine	Eilert *et al.* (1986)
Botrytis sp.	Homogenate	*Papaver somniferum*	Sanguinarine	Eilert *et al.* (1986)
Dendryphion sp.	Extract	*Papaver somniferum*	Sanguinarine	Cline and Coscia (1986)
		Papaver bracteatum	Sanguinarine	
Yeast	Carbohydrate preparation	*Glycine max*	Glyceollin	Brodelius *et al.* (1986)
		Thalictrum rugosum	Berberine	
Aspergillus niger	Homogenate	*Cinchona ledgeriana, Rubia tinctoria,* or *Morinda citrifolia*	Anthraquinones	Wijnsma *et al.* (1986)
Nigeran	—	*Solanum melongena*	Polyacetylenes	Ohta *et al.* (1986)
		Vigna angularis	Isoflavones	

REFERENCES

Amin, M., Kurosaki, F., and Nishi, A. (1986). Extracellular pectinolytic enzymes of fungi elicit phytoalexin accumulation in carrot suspension culture. *J. Gen. Microbiol.* **132,** 771–779.

Ayers, A. R., Ebel, J., Valent, B. S., and Albersheim, P. (1976). Host pathogen interactions. X. Fractionation and biological activity of an elicitor isolated from the mycelial walls of *Phytophthora megasperma* var. *sojae. Plant Physiol.* **57,** 760–765.

Bailey, J. A., and Mansfield, J. W. (1982). "Phytoalexins." Blackie, Glasgow.

Banthorpe, D. V., Bilyard, H. J., and Watson, D. J. (1985). Pigment formation by callus of *Lavandula angustifolia. Phytochemistry* **24,** 2667–2680.

Boller, T. (1983). Ethylene-induced biochemical defenses against pathogenes. *In* "Plant Growth Substances, 1982" (P. F. Wareing, ed.), pp. 303–312. Academic Press, London.

Bolwell, G. P., Robbins, M. P., and Dixon, R. A. (1985). Metabolic changes in elicitor-treated bean cells. Enzymic responses associated with rapid changes in cell wall components. *Eur. J. Biochem.* **148,** 571–578.

Bostock, R. M., Laine, R. A., and Kuć, J. (1982). Factors affecting the elicitation of sesquiterpenoid phytoalexin accumulation by eicosapentaenoic and arachidonic acids in potato. *Plant Physiol.* **70,** 1417–1424.

Brindle, P. A., and Threlfall, D. R. (1982). The metabolism of phytoalexins. *Biochem. Soc. Trans.* **11,** 516–522.

Brindle, P. A., Kuhn, P. J., and Threlfall, D. R. (1983). Accumulation of phytoalexins in potato cell suspension cultures. *Phytochemistry* **22,** 2719–2721.

Brodelius, P., Funk, C., Gügler, K., Haldimann, D., and Umberg, B. (1986a). Elicitor-induced tyrosine decarboxylase in berberine synthesizing suspension cultures of *Thalictrum rugosum. Poster Abstr. Phytochem. Soc. Eur. Symp. Biol. Act. Nat. Prod.*, poster 47.

Brodelius, P., *et al.* (1986b). *In* "Plant Tissue Culture 1986" (D. A. Somer, B. G. Gengenbach, D. D. Biesboer, W. P. Hackett, and C. E. Green, eds.). Univ. of Minneapolis Press, Minneapolis, Minnesota.

Brooks, C. J. W., Watson, D. G., and Freer, I. M. (1986). Elicitation of capsidiol accumulation in suspended callus cultures of *Capsicum annuum. Phytochemistry* **25,** 1089–1092.

Budde, A. D., and Helgeson, J. P. (1981a), Phytoalexins in tobacco callus tissues challenged by zoospores of *Phytophthora parasitica* var. *nicotianae. Phytopathology* **71,** 206.

Budde, A. D., and Helgeson, J. P. (1981b). Chronology of phytoalexin production and histological changes in tobacco callus infected with *Phytophthora parasitica* var. *nicotianae. Phytopathology* **71,** 864.

Callow, J. A., and Dow, J. M. (1980). The isolation and properties of tomato mesophyll cells and their use in elicitor studies. *In* "Tissue Culture Methods for Plant Pathologists" (D. S. Ingram and J. Helgeson, eds.), pp. 197–202. Blackwell, Oxford.

Chappell, J., and Hahlbrock, K. (1984). Transcription of plant defense genes in response to UV-light or fungal elicitor. *Nature (London)* **311,** 76–78.

Chappell, J., Hahlbrock, K., and Boller, T. (1984). Rapid induction of ethylene biosynthesis in cultured parsley cells by fungal elicitor and its relationship to the induction of phenylalanine ammonia-lyase. *Planta* **161,** 475–480.

Cline and Coscia (1986). *In* "Plant Tissue Culture 1986" (D. A. Somer, B. G. Gengenbach, D. D. Biesboer, W. P. Hackett, and C. E. Green, eds.). Univ. of Minneapolis Press, Minneapolis, Minnesota.

Constabel, F., Kurz, W. G. W., and Kutney, J. P. (1982). Variation in cell cultures of periwinkle, *Catharanthus roseus*. *In* "Plant Tissue Culture 1982" (A. Fujiwara, ed.), pp. 301–304. Maruzen, Tokyo.

Cramer, C. L., Ryder, T. B., Bell, J. N., and Lamb, C. J. (1985a). Rapid switching of plant gene expression induced by fungal elicitor. *Science* **227,** 1240–1243.

Cramer, C. L., Bell, J. N., Ryder, T. B., Bailey, J. A., Schuch, W., Bolwell, G. P., Robbins, M. P., and Dixon, R. A. (1985b). Co-ordinated synthesis of phytoalexin biosynthetic enzymes in biologically stressed cells of bean (*Phaseolus vulgaris* L.). EMBO J. **4,** 285–289.

Cruickshank, I. A. M. (1980). Defenses triggered by the invader: Chemical defenses. *In* J. G. Horsfall and E. B. Cowling, eds.), "Plant Disease: An Advanced Treatise" Vol. 5, pp. 247–269. Academic Press, New York.

Darvill, A. G., and Albersheim, P. (1984). Phytoalexins and their elicitors—A defence against microbial infections in plants. *Annu. Rev. Plant Physiol.* **35,** 243–275.

Darvill, A. G., Albersheim, P., McNeil, M., Lau, J. M., York, W. S., Stevenson, T. T., Thomas, J., Doares, S., Gollin, D. J., Chelf, P., and Davis, K. (1985). Structure and function of plant cell wall polysaccharides. *J. Cell Sci., Suppl.* **2,** 203–217.

Davis, K. R., Lyon, G. D., Darvill, A. G., and Albersheim, P. (1984). Host–pathogen interactions. XXV. Endopolygalacturonic acid lyase from *Erwinia carotovora* elicits phytoalexin accumulation by releasing plant cell wall fragments. *Plant Physiol.* **74,** 52.

Davis, K. R., Darvill, A. G., and Albersheim, P. (1986). Several biotic and abiotic elicitors act synergistically in the induction of phytoalexin accumulation in soybean. *Plant Mol. Biol.* **6,** 23–32.

DiCosmo, F., and Misawa, M. (1985). Eliciting secondary metabolism in plant cell cultures. *Trends Biotechnol.* **3,** 318–322.

DiCosmo, F., Norton, R., and Towers, G. H. N. (1982). Fungal culture-filtrate elicits aromatic polyacetylenes in plant tissue culture. *Naturwissenschaften* **69,** 550–551.

Dixon, R. A. (1980). Plant tissue culture methods in the study of phytoalexin induction. *In* "Tissue Culture Methods for Plant Pathologists" (D. S. Ingram and J. P. Helgeson, eds.), pp. 185–196. Blackwell, Oxford.

Dixon, R. A. (1985). "Plant Cell Culture—A Practical Approach." IRL Press Ltd., Oxford.

Dixon, R. A (1986). The phytoalexin response: Elicitation, signalling and control of host gene expression. *Biol. Rev. Cambridge Philos. Soc.* **61**(3), 239–291.

Dixon, R. A., and Bendall, D. S. (1978). Changes in phenolic compounds associated with phaseollin production in cell suspension cultures of *Phaseolus vulgaris. Physiol. Plant Pathol.* **13,** 293–294.

Dixon, R. A., and Fuller, K. W. (1978). Effects of growth substances on non-induced and *Botrytis cinerea* culture filtrate-induced phaseollin production in *Phaseolus vulgaris* cell suspension cultures. *Physiol. Plant Pathol.* **12,** 279–288.

Dixon, R. A., and Lamb, C. J. (1979). Stimulation of de novo synthesis of L-phenylalanine ammonia-lyase in relation to phytoalexin accumulation in *Colletotrichum lindemuthianum* elicitor-treated cell suspension cultures of French bean (*Phaseolus vulgaris*). *Biochim. Biophys. Acta* **586,** 453–463.

Dixon, R. A., Dey, P. M., Murphy, D. L., and Whitehead, I. M. (1981). Dose responses for *Colletotrichum lindemuthianum* elicitor-mediated enzyme induction in French bean cell suspension cultures. *Planta* **151,** 272–280.

Dixon, R. A., Dey, P. M., and Lamb, C. J. (1983b). Phytoalexins, enzymology and molecular biology. *In* "Methods in Enzymology and Related Areas of Molecular Biology" (A. Meister, ed.), pp. 1–136. Wiley, New York.

Dixon, R. A., Gerrish, C., Lamb, C. J., and Robbins, M. P. (1983b). Elicitor-mediated induction of chalcone isomerase in *Phaseolus vulgaris* cell suspension cultures. *Planta* **159,** 561–569.

Dixon, R. A., Prakash, M., Dey, M., Lawton, M. A., and Lamb, C. (1983c). Phytoalexin induction in French bean. Intercellular transmission of elicitation in cell suspension cultures and hypocotyl sections of *Phaseolus vulgaris. Plant Physiol.* **71,** 251–256.

Doke, N., and Furuichi, N. (1982). Response of protoplasts to hyphal cell wall components in relationship to resistance of potato to *Phytophthora infestans. Physiol. Plant Pathol.* **21,** 23–30.

Doke, N., and Tomiyama, K. (1980a). Effect of hyphal wall components from *Phytophthora infestans* on protoplasts of potato tuber tissues. *Physiol. Plant Pathol.* **16,** 169–176.

Doke, N., and Tomiyama, K. (1980b). Suppression of the hypersensitive response of potato tuber protoplasts to hyphal wall components by water-soluble glucans isolated from *Phytophthora infestans. Physiol. Plant Pathol.* **16,** 177–186.

Ebel, J. (1986). Phytoalexin synthesis: The biochemical analysis of the induction process. *Annu. Rev. Phytopathol.* **24,** 235–264.

Ebel, J., Ayers, A. A., and Albersheim, P. (1976). Host pathogen interactions. XII. Response of suspension-cultured soyabean cells to the elicitor isolated from *Phytophthora megasperma* var. *sojae,* a fungal pathogen of soyabean. *Plant Physiol.* **57,** 775–779.

Ebel, J., Schmidt, W. E., and Loyal, R. (1984). Phytoalexin synthesis in soybean cells: Elicitor induction of phenylalanine ammonia-lyase and chalcone synthase mRNAs and correlation with phytoalexin accumulation. *Arch. Biochem. Biophys.* **232,** 240–248.

Edwards, K., Cramer, C. L., Bolwell, G. P., Dixon, R. A., Schuch, W., and Lamb, C. J. (1985). Rapid transient induction of phenylalanine ammonia-lyase mRNA in elicitor-treated bean cells. *Proc. Natl. Acad. Sci. U.S.A.* **82,** 6731–6735.

Eilert, U. and Constabel, F. (1985). Ultrastructure of *Papaver somniferum* cells cultured *in vitro* and treated with fungal homogenate eliciting alkaloid production. *Protoplasma* **128,** 38–42.

Eilert, U., and Constabel, F. (1986). Elicitation of sanguinarine accumulation in *Papaver somniferum* cells by fungal homogenates—An induction process. *J. Plant Physiol.* **125,** 167–173.

Eilert, U., Ehmke, A., and Wolters, B. (1984). Elicitor-induced accumulation of acridone alkaloid epoxides in *Ruta graveolens* suspension cultures. *Planta Med.* **6,** 508–512.

Eilert, U., Kurz, W. G. W., and Constabel, F. (1985). Stimulation of sanguinarine accumulation in *Papaver somniferum* cell cultures by fungal elicitors. *J. Plant Physiol.* **119,** 65–76.

Eilert, U., Constabel, F., and Kurz, W. G. W. (1986a). Elicitor-stimulation of monoterpene indole alkaloid formation in suspension cultures of *Catharanthus roseus. J. Plant Physiol.* **126,** 11–22.

Eilert, U., Kurz, W. G. W., and Constabel, F. (1986b). Elicitor-induction of sanguinarine formation in *Papaver somniferum* cell cultures and semicontinuous sanguinarine production by reelicitation. *Planta Med* **5,** 417–418.

Eilert, U., Wolters, B., and Constabel, F. (1986d). Ultrastructure of acridone idioblasts in roots and cell cultures of *Ruta graveolens. Can. J. Bot.* **64,** 1089–1096.

Eilert, U., DeLuca, V., Constabel, F., and Kurz, W. G. W. (1986e). Elicitor-mediated induction of tryptophan decarboxylase and strictosidine synthase activities in cell suspension cultures of *Catharanthus roseus. Arch. Biochem. Biophys.* (in press).

Eilert, U., *et al.* (1986f). *In* "Plant Tissue Culture 1986" (D. A. Somer, B. G. Gengenbach, D. D. Biesboer, W. P. Hackett, and C. E. Green, eds.). Univ. of Minneapolis Press, Minneapolis, Minnesota.

Eriksson, T. (1965). Studies on the growth requirements and growth measurements of cell cultures of *Haplopappus gracilis*. *Physiol. Plant* **18,** 976–993.
Érsek, T., and Sziraki, I. (1980). Production of sesquiterpene phytoalexins in tissue culture callus of potato tubers. *Phytopathol. Z.* **97,** 364–368.
Esquerre-Tugaye, M. T., Mazau, D., Pellisier, B., Roby, D., Rumeau, D., and Toppan, A. (1985). Induction by elicitors and ethylene of proteins associated with the defense in plants. *UCLA Symp. Mol. Cell. Biol.* [N.S.] **22,** 459–473.
Evans, D. A. (1986). Somaclonal and gametoclonal variation. *In* "Biotechnology for Solving Agricultural Problems," (P. C. Augustine, H. I. Danforth, and M. R. Bakst, eds.), pp. 63–96. Martinus Nijhoff Publ., Dordrecht, The Netherlands.
Farmer, E. E. (1985). Effects of fungal elicitor on lignin biosynthesis in cell suspension cultures of soybean. *Plant Physiol.* **78,** 338–342.
Fujimori, T., Tanaka, H., and Kato, K. (1983). Stress compounds in tobacco callus infiltrated by *Pseudomonas solanacearum*. *Phytochemistry* **22,** 1038.
Fukui, H., Yoshikawa, N., and Tabata, M. (1983). Induction of shikonin formation by agar in *Lithospermum erythrorhizon* cell suspension cultures. *Phytochemistry* **22,** 2451–2453.
Gamborg, O. L., Miller, R. A., and Ojima, K. (1968). Nutrient requirements of suspension cultures of soybean root cells. *Exp. Cell Res.* **50,** 151–158.
Gay, L. (1985). Phytoalexin formation in cell cultures of *Dianthus caryophyllus* treated by an extract from the culture medium of *Phytophthora parasitica*. *Physiol. Plant Pathol.* **26,** 143–150.
Gorelova, O. A., Rzerzabek, J., Korzhenevskaya, T. G., Butenko, R. G., and Gusev, M. V. (1984). Growth and biosynthetic activity of *Solanum laciniatum* cells in mixed cultures with nitrogen fixing cyanobacteria. *Plant Physiol. (Moscow)* **279,** 253–256.
Gossens, J. F. V., and Vendrig, J. C. (1982). Effects of absicisic acid, cytokinins, and light on isoflavonoid phytoalexin accumulation in *Phaseolus vulgaris* L. *Planta* **154,** 441–446.
Grab, D., Loyal, R., and Ebel, J. (1985). Elicitor-induced phytoalexin synthesis in soybean cells: Changes in the activity of chalcone synthase mRNA and the total production of translatable mRNA. *Arch. Biochem. Biophys.* **243,** 523–529.
Gustine, D. L. (1981). Evidence for sulfhydryl involvement in regulation of phytoalexin accumulation in *Trifolium repens* callus tissue cultures. *Plant Physiol.* **68,** 1323–1326.
Gustine, D. L., Sherwood, R. T., and Vance, C. P. (1978). Regulation of phytoalexin synthesis in jackbean callus cultures. Stimulation of phenylalanine ammonia-lyase and *O*-methyltransferase. *Plant Physiol.* **61,** 226–230.
Haberlach, G. T., Budde, A. D., Sequeira, L., and Helgeson, J. P. (1978). Modification of disease resistance of tobacco callus tissues by cytokinins. *Plant Physiol.* **62,** 522–525.
Hadwiger, L. A., and Beckman, J. M. (1980), Chitosan as a component of pea–*Fusarium solani* interactions. *Plant Physiol.* **66,** 205–211.
Hagmann, M., and Grisebach, H. (1984). Enzymic rearrangement of flavanone to isoflavone. *FEBS Lett.* **175,** 199–202.
Hagmann, M. L., Heller, W., and Grisebach, H. (1983). Induction and characterization of a microsomal flavonoid 3′-hydroxylase from parsley cell cultures. *Eur. J. Biochem.* **134,** 547–554.
Hagmann, M. L., Heller, W., and Grisebach, H. (1984). Induction of phytoalexin synthesis in soybean. Stereospecific 3,9-dihydropterocarpan 6α-hydroxylase from elicitor-induced soybean cell cultures. *Eur. J. Biochem.* **142,** 127–131.
Hahlbrock, K. (1981). Flavonoids. *In* "The Biochemistry of Plants" (E. E. Conn, ed.), Vol. 7, pp. 425–455. Academic Press, New York.
Hahlbrock, K., Lamb, C. J., Purwin, C., Ebel, J., Fautz, E., and Schäfer, E. (1981). Rapid

response of suspension-cultured parsley cells to the elicitors from *Phytophthora megasperma* var. *sojae*. *Plant Physiol.* **67,** 768–773.

Hahn, M. G., and Grisebach, H. (1983). Cyclic AMP is not involved as a second messenger in the response of soybean to infection by *Phytophthora megasperma* f. sp. *glycinea*. *Z. Naturforsch., C: Biosci.* **38C,** 578–582.

Hahn, M. G., Darvill, A. G., and Albersheim, P. (1981). Host–pathogen interactions. XIX. The endogenous elicitor, a fragment of a plant cell wall polysaccharide that elicits phytoalexin accumulation in soybeans. *Plant Physiol.* **68,** 1161–1169.

Hargreaves, J. A., and Bailey, J. A. (1978). Phytoalexin production by hypocotyls of *Phaseolus vulgaris* in response to constitutive metabolites released by damaged bean cells. *Physiol. Plant Pathol.* **13,** 89–100.

Hargreaves, J. A., and Selby, C. (1978). Phytoalexin formation in cell suspensions of *Phaseolus vulgaris* in response to an extract of bean hypocotyls. *Phytochemistry* **17,** 1099–1102.

Hattori, T., and Ohta, Y. (1985). Induction of phenylalanine ammonia-lyase activation and isoflavone glucoside accumulation in suspension-cultured cells of read bean, *Vigna angularis,* by phytoalexin elicitors, vanadate, and elevation of medium pH. *Plant Cell Physiol.* **26,** 1101–1110.

Hauffe, K. D., Hahlbrock, K., and Scheel, D. (1986). Elicitor-stimulated furanocoumarin biosynthesis in cultured parsley cells: *S*-adenosyl-L-methionine : bergaptol and *S*-adenosyl-L-methionine : Xanthothoxol *O*-methyltransferases. *Z. Naturforsch., C. Biosci.* **41C,** 228–239.

Heinstein, P. F. (1982). Effect of *Verticillium dahliae* on *Gossypium* cell suspension cultures. *In* "Plant Tissue and Cell Culture 1982" (A. Fujiwara, eds.), pp. 675–676. Maruzen, Tokyo.

Heinstein, P. F. (1985a). Future approaches to the formation of secondary natural products in plant cell suspension cultures. *J. Nat. Prod.* **48,** 1–9.

Heinstein, P. F. (1985b). Stimulation of sesquiterpene aldehyde formation in *Gossypium arboreum* cell suspension cultures by conidia of *Verticillium dahliae*. *J. Nat. Prod.* **48,** 907–915.

Helgeson, J. P. (1983). Studies of host–pathogen interactions *in vitro*. *In* "Use of Tissue Culture and Protoplasts in Plant Pathology" (J. P. Helgeson and B. J. Deverall, ed.), pp. 9–38. Academic Press, New York.

Helgeson, J. P., Budde, A. D., and Haberlach, G. T. (1978). Capsidiol. A phytoalexin produced by tobacco callus tissues. *Plant Physiol.* **61,** Suppl. 58.

Heller, W., and Kühnl, T. (1985). Elicitor induction of a microsomal 5-*O*-(4-coumaroyl)shikimate 3′-hydroxylase in parsley cell suspension cultures. *Arch. Biochem. Biophys.* **241,** 453–460.

Hille, A., Purwin, C., and Ebel, J. (1982). Induction of enzymes of phytoalexin synthesis in cultured soybean cells by an elicitor from *Phytophthora megasperma* f. sp. *glycinea*. *Plant Cell Rep.* **1,** 123–127.

Holliday, M. J., and Klarmann, W. L. (1979). Expression of disease reaction types in soybean callus from resistant and susceptible plants. *Phytopathology* **69,** 576–578.

Ingram, D. S., and Helgeson, J. P., eds. (1980). "Tissue Culture Methods for Plant Pathologists." Blackwell, Oxford.

Jones, D. H. (1984). Phenylalanine ammonia lyase: Regulation of its induction, and its role in plant development. *Phytochemistry* **23,** 1349–1359.

Keen, N. T. (1986). Phytoalexins and their involvement in plant disease resistance. *Iowa State J. Res.* **60** (4), 477–499.

Keen, N. T., and Horsch, R. (1972). Hydroxyphaseollin production by various soyabean tissues: A warning against the use of unnatural host–parasite systems. *Phytopathology* **62,** 439–442.

Keen, N. T., and Yoshikawa, M. (1983). β-1,3-Endoglucanase from soybean releases elicitor-active carbohydrates from fungus cell walls. *Plant Physiol.* **71,** 460–465.

Keen, N. T., Partridge, J. E., and Zaki, A. I. (1972). Pathogen-produced elicitor of a chemical defense mechanism in soybean mono-genically resistant to *Phytophthora megasperma* var. *sojae*. *Phytopathology* **62,** 768.

Keen, N. T., Yoshikawa, M., and Wang, M. C. (1983). Phytoalexin elicitor activity of carbohydrates from *Phytophthora megasperma* f. sp. *glycinea* and other sources. *Plant Physiol.* **71,** 466–471.

Köhle, H., Young, D. H., and Kauss, H. (1984). Physiological changes in suspension-cultured soybean cells elicited by treatment with chitosan. *Plant Sci. Lett.* **33,** 221–230.

Köhle, H., Jeblick, W., Poten, F., Blaschek, W., and Kauss, H. (1985). Chitosan-elicited callose synthesis in soybean cells as a Ca^{2+}-dependent process. *Plant Physiol.* **77,** 544–551.

Kombrink, E., and Hahlbrock, K. (1985). Dependence of the level of phytoalexin and enzyme induction by fungal elicitor on the growth stage of *Petroselinum crispum* cell cultures. *Plant Cell Rep.* **4,** 277–280.

Kombrink, E., and Hahlbrock, K. (1986). Responses of cultured parsley cells to elicitors from phytopathogenic fungi. *Plant Physiol.* **81,** 216–221.

Krasnuk, M., Witham, F. J., and Tegley, J. R. (1971). Cytotoxins extracted from pinto bean fruit. *Plant Physiol.* **48,** 320–324.

Kuć, J. (1972). Phytoalexins. *Annu. Rev. Phytopathol.* **10,** 207–232.

Kuhn, D. N., Chappell, J., Boudet, A., and Hahlbrock, K. (1984). Induction of phenylalanine ammonia-lyase and 4-coumarate : CoA ligase mRNAs in cultured plant cells by UV-light or fungal elicitor. *Proc. Natl. Acad. Sci. U.S.A.* **81,** 1102–1106.

Kurosaki, F., and Nishi, A. (1983). Isolation and antimicrobial activity of the phytoalexin 6-methoxymellein from cultured carrot cells. *Phytochemistry* **22,** 669–672.

Kurosaki, F., Futamura, K., and Nishi, A. (1985a). Factors affecting phytoalexin production in cultured carrot cells. *Plant Cell Physiol.* **24,** 693–700.

Kurosaki, F., Yutaka, T., and Nishi, A. (1985b). Phytoalexin production in cultured carrot cells treated with pectinolytic enzymes. *Phytochemistry* **24,** 1479–1480.

Lamb, C. J., and Dixon, R. A. (1978). Stimulation of de novo synthesis of L-phenylalanine ammonia-lyase during induction of phytoalexin biosynthesis in cell suspension cultures of *Phaseolus vulgaris*. *FEBS Lett.* **94,** 277–280.

Lamb, C. J., Lawton, M. A., Taylor, S. J., and Dixon, R. A. (1980). Elicitor modulation of phenylalanine ammonia-lyase in Phaseolus vulgaris. Ann. Phytopathol. **12,** 423–433.

Latunde-Dada, A. O., and Lucas, J. A. (1985). Involvement of the phytoalexin medicarpin in the differential response of callus lines of lucerne (*Medicago sativa*) to infection by *Verticillium albo-atrum*. *Physiol. Plant Pathol.* **26,** 31–42.

Lawton, M. A., Dixon, R. A., and Lamb, C. J. (1980). Elicitor modulation of the turnover of L-phenylalanine ammonia-lyase in French bean cell suspension cultures. *Biochim. Biophys. Acta* **633,** 162–175.

Lawton, M. A., Dixon, R. A., Hahlbrock, K., and Lamb, C. J. (1983a). Rapid induction of the synthesis of phenylalanine ammonia-lyase and chalcone synthase in elicitor-treated plant cells. *Eur. J. Biochem.* **129,** 593–601.

Lawton, M. A., Dixon, R. A., Hahlbrock, K., and Lamb, C. J. (1983b). Elicitor induction of mRNA activity. Rapid effects of elicitor on phenylalanine ammonia-lyase and chalcone synthase mRNA activities in bean cells. *Eur. J. Biochem.* **130,** 131–139.

Lee, S. C., and West, C. A. (1981). Polygalacturonase from *Rhizopus stolonifer,* an elicitor of casbene synthetase activity in castor bean (*Ricinus communis* L.) seedlings. *Plant Physiol.* **67,** 33–39.

Leube, J., and Grisebach, H. (1983). Further studies on induction of enzymes of phytoalexin synthesis in soybean and cultured plant cells. *Z. Naturforsch., C: Biosci.* **38C.** 730–735.

Linsmaier, E. M., and Skoog, F. (1965). Organic growth factor requirements of tobacco tissue cultures. *Physiol. Plant* **18,** 100–112.

Mieth, H., Speth, V., and Ebel, J. (1986). Phytoalexin production by isolated soybean protoplasts. *Z. Naturforsch., C: Biosci.* **41C,** 193–201.

Miller, S. A., and Maxwell, D. P. (1983). Evaluation of disease resistance. *In* "Handbook of Plant Cell Culture" (D. A. Evans, R. Sharp, D. Ammirato, and Y. Yamada, eds.), Vol. 1, pp. 853–879. Macmillan, New York.

Moesta, P., and Grisebach, H. (1980). Effects of biotic and abiotic elicitors on phytoalexin metabolism in soybean. *Nature (London)* **286,** 710–711.

Moesta, P., and Grisebach, H. (1981). Investigation of the mechanism of phytoalexin accumulation in soybean induced by glucan or mercuric chloride. *Arch. Biochem. Biophys.* **211,** 39–43.

Müller, K. O. (1956). Einige einfache Versuche zum Nachweis von Phytoalexinen. *Phytopathol. Z.* **27,** 237–254.

Murashige, T., and Skoog, F. (1962). A revised medium for rapid growth and bioassays with tobacco tissue cultures. *Physiol. Plant* **15,** 473–497.

Oba, K., and Uritani, I. (1979). Biosynthesis of furano-terpenes by sweet potato cell culture. *Plant Cell Physiol.* **20,** 819–826.

Ohta, Y. *et al.* (1986). *In* "Plant Tissue Culture 1986" (D. A. Somer, B. G. Gengenbach, D. D. Biesboer, W. P. Hackett, and C. E. Green, eds.), Univ. of Minneapolis Press, Minneapolis, Minnesota.

Okazawa, Y., Katsura, N., Tagawa, T. (1967). Effects of auxin and kinetin on the development and differentiation of potato tissue cultured *in vitro. Physiol Plant.* **20,** 862–869.

Paradies, I., Konze, J. R., and Elstner, E. F. (1980). Ethylene: Indicator but not inducer of phytoalexin synthesis in soybean. *Plant Physiol.* **66,** 1106–1109.

Peters, B. M., Cribbs, D. H., and Stelzig, D. A. (1978). Agglutination of plant protoplasts by fungal cell wall glucans. *Science* **201,** 364–365.

Robbins, M. P., and Dixon, R. A. (1984). Induction of chalcone isomerase in elicitor-treated bean cells. Comparison of rates of synthesis and appearance of immunodetectable enzyme. *Eur. J. Biochem.* **145,** 195–202.

Robbins, M. P., Bolwell, G. P., and Dixon, A. R. (1985a). Metabolic changes in elicitor-treated bean cells. Selectivity of enzyme induction in relation to phytoalexin accumulation. *Eur. J. Biochem.* **148,** 463–569.

Robertson, N. F., Friend, J., Aveyard, M., Brown, J., Huffee, M., and Homas, A. L. (1968). Accumulation of phenolic acids in tissue culture pathogen combinations of *Solanum tuberosum* and *Phytophthora infestans. J. Gen. Microbiol.* **54,** 261–268.

Rokem, J. S., Schwarzberg, J., and Goldberg, I. (1984). Autoclaved fungal mycelia increase diosgenin production in cell suspension cultures of *Dioscorea deltoidea. Plant Cell Rep.* **3,** 159–160.

Ryder, T. B., Cramer, C. L., Bell, J. N., Robbins, M. P., Dixon, R. A., and Lamb, C. J. (1984). Elicitor rapidly induces chalcone synthase mRNA in *Phaseolus vulgaris* cells at the onset of the phytoalexin defense reaction. *Proc. Natl. Acad. Sci. U.S.A.* **81,** 5724–5728.

Schenk, R. U., and Hildebrandt, A. C. (1972). Medium and techniques for induction and growth of monocotyledonous and dicotyledonous plant cell cultures. *Can. J. Bot.* **50,** 199–204.

Schmelzer, E., Boerner, H., Grisebach, H., Ebel, J., and Hahlbrock, K. (1984). Phytoalexin synthesis in soybean (*Glycine max*). Similar time courses of mRNA induction in hypocotyls infected with a fungal pathogen and in cell cultures treated with fungal elicitor. *FEBS Lett.* **172,** 59–63.

Schmelzer, E., Somssich, I., and Hahlbrock, K. (1985). Coordinated changes in transcription and translation rates of phenylalanine ammonia-lyase and 4-coumarate : CoA ligase mRNAs in elicitor-treated *Petroselinum crispum* cells. *Plant Cell Rep.* **4,** 293–296.

Somssich, I. E., Schmelzer, E., Bollmann, J., and Hahlbrock, K. (1986). Rapid activation by fungal elicitor of gene encoding "pathogenesis-related" proteins in cultured parlsey cells. *Proc. Natl. Acad. Sci. U.S.A.* **83,** 2427–2430.

Strasser, H., and Matern, U. (1986). Minimal time requirement for lasting elicitor effects in cultured parsley cells. *Z. Naturforsch., C: Biosci.* **41C,** 222–227.

Strasser, H., Tietjen, K. G., Himmelspach, K., and Matern, U. (1983). Rapid effect of an elicitor on uptake and intracellular distribution of phosphate in cultured parsley cells. *Plant Cell Rep.* **2,** 140–143.

Tallevi, *et al.* (1986). *In* "Plant Tissue Culture 1986" (D. A. Somer, B. G. Gengenbach, D. D. Biesboer, W. P. Hackett, and C. E. Green, eds.), Univ. of Minneapolis Press, Minneapolis, Minnesota.

Tanaka, H., and Fujimori, T. (1985). Accumulation of phytuberin and phytuberol in tobacco callus inoculated with *Pseudomonas solanacearum* or *Pseudomonas syringae* pv. *tabaci*. *Phytochemistry* **24,** 1193–1195.

Tietjen, K. G., and Matern, U. (1983). Differential response of cultured parsley cells to elicitors from two non-pathogenic strains of fungi. 2. Effects on enzyme activities. *Eur. J. Biochem.* **131,** 409–413.

Tietjen, K. G., and Matern, U. (1984). Induction and suppression of phytoalexin biosynthesis in cultured cells of safflower, *Carthamus tinctorius* L. by metabolites of *Alternaria carthami* Chowdhury. *Arch. Biochem. Biophys.* **229,** 136–144.

Tietjen, K. G., Hunkler, D., and Matern, U. (1983). Differential response of cultured parsley cells to elicitors from two non-pathogenic strains of fungi. 1. Identification of induced products as coumarin derivatives. *Eur. J. Biochem.* **131,** 401–407.

Walker-Simmons, M., and Ryan, C. A. (1986). Proteinase inhibitor I accumulation in tomato suspension cultures. *Plant Physiol.* **80,** 68–71.

Watson, D. G., Rycroft, D. S., Freer, I. M., and Brooks, C. J. W. (1985). Sesquiterpenoid phytoalexins from suspended callus cultures of *Nicotiana tabacum*. *Phytochemistry* **24,** 2195–2200.

West, C. A. (1981). Fungal elicitors of the phytoalexin response in higher plants. *Naturwissenschaften* **68,** 447–457.

Wijnsma, R., Go, J. T. K. A., van Weerden, I. N., Harkes, P. A. A., Verpoorte, R., and Baerheim-Svendsen, A. (1985a). Anthroquinones as phytoalexins in cell tissue cultures of *Cinchona* spec. *Plant Cell Rep.* **4,** 241–244.

Wijnsma, R., van Weerden, I. N., Verpoorte, R., Harkes, P. A. A., Lugt, B., Scheffer, J. J. C., and Baerheim-Svendsen, A. (1985b). Anthraquinones in *Cinchona ledgeriana* bark infected with *Phytophthora cinnamomi*. *Planta Med.*, pp. 211–212.

Wijnsma, R. *et al.* (1986). *In* "Plant Tissue Culture 1986" D. A. Somer, B. G. Gengenbach, D. D. Biesboer, W. P. Hackett, and C. E. Green, eds.). Univ. of Minneapolis Press, Minneapolis, Minnesota.

Wolters, B., and Eilert, U. (1982). Acridonepoxidgehalte in Kalluskulturen von *Ruta grav-*

eolens und ihre Steigerung durch Mischkultur mit Pilzen. *Z. Naturforsch., C: Biosci.* **37C,** 575–583.

Wolters, B., and Eilert, U. (1983). Elicitoren-Auslöser der Akkumulation von Pflanzenstoffen. Ihre Anwendung zur Produktionssteigerung in Zellkulturen. *Dtsch. Apoth. Ztg.* **123,** 659–667.

Yoshikawa, M. (1978). Diverse modes of action of biotic and abiotic phytoalexin elicitors. *Nature (London)* **275,** 546–547.

Yoshikawa, M., Keen, N. T., and Wang, M. C. (1983). A receptor on soybean membranes for a fungal elicitor of phytoalexin accumulation. *Plant Physiol.* **73,** 497–506.

Zenk, M. H., El-Shagi, H., Arens, H., Stöckigt, J., Weiler, E. W., and Deus, B. (1977). Formation of the indole alkaloids serpentine and ajmalicine in cell suspension cultures of *Catharanthus roseus. In* "Plant Tissue Culture and Its Biotechnological Application" (W. Barz, E. Reinhard, and M. H. Zenk, eds.), pp. 27–43. Springer-Verlag, Berlin and New York.

CHAPTER **10**

Techniques, Characteristics, Properties, and Commercial Potential of Immobilized Plant Cells

Michael M. Yeoman

Department of Botany
University of Edinburgh
Edinburgh, Scotland

I. INTRODUCTION

A fully differentiated multicellular plant in which there is a high degree of cellular organization and division of labor between the component parts is a perfect example of the immobilized state. In this "state" there is an exchange of molecular information between adjacent cells and the provision of nutrients, metabolic intermediates, and regulatory substances by one part of a plant to another ensuring a controlled sequence of development (Davidson *et al.*, 1976). It therefore appeared to me back in the early 1970s (see Yeoman *et al.*, 1978) that if the full synthetic potential of cultured plant cells was to be harnessed to the commercial production of secondary metabolites, i.e., the metabolic activities of the plant were to be mimicked, then one possible route was to design a culture system in which the cells would remain in close contact with one another and at least exhibit the necessary biochemical differentiation. The achievement of the expression of the various biosynthetic pathways was indeed a major undertaking still unfulfilled and necessitated the development of a fresh approach to plant cell culture: the immobilization of cultured plant cells.

II. EARLY RESEARCH ON IMMOBILIZATION SYSTEMS

The development of plant cell culture techniques (Street, 1977; Volumes 1 and 2, this treatise), owes much to the early achievements of microbiologists. Our basic procedures of culture on an agar surface, in liquid suspension, in fermenters, and in turbidostats and chemostats are borrowed and modified from those used routinely by microbiologists. This has, however, conditioned some plant tissue culturists into believing, at least subconsciously, that multicellular plants behave in culture like microorganisms. Indeed many have sought to establish this supposition. Careful consideration after some experience with the culture of plant cells and tissues inevitably leads to the conclusion that while there are superficial similarities, there are fundamental differences between a culture of yeast and a plant cell suspension which relate to the basic organization of the organism from which the culture is derived. A yeast, e.g., *Saccharomyces,* is the organism, while a plant cell culture, e.g. from *Atropa,* is only a collection of cells from a highly complex multicellular organism, the *Atropa* plant.

It is therefore not surprising that a collection of cultured plant cells in a rich nutrient medium, shaken or agitated at relatively high speed, in a system specifically designed to eliminate or minimize gas and nutrient gradients will respond quite differently than the intact plant in terms of the pattern and extent of differentiation and development. Indeed if there is a link between differentiation and the expression of secondary metabolism leading to the accumulation of the metabolites, and this seems highly probable in most cases (Yeoman *et al.,* 1980, 1982a), then conventional cell culture techniques will generally fail to produce the secondary products typical of the species. Of course there is always the possibility that cells more or less dedicated to the production of a particular substance will appear in these cultures and can then be cloned and exploited, e.g., shikonin production by *Lithospermum* (see Curtin, 1983), but this is a comparative rarity and even here a two-stage fermentation is involved in which growth (biomass production) is followed by a production phase in which the expression of the shikonin pathway is achieved (i.e., biochemical differentiation). Before considering the characteristics of immobilized cells and the advantages they may have over free cell suspensions, it is necessary to examine the techniques of cell immobilization.

III. TECHNIQUES OF CELL IMMOBILIZATION

A. Entrapment in Gels

The methods and procedures currently employed for the immobilization of cultured plant cells are similar to those used successfully with microorganisms (e.g., Kierstan and Bucke, 1977; see also Chapter 60, Volume 1, this treatise). The most commonly used techniques involve the entrapment of cells or protoplasts in some kind of gel or combination of gels which are allowed to polymerize around them (active polymerization). This technique has been used for the immobilization of microbial cells for a number of years but has only relatively recently been employed for plant cells. The first report using this approach was described by Brodelius *et al.* (1979), in which they described the entrapment of cells of *Catharanthus roseus, Digitalis lanata,* and *Morinda citrifolia* in a matrix of calcium alginate. A similar method was simultaneously developed at Edinburgh (see Lindsey and Yeoman, 1983a) for the immobilization of cultured cells of *Datura innoxia, Solanum nigrum,* and *Capsicum frutescens,* and the number of reports of the application of this technique has rapidly increased since 1980 (Table I). Other gels have also been tried, with varying success. These include agar, polyacrylamide, agarose, gelatin, and carrageenan, but of the gels alginate remains the most widely used, because of its relative lack of toxicity and simplicity in use, although its instability in the presence of dissolved phosphate can be troublesome. The full experimental details of a number of gel entrapment techniques can be found in Lindsey and Yeoman (1986) and in Chapter 60, Volume 1, of this treatise.

B. Entrapment in Nets or Foam

Gel entrapment techniques frequently involve the active polymerization of monomers or cross-linking of polymer chains around cells to immobilize them in an inert gel. An alternative approach to entrapping cells exploits the ability of cells suspended in liquid to become passively entrapped and be retained within the interstices of pre-formed polymers or meshes added to the cell suspensions. Atkinson *et al.* (1979) have developed the use of such biomass support particles (BSP) constructed, for example, from stainless steel mesh and folded perforated aluminum sheets, for the immobilization of yeasts and fungi.

TABLE I

A Summary of Techniques Used for the Immobilization of Cultured Plant Cells

Species	Immobilized substratum	Reference
Umbilicaria pustulata	Polyacrylamide	Mosbach and Mosbach (1966)
Catharanthus roseus, Morinda citrifolia, Digitalis lanata	Calcium alginate	Brodelius *et al.* (1979)
Digitalis lanata	Calcium alginate	Alfermann *et al.* (1980)
Catharanthus roseus	Calcium alginate, agarose agar, carrageenan, gelatin, polyacrylamide, alginate + gelatin, agarose + gelatin	Brodelius and Nilsson (1980)
Solanum aviculare	Polyphenyleneoxide	Jirku *et al.* (1981)
Cannabis sativa, Daucus carota, Ipomoea sp.	Calcium alginate	Jones and Veliky (1981)
Glycine max	Hollow fibers	Shuler (1981)
Catharanthus roseus	Hypol 3000	Felix and Mosbach (1982)
Catharanthus roseus	Polyacrylamide + alginate	Lambe and Rosevear (1982)
Capsicum frutescens, Datura innoxia, Solanum nigrum	Fibrous polypropylene matting, calcium alginate + nylon, agar + nylon	Lindsey and Yeoman (1983a)
Capsicum frutescens, Daucus carota	Reticulate polyurethane	Lindsey *et al.* (1983)
Daucus carota, Petunia hybrida	Hollow fibers	Prenosil and Pedersen (1983)
Mucuna pruriens	Calcium alginate	Wichers *et al.* (1983)
Papaver somniferum	Calcium alginate	Furuya *et al.* (1984)
Amaranthus tricolor	Chitosan gel	Knorr and Teutonico (1986)
Capsicum frutescens	Reticulate polyurethane	Mavituna and Park (1985)
Humulus lupulus	Reticulate polyurethane	Rhodes *et al.* (1985)
Capsicum frutescens	Reticulate polyurethane	Mavituna *et al.* (1987)
Hyoscyamus muticus, Atropa belladonna	Reticulate polyurethane	Collinge and Yeoman (1986)

In this laboratory (Lindsey *et al.*, 1983) we have shown that suspended cells of various species (e.g., *Daucus carota, Capsicum frutescens, Zingiber officinale, Hyoscyamus muticus, Atropa belladonna, Digitalis lanata, D. purpurea,* and *Mentha piperita*) growing actively in a liquid nutrient medium can be immobilized readily in blocks of polyurethane foam in a one-step process. Although cell suspensions of almost any degree of aggregation can usually be accommodated by using reticulate polyurethane of different pore sizes, very fine suspensions may be difficult to immobilize. On the other hand, fairly lumpy suspensions which are totally unsuitable for the inoculation of stirred and air-lift fermenters will immobilize readily using standard conditions (Lindsey *et al.*, 1983). This is an advantage of this and some other immobilization techniques because it eliminates the lengthy procedures which must be followed to obtain fine suspensions of cells and, in addition, facilitates cell immobilization from cultures which cannot be persuaded to form finely divided suspensions. This can provide an extra bonus, because aggregated cultures often exhibit a high potential for the synthesis and accumulation of a particular secondary metabolite, a feature not typical of fine cell suspensions which are selected for their mode of growth (Yeoman *et al.*, 1980; Lindsey and Yeoman, 1983b).

The techniques of foam preparation and cell immobilization are quite simple. First the foam is washed with ethanol and then water to remove any materials which may have remained after manufacture, then sterilized by autoclaving, and introduced into a suspension of plant cells in the early stages of culture (Lindsey *et al.*, 1983). Under laboratory conditions the cultures are raised in 250-ml Erlenmeyer flasks containing 60 ml of medium, however, the whole process can be scaled up so that immobilization takes place in small bioreactors before transfer to a larger vessel or in the bioreactor in which the production of the secondary metabolite is to take place (Mavituna *et al.*, 1987). In all cases, irrespective of scale, the ratio of medium volume to the weight of immobilized biomass is important. If the ratio of medium volume to cell mass is too high, "washout" will occur; essential substances important to the growth of the cells will be lost, and the culture will be retarded or even rendered nonviable. A deficiency in culture growth due to the inability of cells to achieve an equilibrium with the nutrient medium in which they are placed is well known to most plant tissue culturists and is most obvious when protoplasts or single cells are plated onto agar (Reinert and Yeoman, 1982). In these cases the use of conditioned media or nurse cultures can help to preserve the viability of single cells and ensure the establishment of multicellular colonies (Street, 1977).

The foam (polyurethane or polyester, Declon, Corby, Northants, England) may be used as blocks of various dimensions or as sheets. For most experimental purposes at the laboratory scale, blocks of 1 cm^3 have proved convenient. However, it is likely that in fixed-bed bioreactors sheets will be most effective (F. Mavituna, personal communication). As soon as the blocks are immersed in the cell suspension, and the flasks are agitated on a rotary shaker, the process of passive immobilization begins. At first the free cells and cell aggregates are washed in and out of the blocks, but soon the aggregates become trapped deep within the inert matrix where the component cells divide, expand, and form much larger aggregates within the compartments of the foam. After approximately 14 days the whole of the block has become filled (Lindsey *et al.*, 1983) with cells, and, if the process is allowed to continue, the cells, still attached, grow out beyond the faces of the cube until a point is reached when the foam cube is completely hidden. Even at this stage only small numbers of cells become detached and remain in the medium. Often all of the cells are removed from the culture medium by the immobilization process leaving the medium clear. However, this "overimmobilization" can be prevented by simply removing the loaded cubes from the medium in which they were immobilized and placing them in a non- or low-growth medium which will support metabolic activity but not active cell division (by removal or lowering the concentration of nitrate or phosphate). Therefore immobilized cells placed in a nongrowth medium do not proliferate further to any marked extent, and the cells are not released into the medium, a problem frequently encountered with cells entrapped in alginate beads.

Immobilization in a polyurethane matrix does not appear to affect cell viability (Lindsey and Yeoman, 1984a) to any marked extent. Indeed it has been demonstrated that immobilized cells of *Capsicum frutescens* produce much higher yields of the secondary metabolite capsaicin than do freely suspended cells cultured under similar conditions. In both cases the product is released into the medium rather than accumulated intracellularly (Lindsey and Yeoman, 1984b). Also, since the reticulate polyurethane is approximately 97% void, it offers no significant barrier to the inward and outward diffusion of metabolites.

C. Immobilization in Hollow-Fiber Membranes

Recently the use of hollow-fiber membranes has been described as a method for the immobilization of intact plant cells. Tubular fibers of, for example, cellulose acetate silicone polycarbonate, organized in parallel bundles within a reactor vessel, have been used for the culture of animal cells (Knazek *et al.*, 1972, 1974) and microorganisms (Kan and Shuler, 1978; Vick Roy *et al.*, 1982), and Shuler (1981) has described the entrapment of cells of *Glycine max* in a similar structure. Prenosil and Pedersen (1983) have also described the immobilization of cells of *Daucus carota* and *Petunia hybrida* in hollow-fiber cartridges. Cells are entrapped in the spaces between the fiber membranes, which are permeable to, and through which are supplied, nutrients and precursors to specific secondary products.

IV. CHARACTERISTICS AND PROPERTIES OF IMMOBILIZED PLANT CELLS WHICH MAKE THEM SUITABLE FOR THE PRODUCTION OF CHEMICALS

What then do the immobilized systems offer to those seeking to produce secondary metabolites by cultured plant cells? The major advantage of immobilizing microorganisms is to produce a high density culture in which the metabolite is synthesized and then released to the circulating medium from which it can be removed, providing a process which can be worked more or less continuously. The facility with which substances can be added to, and removed from, the medium is also important. The advantages of immobilized plant cells for the production of metabolites are much more complex, particularly in view of the relationship between differentiation and metabolite synthesis (Yeoman *et al.*, 1982b). To understand this we must examine the characteristics and properties of entrapped cells.

A. Cell–Cell Contact

The highly aggregated cell masses characteristic of all immobilized cultures exhibit high cell-to-cell contact and a much greater degree of cell diversity and differentiation than is commonly observed in a cell suspension culture (Lindsey and Yeoman, 1985). The state of cellular organization approximates that observed in a partially differentiated callus mass, and the range of cell sizes is very wide, from 20 μm in diameter to 200 μm in, for example, a pepper culture immobilized in polyurethane foam (Lindsey *et al.*, 1983). There are also often signs of organization into identifiable structures such as roots, shoots, proembryoids, and embryoids. In a number of cultures, changes to the auxin–cytokinin balance result in massive structural differentiation *in situ*. Cells of *Zingiber officinale* form massive root clusters when immobilized in polyurethane foam (Aitken, personal communication), and *Atropa belladonna* and *Hyoscyamus muticus* (Collinge and Yeoman, 1986) form roots and shoots while immobilized. Indeed an ultimate objective of the research at Edinburgh is to develop an immobilized embryoid or plantlet system to increase yields of various substances. Therefore it would appear that the high degree of cell-to-cell contact associated with aggregation and higher states of structural differentiation is promoted in immobilized cultures and this in turn leads to the expression of synthetic pathways and the formation of secondary metabolites characteristic of the species (Lindsey and Yeoman, 1985).

B. Nutrient, Hormone, and Gaseous Gradients

The central aim of growing cells in liquid suspension is to create, as far as possible, a homogeneous environment in which gradients are eliminated or minimized. The agitation of the culture by shaking or stirring produces a condition in which the cells grow quickly and behave apparently like partially aggregated microorganisms. However, despite the efforts of many plant tissue culturists, the single cell suspension has not been attained, and the "best" cultures consist mainly of small aggregates of cells (2–50) with some single cells (Street, 1977; Reinert and Yeoman, 1982). Clearly the tendency of plant cells to divide and remain together eventually forming a large aggregate or a plant (!) is at least partially frustrated. The removal or reduction of gradients and the "neutralization" of gravity and other environmental stimuli, e.g., unilateral light, radically changes the developmental pattern of the organism, pressurizing it toward forming uniform nondifferentiated clumps. This inhibits or curtails the expression of the biosynthetic pathways essential to secondary metabolite synthesis and accumulation. A simple, perhaps naive, view is that if you wish to encourage a cell culture to behave like a plant then you should treat it like a plant, exposing it to the gradients which are all part of plant development. The immobilized state coupled with a bioreactor design in which the cells are held stationary goes some way toward the achievement of this objective.

C. Manipulation of the Cultures

A severe limitation of batch cultures in suspension is that substances may be added during the course of the growth cycle but changes to the medium involving the removal of one or more compounds, including the product, is impossible without terminating the culture. It is of course possible with chemostats and turbidostats to regulate levels of nutrients and growth regulators during continuous culture, but such complex culture systems can only be used with highly dispersed cell suspensions which usually have lost the ability to produce any useful secondary compound. On the other hand, immobilized cells present the investigator with a system that can be manipulated with ease, and complex alterations to the medium can be made without terminating the culture period. For example, the cells can be given short pulse treatments with precursors to the desired product, or growth regulators and can be changed from a growth state to a nongrowth state rapidly without disturbing the culture (Lindsey and Yeoman, 1983b, 1986).

This brings us to another important advantage of immobilized cells which relates to the ease of manipulation of such cultures. In batch culture a considerable length of time is taken to build up the level of biomass from which the secondary metabolite is synthesized and accumulates. Even in a one-stage process of, say, 50–100 liters in which the product is usually produced after cell division has ceased, the time taken to produce the synthesizing cell population may be several weeks while the time over which the product accumulates is only a few days or a week. As the scale of the process is increased the problem increases and the cost of producing a particular substance rises steeply. This means that over and above the problem of yield, which is normally quite modest (Fowler, 1983, 1986), the time taken to produce the metabolite is excessively long, and unless the product has a high commercial value the process is far from economic. Even in instances where the yield of the product is very high and the price of the product is significant, such as the one-step biotransformation of β-methyldigitoxin to β-methyldigoxin (Alfermann *et al.*, 1980), the time taken to produce the biomass, which is not reusable, makes the cost of production prohibitively expensive. The relatively slow growth of plant cells compared to yeasts and bacteria means that *apart* from high value substances production of secondary metabolites or indeed the biotransformation of one compound to another utilizing the wide biochemical repertoire of plant cells is not commercially viable using a batch approach.

Even the marvels of genetic manipulation will not increase the growth rate of plant cells to that of yeasts, so another solution must be sought. Clearly the alternative is reuse of the biomass so that the production phase for the metabolite forms a relatively large proportion of the total time the cells are in the bioreactor. One means of achieving this end is with immobilized cells where, once the initial cell population has been produced, long production phases can alternate with short regrowth phases to rejuvenate the biomass (Yeoman and Lindsey, 1985). This means that immobilized plant cells in a fixed-bed reactor can be used in a continuous or semicontinuous process lasting 6–12 months. There is, however, a problem which is that the product must be released to the culture medium and be harvested continuously. A number of metabolites, e.g., capsaicin, gingerol, zingiberone, and L-dihydroxyphenylalanine (L-dopa), are released from the cells while others such as atropine and the tropane alkaloids are not. Current research in Edinburgh and elsewhere (e.g., Parr *et al.*, 1987) is aimed at developing techniques and approaches which allow release of metabolites without impairing the synthetic capabilities of the cells. Indeed if immobilized cells are to be used to their maximum advantage, it is necessary that the product be released into the surrounding medium from which it can be continuously removed leaving the cell population intact. Clearly the problems associated with the removal of relatively small amounts of metabolites continuously from large amounts of an aqueous medium are great but are not confined to the immobilized cell systems. It is important to point out here that the presence of the product in the medium even at quite low concentrations will inhibit synthesis of the compound by the cells, and yields will remain low (Lindsey and Yeoman, 1984a,b).

V. BIOREACTOR CONFIGURATIONS FOR USE WITH IMMOBILIZED CELLS

The introduction and development of techniques for plant cell immobilization have led to a reappraisal of the types and design of bioreactors which can be used most effectively with entrapped cells (Shuler *et al.*, 1984; Mavituna and Park, 1985; Mavituna *et al.*, 1987). Traditionally, free plant cells in suspension are most usually cultured in bioreactors similar to, and developed from, fermenters used for the culture of microorganisms (Fowler, 1984). As already emphasized earlier in this chapter, the considerable differences that exist between the structure, size, and robustness of microorganisms and plant cells have presented those attempting to grow plant cell suspensions on a large scale with many problems (Fowler, 1986). In trying to resolve these problems they have tended to move away from more conventional stirred-tank bioreactors toward air-lift systems principally to avoid the unfavorable shear characteristics associated with the impeller devices used in stirred tanks. However, air-lift bioreactors also have their own intrinsic problems when used with plant cells; high rates of air flow through the cultures affect biomass yield (Tanaka, 1981), lowering substantially the biomass per volume of culture attainable with stirred tanks. Also the effects of reducing the CO_2 level in the cultures almost to zero (e.g., CO_2 stripping) produces adverse changes in metabolism.

Therefore, despite the successful application of both stirred-tank and air-lift fermenters to the culture of some plant cells, the problems remain and no bioreactor configuration suitable for a wide range of plant cells is available. Currently the choice appears to be the best compromise between adequate mass transfer of nutrients (especially oxygen) and mixing, and the reduction of cell breakage and lysis due to shear by using "hybrids" of air-lift and stirred-tank bioreactors. The problems of shear, foaming, overaeration leading to CO_2 stripping, and perhaps oxygen poisoning are all overcome when a bioreactor using immobilized cells is adopted. However, as might be expected, other problems arise, and their solution requires careful bioreactor design based on a comprehensive knowledge of the physiology and metabolism of entrapped cells.

Basically, bioreactors currently used with immobilized plant cells may be placed in one of two categories, (1) fluidized bed and (2) fixed bed. In a *fluidized-bed bioreactor* the cells are entrapped in a gel as beads (Brodelius *et al.*, 1979) or in metal or foam particles (Lindsey *et al.*, 1982). The immobilized entities are then agitated either by a flow of air (Alfermann *et al.*, 1980; Mantell and Smith, 1983; Mavituna *et al.*, 1987; also see Fig. 1) or by medium pumped through the bioreactor, or both (Atkinson *et al.*, 1979). In a *fixed-bed bioreactor* the cells immobilized within gel beads (Brodelius *et al.*, 1979), foam, metal, or other particles or in a continuous matrix (Lindsey and Yeoman, 1983a; Rhodes and Kirsop, 1982; Yeoman *et al.*, 1980) are held stationary and perfused with an aerated liquid culture medium at a relatively slow rate. The medium may then be recirculated unchanged or may be modified, by either removal or addition of constituents. With both types of bioreactor the cells are protected from shear; however, fluid mixing is probably more effective with the fluidized bed (Shuler *et al.*, 1984). There are also other major advantages of both systems over freely suspended cells, in particular the ease of separation of culture growth and production of the metabolite by the precise manipulation of the chemical environment, allowing continuous or at least semicontinuous operation (Lindsey and Yeoman, 1984a, 1986; Yeoman and Lindsey, 1985).

Another example of a fixed-bed bioreactor already referred to earlier is the *membrane reactor* (Shuler *et al.*, 1984). The entrapment of plant cells between or within membranes has been used successfully by Shuler and colleagues (see Shuler, 1981). It is generally more expensive than gel or foam entrapment systems but is mechanically stable and, unlike some

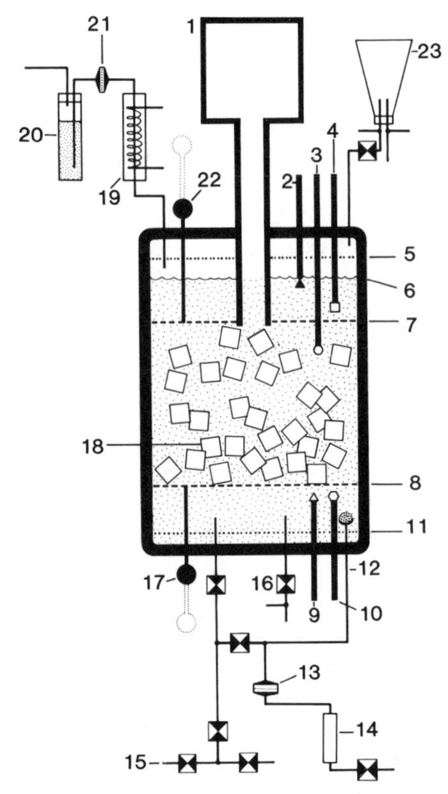

Fig. 1. Schematic representation of a fluidized-bed bioreactor for use with cultured plant cells immobilized in blocks of polyurethane foam. (Redrawn and modified from Mavituna *et al.*, 1987.) The vessel is an adaptation of a 5-liter LH fermenter. The two movable stainless steel grids cover the complete horizontal cross section of the reactor and allow the foam blocks (18) to be held in the center of the reactor as a packed bed during the immobilization process. After immobilization the blocks are circulated in the medium with sterile air (12) after the retraction of the grids.

Key:

1 Biomass sampler vessel
2 Antifoam probe
3 pH probe
4 Oxygen probe
5 Grid (raised position)
6 Liquid level
7 Grid (lowered position)
8 Grid (raised position)
9 Héater
10 Temperature sensor
11 Grid (lowered position)
12 Air input through sparge
13 Filter
14 Rotameter
15 Steam point
16 Liquid sample point
17 Handle to move grid
18 Foam particles
19 Grid (raised position)
20 Condenser
21 Filter
22 Handle to move grid
23 Inoculation vessel

gel systems which become seriously compacted in large columns, offers better control of fluid dynamics and flow distribution. Here there is great potential for scale-up to an industrial level especially if the expensive membrane systems can be reused.

Despite the apparent advantages of the fluidized bed and its appeal to the engineer, it is in my view inferior to the fixed bed for the production of secondary metabolites from plant cells. Although the continuous movement of the particles within the medium ensures good mixing it does present problems, and in order to fully appreciate and understand what is involved it is necessary to refer back to the general points raised earlier in this chapter on the characteristics of immobilized cells and how their biosynthetic potential may be exploited. It has already been established (Yeoman *et al.*, 1980, 1982b; Lindsey and Yeoman, 1985) that there is a positive correlation between differentiation and the ability of a culture to synthesize and accumulate a particular secondary metabolite. It follows that to enable a plant culture to differentiate to a state in which secondary pathways are expressed it is necessary to establish conditions similar to those which exist within the plant. Immobilization goes some way toward this end but does not in itself ensure that the cells remain stationary and are subjected to nutrient and gaseous gradients within the bioreactor chamber similar to those in the plant. This could be important. However, something perhaps more immediately obvious is the importance of the ratio of circulating medium to biomass. In a fluidized bed the ratio is very high, and this leads to washout of nutrients and other solutes, which affects the metabolism of the cells. In a fixed bed in which the medium permeates slowly, much smaller volumes of liquid can be used, and this appears to be beneficial. Last, the use of a continuous matrix, which is not possible with a fluidized bed, is certainly advantageous when scale-up is contemplated. On balance it would appear that fixed-bed bioreactors in which a continuous inert matrix can be used and in which rapid immobilization of cells can be achieved offers the most attractive proposition (Yeoman and Lindsey, 1985; Mavituna *et al.*, 1987).

VI. FUTURE COMMERCIAL EXPLOITATION

Although the production of secondary metabolites from cultured plant cells has been the goal of many plant tissue culturists since the early 1950s, it was only comparatively recently that a process using plant cells was commercialized. In 1983 the Mitsui Petrochemical Company of Japan, with an industrial plant originally designed for use with microorganisms, successfully produced shikonin, a red pigment with mild antiseptic properties, from suspension cultures of *Lithospermum erythrorhizon* (Curtin, 1983). They used a two-stage fermentation in stirred-tank bioreactors. So far there is no process in which immobilized plant cells have been exploited commercially to produce a marketable product, but several are in prospect. It is now appropriate to consider here the criteria which must be satisfied before commercial viability becomes a reality and to decide if and how cell immobilization may be involved.

A. Cost of Product

The products must have a relatively high intrinsic value, which usually reflects the price of the substance extracted from the natural source, the intact plant. It is generally agreed (see, for example, Goldstein *et al.*, 1980; Sahai and Knuth, 1985) that at 1986 prices the lower limit of product value for the process to be commercial should be of the order of £300 per kilogram. These calculations are based on a series of assumptions and formulae and must be considered with some caution. The stated market value of shikonin is about £3000/kg (Sahai and Knuth, 1985), but just how realistic this is remains to be seen. Saffron, usually considered to be the most expensive flavor product on the market, is about £5000/kg. The most valuable secondary metabolites from a plant source, however, are vinblastine and vincristine, two alkaloids used in the treatment of cancer, which are sold at around £100,000/kg. Clearly the high intrinsic value of these alkaloids has made them a popular target for those wishing to set up a commercially viable process for secondary metabolite production. So far these alkaloids have not been produced by plant cells in culture.

B. Cost of Process

It is extremely difficult to estimate the capital investment required for a process based on immobilized cells although it is likely that the cost will be considerably lower than comparable costs for more conventional bioreactors based on microbiological practice, e.g., large stirred-tank or air-lift fermenters. Sahai and Knuth (1985) have made a comparison of production costs and fixed capital investment between batch fermentation processes with three products, diosgenin, anthraquinones, and rosmarinic acid, and an immobilized cell process for a hypothetical product, all at 200,000 kg per year, and concluded that a process based on cells entrapped in alginate beads in a column bioreactor can be more efficient both in terms of production costs and capital investment than a batch fermentation process. However, these figures must be treated with extreme caution due to lack of any hard evidence of how much it would cost to set up column bioreactors for use with entrapped cells. Also the assumption is made that yields will be much higher with immobilized cells and efficient processes will be available to remove the product from the medium, preventing its accumulation. Lindsey and Yeoman (1984a) have compared the yields of capsaicin from suspended and polyurethane-entrapped cells of *Capsicum frutescens* and shown a difference of at least one order of magnitude in favor of immobilized cells. However, whether these yields would be maintained when the process is scaled up from a laboratory scale to a large scale and how much the plant would cost remain unknown.

The very nature of a process in which cells are entrapped in an inert matrix and the product is released into the medium facilitates continuous operation over long periods of up to a year. As has already been discussed this would ensure long periods of metabolite production alternating with shorter phases of regrowth without disturbing the biomass. Such an arrangement reduces the risk of contamination and cuts down the time involved in setting up and dismantling the bioreactors.

C. Range of Products

A wide range of products is required so that the manufacturing plant can be used to full effect. The attraction of one set of hardware for many processes is extremely important. It is also advantageous to design the bioreactors for both multistep syntheses, e.g., phenylalanine to capsaicin, or one-step biotransformations, e.g., tyrosine to L-dopa or β-methyldigitoxin to β-methyldigoxin. Indeed the higher yields of product from precursor in one-step biotransformations in a process carried out continuously could prove crucial in the development of a commercial process.

D. A Possible Process

An outline flow diagram of the projected production of a secondary metabolite by cultured immobilized plant cells is presented in Fig. 2,

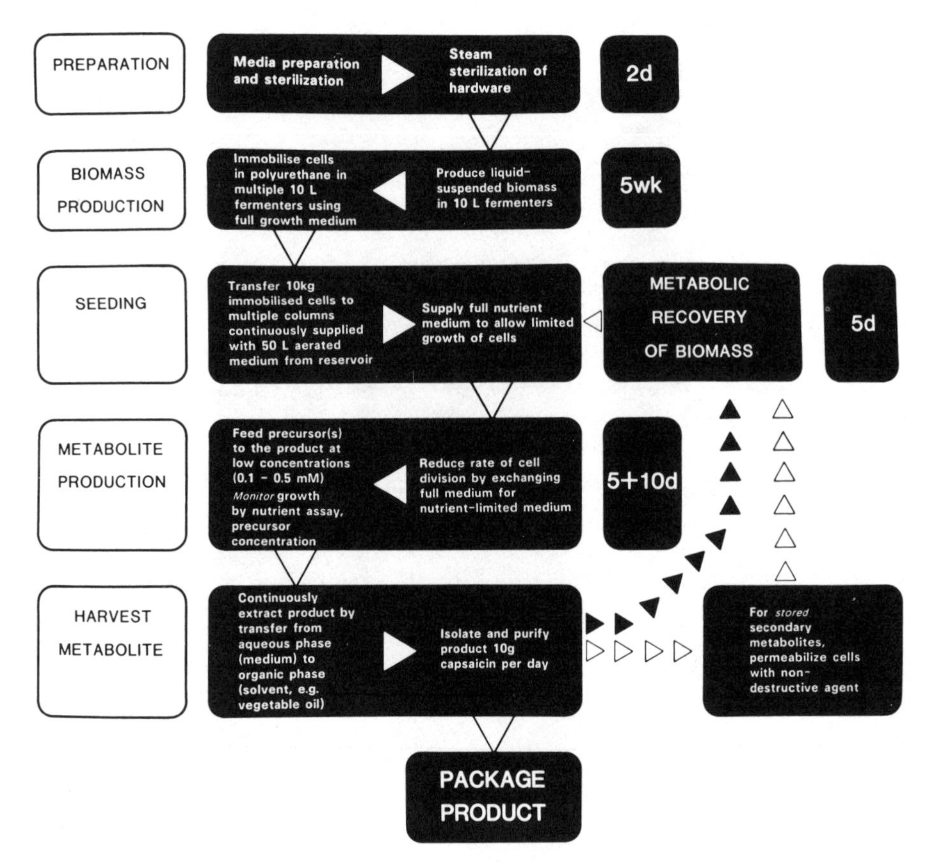

Fig. 2. Flow diagram of the prospective production of a secondary metabolite by immobilized plant cells.

from the preparation of the media and production of the biomass to the extraction and packaging of the product. The assumption has been made that a small-scale laboratory process for the production of capsaicin from entrapped cells of *C. frutescens* can be scaled up without a significant loss in yield and that a suitable extraction technique can be developed. The figures for yield are taken from Lindsey and Yeoman (1984b). From the published values it is calculated that 10 kg fresh weight of cells (1 kg dry weight) will produce approximately 20 g of capsaicin per day (allowing for nonproductive growth time). Over a 6-month period the yield will be 3.6 kg per kilogram dry weight. If capsaicin (in a crude form) has a value of £300/kg, then the total value of the product equals £1,080. As stated earlier £300/per kg represents the lower limit of product value based on a batch fermentation process; it could be that cultures producing capsaicin at the above yield could form the basis of a commercial enterprise together with another 10 products of equal or higher value. These calculations, although based on actual yields, are based on as yet untested assumptions and must be taken at face value. However, the general trends which are emerging give some reason to hope that a process or processes involving immobilized cells can provide a commercial possibility.

ACKNOWLEDGMENT

I wish to thank Mrs. E. A. Raeburn and Mrs. J. Summers for typing and processing the manuscript. I also wish to express my gratitude to Dr. C. E. Jeffree for his assistance with the preparation of the figures.

REFERENCES

Alfermann, A. W., Schuller, I., and Reinhard, E. (1980). Biotransformation of cardiac glycosides by immobilised cells of *Digitalis lanata*. *Planta Med.* **40,** 218–223.

Atkinson, B., Black, G. M., Lewis, P. J. S., and Pinches, A. (1979). Biological particles of given size, shape and density for use in biological reactors. *Biotechnol. Bioeng.* **21,** 193–200.

Brodelius, P., and Nilsson, K. (1980). Entrapment of plant cells in different matrices. *FEBS Lett.* **122,** 312–316.

Brodelius, P., Deus, B., Mosbach, K., and Zenk, M. H. (1979). Immobilised plant cells for the production and transformation of natural products. *FEBS Lett.* **103,** 93–97.

Collinge, M. A., and Yeoman, M. M. (1986). The relationship between tropane alkaloid production and structural differentiation in plant cell cultures of *Atropa belladonna* and *Hyoscyamus muticus*. *In* "Secondary Metabolism in Plant Cell Cultures" (P. Morris, A. H. Scragg, A. Stafford, and M. W. Fowler, eds.). Cambridge Univ. Press, London and New York.

Curtin, M. E. (1983). Harvesting profitable products from plant tissue culture. *Bio/Technology* **1,** 649–657.

Davidson, A. W., Aitchison, P. A., and Yeoman, M. M. (1976). Disorganised systems. *In* "Cell Division in Higher Plants" (M. M. Yeoman, ed.), pp. 407–438. Academic Press, New York.

Felix, H. R., and Mosbach, K. (1982). Enhanced stability of enzymes in permeabilized and immobilized cells. *Biotechnol. Lett.* **4,** 181–186.

Fowler, M. W. (1983). Commercial applications and economic aspects of mass plant and cell culture. *In* "Plant Biotechnology" (S. H. Mantell and H. Smith, eds.), pp. 3–38. Cambridge Univ. Press, London and New York.

Fowler, M. W. (1984). Plant cell culture: Natural products and industrial application. *Biotechnol. Genet. Eng. Rev.* **2,** 41–67.
Fowler, M. W. (1986). Industrial applications of plant cell culture. *Bot. Monogr. (Oxford).* **23,** 202–227.
Furuya, T., Yoshikawa, T., and Taira, M. (1984). Biotransformation of codeinone to codeine by immobilised cells of *Papaver somniferum. Phytochemistry* **23,** 999–1002.
Goldstein, W. E., Ingle, M. B., and Lasure, L. (1980). Product cost analysis. *In* "Plant Tissue Culture as a Source of Biochemicals" (E. J. Staba, ed.), pp. 191–234. CRC Press, Boca Raton, Florida.
Jirku, V., Macek, T., Vanek, T., Krumphanzl, V., and Kubanek, V. (1981). Continuous production of steroid glycoalkaloids by immobilised plant cell. *Biotechnol. Lett.* **3,** 447–450.
Jones, A., and Veliky, I. A. (1981). Effect of medium constituents on the viability of immobilised plant cells. *Can. J. Bot.* **59,** 2095–2101.
Kan, J. K., and Shuler, M. L. (1978). Urocanic acid production using whole cells immobilised in a hollow fibre reactor. *Biotechnol. Bioeng.* **20,** 217–230.
Kierstan, M., and Bucke, C. (1977). The immobilisation of microbial cells, sub-cellular organelles and enzymes in calcium alginate gels. *Biotechnol. Bioeng.* **14,** 387–397.
Knazek, R. A., Gullino, P. M., Kohler, P. O., and Dedrick, R. L. (1972). Cell culture on artificial capillaries: An approach to tissue growth *in vitro. Science* **178,** 65–66.
Knazek, R. A., Kohler, P. O., and Gullino, P. M. (1974). Hormone production by cells grown *in vitro* on artificial capillaries. *Exp. Cell Res.* **84,** 251–254.
Knorr, D., and Teutonico, R. A. (1986). Chitosan immobilization and permeabilization of *Amaranthus tricolor* cells. *J. Agric. Food Chem.* **34,** 96–97.
Lambe, C. A., and Rosevear, A. (1982). Production of chemical compounds from viable cells. U.K. Pat. No. GB2 096 169 A.
Lindsey, K., and Yeoman, M. M. (1983a). Novel experimental systems for studying the production of secondary plant metabolites by plant tissue cultures. *Semin. Ser.—Soc. Exp. Biol.* **18,** 39–66.
Lindsey, K., and Yeoman, M. M. (1983b). The relationship between growth rate, differentiation and alkaloid accumulation in cell culture. *J. Exp. Bot.* **34,** 1055–1065.
Lindsey, K., and Yeoman, M. M. (1984a). The viability and biosynthetic activity of cells of *Capsicum frutescens* Mill. cv. *annuum* immobilised in reticulate polyurethane foam. *J. Exp. Bot.* **35,** 1684–1696.
Lindsey, K., and Yeoman, M. M. (1984b). The synthetic potential of immobilised cells of *Capsicum frutescens* Mill. cf. *annum. Planta* **162,** 495–501.
Lindsey, K., and Yeoman, M. M. (1985). Dynamics of plant cell cultures. *In* "Cell Culture and Somatic Cell Genetics of Plants" (I. K. Vasil, ed.), Vol. 2, pp. 61–101. Academic Press, New York.
Lindsey, K., and Yeoman, M. M. (1986). Immobilized plant cells. *Bot. Monogr. (Oxford)* **23,** 228–267.
Lindsey, K., Yeoman, M. M., Black, G. M., and Mavituna, F. (1983). A novel method for the immobilisation and culture of plant cells. *FEBS Lett.* **155,** 143–149.
Mantell, S. H., and Smith, H. (1983). Cultural factors that influence secondary metabolite accumulations in plant cell and tissue cultures. *Semin. Ser.—Soc. Exp. Biol.* **18,** 75–108.
Mavituna, F., and Park, J. M. (1985). Growth of immobilised plant cells in reticulate polyurethane foam matrices. *Biotechnol. Lett.* **7,** 637–640.
Mavituna, F., Park, J. M., Williams, P. D., and Wilkinson, A. K. (1987). Characteristics of immobilised plant cell reactors. *In* "Process Possibilities for Plant and Animal Cell

Cultures" (C. Webb, F. Mavituna, and J. J. Faria, eds.). Inst. Chem. Eng., London.

Mosbach, K., and Mosbach, R. (1966). Entrapment of enzymes and microorganisms in synthetic cross-linked polymers and their application in column techniques. *Acta Chem. Scand.* **20,** 2807–2810.

Parr, A. J., Robins, R. J., and Rhodes, M. J. C. (1987). Release of secondary products by plant cell cultures. *In* "Process Possibilities for Plant and Animal Cell Cultures" (C. Webb, F. Mavituna, and J. J. Fario. eds.), in press. Inst. Chem. Eng., London.

Prenosil, J. E., and Pedersen, H. (1983). Immobilised plant cell reactors. *Enzyme Microbial Technol.* **5.** 323–331.

Reinert, J., and Yeoman, M. M. (1982). "Plant Cell and Tissue Culture." Springer-Verlag, Berlin and New York.

Rhodes, M. J. C., and Kirsop, B. H. (1982). Plant cell cultures as sources of valuable secondary products. *Biologist* **29,** 134–140.

Rhodes, M. J. C., Robins, R. J., Turner, R. J., and Smith, J. I. (1985). Mucilagainous film production by plant cells immobilised in a polyurethane or nylon matrix. *Can. J. Bot.* **63,** 2357–2363.

Sahai, O., and Knuth, M. (1985). The technology of phytoproduction in plant tissue culture and process economics. *Biotechnol. Prog.* **1,** 1–9.

Shuler, M. L. (1981). Production of secondary metabolites from plant tissue culture—Problems and Prospects. *Ann. N.Y. Acad. Sci.* **369,** 65–79.

Shuler, M. L., Pyne, J. W., and Hallsby, G. A. (1984). Prospects and problems in the large scale production of metabolites from plant cell tissue cultures. *J. Am. Oil Chem. Soc.* **61,** 1724–1728.

Street, H. E., ed. (1977). "Plant Tissue and Cell Culture," 2nd ed. Blackwell, Oxford.

Tanaka, H. (1981). Technological problems in cultivation of plant cells at high density. *Biotechnol. Bioeng.* **23,** 1203–1218.

Vick Roy, T. B., Blanch, H. W., and Wilke, C. R. (1982). Lactic acid production by *Lactobacillus delbrenkii* in a hollow fibre reactor. *Biotechnol. Lett.* **4,** 483–488.

Wichers, H. J., Malingre, T. M., and Huizing, H. J. (1983). The effect of some environmental factors on the production of L-DOPA by alginate-entrapped cells of *Mucuna pruriens. Planta* **158,** 482–486.

Yeoman, M. M., and Lindsey, K. (1985). The scientific and commercial potential of immobilized plant cells. *In* "Advances in Fermentation 1985", pp. 87–92. Turret-Wheatland, U.K.

Yeoman, M. M., Fagandini, D. A. A., and Childs, A. F. (1978). Callus culture in nutrient flow. British Patent No. BP 32185/78.

Yeoman, M. M., Miedzybrodzka, M. B., Lindsey, K., and McLauchlan, W. R. (1980). The synthetic potential of cultured plant cells. *In* "Plant Cell Cultures: Results and Perspectives" (F. Sala, B. Parisi, R. Cella, and O. Cifferi, eds.), pp. 327–343. Elsevier/North-Holland, Amsterdam.

Yeoman, M. M., Lindsey, K., Miedzybrodzka, M. B., and McLauchlan, W. R. (1982a). *Symp. Br. Soc. Cell Biol.* **4.**

Yeoman, M. M., Lindsey, K., and Hall, R. (1982b). Differentiation as a prerequisite for the production of secondary metabolites. *In* "Proceedings of the Plant Cell Culture Conference," pp. 1–7. Oyez Scientific and Technical Services Limited, Sudbury House, London.

CHAPTER **11**

Cryopreservation of Secondary Metabolite-Producing Plant Cell Cultures*

K. K. Kartha

Plant Biotechnology Institute
National Research Council
Saskatoon, Saskatchewan, Canada S7N 0W9

I. INTRODUCTION

Plant cells cultured *in vitro* have been shown to synthesize and accumulate a range of primary and secondary metabolites. The latter category which includes compounds of economic importance are pigments, vitamins, alkaloids, and steroids (Yamada, 1984). Maintenance of cell or callus cultures in a continued state of division through repeated subculture on appropriate nutrient media often results in increased ploidy; accumulation of spontaneous mutations; decline and/or loss of morphogenetic potential, biosynthetic capacity for product formation; reversion of selected lines or mutants to wild types; and, most importantly, unintentional selection of undesirable phenotypes. Any or all of these factors could severely impede the exploitation of cell culture systems for industrial production of valuable compounds envisaged by a number of authors (Zenk, 1978; Constabel, 1981). In fact, loss of ability to synthesize alkaloids over a period of time has been observed with cultured cells of several species (Dhoot and Henshaw, 1977; Barz and Ellis, 1981). Various means of preserving biosynthetic capacity in cell cultures have been contemplated, and cryopreservation of desirable cell lines is regarded as the most logical and effective way.

The general area of cryopreservation of plant cells and organs has

*NRCC No. 26776.

received extensive treatment in several review articles (Kartha, 1985; Withers, 1985; see also Chapter 7, Volume 2, this treatise). This chapter, therefore, is intended to focus specifically on the progress made in the cryopreservation of cells which produce secondary metabolites.

II. CRYOPRESERVATION COMPONENTS

Development of a successful cryopreservation strategy involves identification and optimization of various critical components of two major yet different scientific disciplines, cell technology and cryobiology, so that they interact in concert to inflict minimal freezing and/or other associated stress thereby resulting in maximal recovery of viable cells. The following are some of the critical components to be considered while developing a cryopreservation protocol.

A. Experimental Material

Plant cells in culture exhibit an array of heterogeneity with respect to growth rate, doubling time, mitotic index, cell synchrony, nuclear–cytoplasmic ratio, extent of vacuolation, etc. In addition to being in various stages of cell cycle, cells present in any given sample of suspension and callus culture also exhibit a variety of physiological and morphological variations. The ability of these cells to withstand freezing and thawing stress would vary and hence the reason why all the cells do not survive cryopreservation. Although extensive studies have not been carried out to examine the role of these variables, evidence available so far indicates that for best results cells have to be maintained in a rapid rate of cell division often accompanied by frequent subculture.

A typical growth curve of plant cells has four phases: (1) a lag phase involving changes in cell structure accompanied by synthesis of cytoplasmic components and decrease in vacuolar volume; (2) a growth phase which is mostly exponential followed by a decline in growth resulting from depletion of nutrients; (3) a decline in rate of cell division toward zero; and (4) a stationary phase leading to increase in cellular and vacuolar volume and senescence (Street, 1977). In the case of alkaloid-producing and -nonproducing cell cultures of *Catharanthus roseus,* cells sampled 3–4 days after subculture to fresh media and which had undergone active cell division were particularly resistant to freezing and thawing injury than those taken several days after subculture (Kartha *et al.*, 1982; Chen *et al.*, 1984b). The dense cytoplasm present in the predominantly small-sized cell population may be related to enhanced survival. Also worth examining is the level of resistance of these cells to dehydration stress imposed on them, during pregrowth stages prior to, and during, freezing. Careful manipulation of the cell culture regime should facilitate identification of stages in cell growth ideal for successful cryopreservation provided the freezing conditions employed are optimal to permit maximum recovery of viable cells.

Postfreezing viability in relation to the growth phase of the cells that produce secondary metabolites has not been studied in any great detail; however, results from another type of cell culture would serve to illustrate its effect. For example, Sugawara and Sakai (1974) obtained high viability of *Acer pseudoplatanus* cells when cells at the late lag phase or early division phase were used for cryopreservation. In a subsequent study with cell cultures of the same species, Withers (1978) obtained enhanced postfreezing survival and mitotic index from the cells newly entered into G_1 phase of the cell cycle. Similarly, Sala *et al.* (1979) and Withers and Street (1977) provided evidence to indicate that the viability and regrowth potential of cryopreserved cells are enhanced when cells in late lag and early-to-mid exponential phases are selected for experimentation. The sensitivity of cells in early lag phase or stationary phase to freeze–thaw stress is attributable to the increase in cell size, vacuolar volume, and water content.

B. Pretreatments, Additives, and Cryoprotection

Freezing of cells is always preceded by the application of cryoprotectants. A number of compounds such as dimethyl sulfoxide (DMSO), glycerol, polyethylene glycol, sugars, and sugar alcohols protect living cells against damage during freezing and thawing. Such compounds can lower the temperature at which freezing first occurs (freezing point depression) so that the cells can withstand an otherwise lethal temperature. High water solubility and low toxicity are the two essential characteristics these compounds should possess. The cryoprotectants generally used for freezing biological specimens fall into two categories, permeating and nonpermeating. The most commonly used permeating additives are DMSO and glycerol. Dimethyl sulfoxide permeates into the cells more rapidly than glycerol and, therefore, requires shorter treatment duration. The nonpermeating compounds include sugars, sugar alcohols, and such high molecular weight additives as polyethylene glycol, polyvinylpyrrolidone, dextran, and hydroxyethyl starch. Most of the cryoprotectants exhibit varying degrees of cytotoxicity at higher concentrations. Generally a concentration of 5–10% for DMSO and 10–20% for glycerol is adequate for most material. In instances where application of a single cryoprotectant does not result in high survival, a mixture of cryoprotectants has been beneficial (Diettrich *et al.*, 1982, 1985; Seitz *et al.*, 1983; Watanabe *et al.*, 1983; Butenko *et al.*, 1984; Chen *et al.*, 1984a,b).

In addition to the application of cryoprotectants, certain physiological modifications become necessary for cell culture with biosynthetic capacities for secondary metabolite production. Preconditioning of cells prior to cryopreservation has been beneficial in increasing the freezing resistance of cells. Supplementation of the culture medium with such osmotically active compounds as mannitol, sorbitol, sucrose, and proline has resulted in increasing the freezing resistance. However, preculturing in mannitol-supplemented medium did not enhance the survival of cryopreserved cells of the alkaloid-producing cell cultures of *Catharanthus roseus* (Chen *et al.*, 1984b). Preculturing the cells in 5% DMSO-supplemented medium for periods of 24–48 hr prior to freezing has resulted in better survival of the alkaloid-nonproducing line of *C. roseus* but not the producing line (Kartha *et al.*, 1982).

Sorbitol has been used as an osmotic agent in the preculture medium followed by the utilization of sorbitol and DMSO as cryoprotectants for the successful cryopreservation of the alkaloid-producing cell cultures of *C. roseus* (Chen *et al.*, 1984b). The preculture method included incubation of *C. roseus* cells in 1 *M* sorbitol for 6–20 hr followed by freezing the cells using 1 *M* sorbitol plus 5% DMSO. Equimolar concentrations of $CaCl_2$ and KCl, in addition to being toxic to cells, were most ineffective in their cryoprotective efficacy. Although considerable survival was noticed subsequent to preculturing the cells in media supplemented with 1 *M* concentrations of glucose, trehalose, and sucrose, the overall survival and recovery was much superior when sorbitol was used as the preculturing agent (Chen *et al.*, 1984b).

In order to understand the mode of action of the cryoprotectants, the freezing behavior of DMSO and sorbitol solutions and alkaloid-producing *C. roseus* cells treated with DMSO and sorbitol alone and in combination was examined by nuclear magnetic resonance (NMR) and differential thermal analysis (DTA). Incorporation of DMSO or sorbitol into the liquid medium had a significant effect on the temperature range for initiation to completion of ice crystallization. Compared to control, less water crystallized at temperatures below −30°C in DMSO-treated cells. Similar results were obtained with sorbitol-treated cells, except sorbitol had less effect on the amount of water crystallized below −25°C. There was a close association between the percent unfrozen water at −40°C and percent cell survival after freezing in liquid nitrogen. These studies indicated that in alkaloid-producing *C. roseus* cells, the amount of liquid water at −40°C is critical for successful cryopreservation, and the combination of DMSO and sorbitol was the most effective in preventing water from freezing (Chen *et al.*, 1984a). These results may explain the cryoprotective properties of DMSO and sorbitol and why DMSO and sorbitol in combination are more effective than when used alone. In instances where a synergistic effect has been noticed in cryoprotection using a mixture of cryoprotectants, similar events may be occurring.

The sensitivity of cultured plant cells to sorbitol differs markedly between species. In the author's laboratory, it was observed that many culture systems such as cell cultures and somatic embryos of alfalfa, apple cell cultures, haploid cell cultures and auxotrophic mutants of *Datura innoxia*, and cell cultures of white spruce could be successfully cryopreserved by paying careful attention to the concentration and preculture duration of the experimental material in the sorbitol-enriched media as well as by strictly controlling the cooling program. The studies revealed that a universal cryopreservation strategy for cell cultures could be evolved using sorbitol preculturing followed by controlled slow freezing in sorbitol plus DMSO solution.

Diettrich *et al.* (1982, 1985) employed 0.17 *M* mannitol in their preculture medium while Seitz *et al.* (1983) used 6% for cryopreservation of *Digitalis lanata* cell cultures. The amino acids asparagine, alanine, proline, and serine have also been employed as additives to the preculture media for the cryopreservation of *Dioscorea deltoidea*. Similarly, cold hardening of *Panax ginseng* cells at 2–10°C in the presence of 7.0–25.0% sucrose was also found to be beneficial (Butenko *et al.*, 1984). From the above examples, it becomes evident that a preculture requirement is essential for successful cryopreservation of cell cultures, especially those which produce secondary metabolites.

C. Freezing and Thawing Methods

Several types of freezing methods are available for the cryopreservation of biological material. However, slow freezing is the most suitable method for all types of cell cultures. Freezing methods are based on the physicochemical events occurring during freezing. Mazur (1969) identified the events to which a cell is subjected during freezing and thawing. With temperature reduction, the cell and its external medium initially supercool followed by ice formation in the medium. The cell membrane/wall acts as a physical barrier and prevents the ice from seeding the cell interior at temperatures above approximately −10°C, and thus the cell remains unfrozen but supercooled. As the temperature is further lowered, an increasing fraction of extracellular solution is converted to ice, resulting in the concentration of extracellular solutes. Since the cell remains supercooled and its aqueous vapor pressure exceeds that of frozen exterior, the cell equilibrates by loss of water to external ice (dehydration). Slowly cooled cells reach equilibrium with the external ice by efflux of water and will remain shrunk provided the cell is sufficiently permeable to water. In such cases intracellular ice formation, generally considered to be one of the factors responsible for causing freezing injury, will not occur. This phenomenon is utilized in devising slow freezing cryopreservation techniques.

The success of the slow freezing method depends on a number of factors such as cooling rates, pretreatments and cryoprotection, type and physiological state of the experimental material, and the temperature at which controlled freezing is arrested prior to immersion in liquid nitrogen. The most commonly used method of freezing cell cultures is by regulated slow cooling at a rate of 0.5–1°C/min down to either −30, −35, or −40°C followed by storage in liquid nitrogen. A brief summary of cryopreservation methodology used for the secondary metabolite-producing cells is given in Table I. In some cases a modification to slow cooling by holding the cell samples for a predetermined period of time in the temperature region of −30 to −40°C is also practiced. The optimal cooling rate would depend on the type of cells, the cryoprotectant used, and also the preculturing methods employed. Freezing is carried out either in cryogenic glass or plastic ampules with cells concentrated to an adequate cell density (~10^6 cells/ml) and dispersed in 1.0 ml of the cryoprotectant solution of optimal concentration. The cooling rate is monitored by a temperature probe inserted into one of the ampules. Although some plant organs such as meristems, pollen, and seeds have been successfully cryopreserved by ultra-rapid freezing by direct immersion of the cryoprotectant-treated material in liquid nitrogen, this approach may not be applicable for cell cultures. Even in a controlled freezing protocol, a proportionate loss of viability could be observed at cooling rates above or below the optimal which in turn is attributable to the degree of over or under dehydration attained by the cells.

Thawing is generally carried out rapidly, in order to avoid recrystallization of minute ice crystals, by immersing the ampules for 1–2

TABLE I

A Brief Outline of Protocol Used for the Cryopreservation of Secondary Metabolite-Producing Cell Cultures

Species	Methodology	Reference
Catharanthus roseus	Four-day-old cultures precultured in liquid medium with 1 *M* sorbitol for 6–2 hr. Frozen using 1 *M* sorbitol + 5% DMSO. Optimal cooling rate 0.5°C/min till −35°C; stored in liquid nitrogen (LN). Regrowth on filter paper placed over nutrient medium. No postthaw wash.	Chen *et al.* (1984)
Daucus carota	Two-week-old cultures treated with equal volume of medium containing 10% DMSO and frozen at 1°C/min to −70°C followed by storage in LN. Cells recultured after washing.	Dougall and Whitten (1980)
Digitalis lanata	Cells precultured for 1 week in medium with 3% mannitol; cryoprotectants: sucrose, glycerol, and DMSO. Cooling rate 0.5–2.0°C/min till −60°C followed by storage in LN. Regrowth on solid medium. No postthaw wash.	Diettrich *et al.* (1982, 1985)
D. lanata	Six-day-old cultures pregrown in medium with 6% mannitol for 3 days. Cryoprotectants: DMSO, glycerol, and sucrose. Cooling rate 1°C/min to −35°C and stored in LN. Regrowth on solid medium. No postthaw wash.	Seitz *et al.* (1983)
Dioscorea deltoidea	Cells precultured in medium containing asparagine (20 m*M*), or alanine (50 m*M*), or proline (20 m*M*), or serine (10 m*M*). Cryoprotectant: 7% DMSO. Cooling rate 0.5°C/min to −30°C, 9°C/min from −30 to −70°C, followed by storage in LN. Postthaw wash.	Butenko *et al.* (1984)
Lavandula vera (callus)	Green callus pieces suspended in liquid medium. Cryoprotectants: 5% DMSO and 10% glucose. Cooling rate 1°C/min to −40°C followed by storage in LN. Reculture on solid medium after washing.	Watanabe *et al.* (1983)
Panax ginseng	Cold hardening of cells by gradually increasing the sucrose concentration from 3 to 25% and simultaneous decrease of culture temperature to 2°C. Cryoprotectants: sucrose–glycerine–DMSO. Cooling rate 0.5°C/min to −30°C, 9°C/min from −30 to −70°C, followed by storage in LN. Reculture after postthaw wash.	Butenko *et al.* (1984)

min in a water bath held at 40°C. Slow thawing of slowly frozen cells invariably results in loss of viability.

D. Regrowth and Viability Assays

Conventionally, prior to assessing viability the thawed cells are gradually washed with chilled nutrient medium and returned to culture either in liquid or on semisolid medium as was practiced for alkaloid-nonproducing *C. roseus* cells (Kartha *et al.*, 1982). This technique proved to be unsatisfactory for a number of cell cultures (see Withers, 1985). The washing step has been found to be deleterious to a few cell cultures including alkaloid-producing *C. roseus* (Chen *et al.*, 1984b). Furthermore, a special technique had to be developed to induce regrowth of the alkaloid-producing cultures of *C. roseus.* This technique consisted of transferring the cells without post-thaw wash onto filter paper disks over nutrient media solidified with agar for a period of 4–5 hr. The filter paper disks with the cells were then transferred to fresh media of the same composition where the cells resumed growth (Chen *et al.*, 1984b). Culturing the cells, postthawing, without washing has been practiced with *Digitalis lanata* cells as well (Seitz *et al.*, 1983; Diettrich *et al.*, 1982, 1985).

The most accurate test for assessing viability of cryopreserved cells should be based on regrowth. Triphenyltetrazolium chloride (TTC) reduction assay, other staining methods with vital stains, or fluorescent stains such as fluorescein diacetate could also be used to obtain a rough estimate of viability. A reliable estimate should combine several viability assay techniques.

E. Stability and Biosynthetic Capability of Cryopreserved Cells

Stability in terms of ploidy, growth rate, mitotic index, and, most importantly, biosynthetic capability is to be preserved in cells postcryopreservation. The alkaloid-nonproducing cell cultures of *C. roseus* maintained identical levels of ploidy and mitotic indices before and after cryogenic storage (Kartha *et al.*, 1982). Studies carried out in the authors' laboratory with three lines of *C. roseus* also revealed that all the cryopreserved cultures retained the capacity for alkaloid biosynthesis (Chen *et al.*, 1984b). Twenty-five different anthocyanin-producing cultures of wild carrot produced approximately the same amount of anthocyanin after cryogenic storage as did the unfrozen controls (Dougall and Whitten, 1980). The cryopreserved cells of *Digitalis lanata* also retained the biochemical activity relative to the transformation of cardenolides (Diettrich *et al.*, 1982, 1985; Seitz *et al.*, 1983). *Lavandula vera* callus recovered after cryopreservation retained not only the biosynthetic capability for biotin but also the plant regeneration potential, as did the original green callus cultures (Watanabe *et al.*, 1983). The growth rate and viability of cryopreserved *Dioscorea deltoidea* cells did not differ from those of unfrozen controls. It was also shown that the *Dioscorea* cells, after cryogenic storage, produced the same spectrum and amount of steroid compounds, e.g., diosgenine, as did controls. From these examples, although limited in number, it could be concluded that cryopreservation per se does not induce major variation with respect to biosynthetic capacities of these cells.

III. PROSPECTS

The evidence available so far indicates that cryopreservation is the most reliable approach for the long-term preservation of cell cultures which possess biosynthetic capacity for the synthesis and accumulation of secondary metabolites. In relation to that of other types of cells, cryostorage of these special cells poses problems in regard to optimizing the cryobiological components. Since accumulation of secondary metabolites in itself requires a certain degree of structural differentiation often accompanied by increases of cellular and in particular vacuolar volume and lengthening of the generation period, the identification of appropriate growth stages most conducive to resisting freezing injury becomes very critical in devising an appropriate cryopreservation strategy. With the recent developments in preculture techniques using various additives such as sorbitol, mannitol, sucrose, and amino acids coupled with the judicious use of cryoprotectants, either alone or in combination, and regrowth techniques, it is envisaged that many more secondary metabolite-producing cell cultures could be successfully cryopreserved. Demonstration of postfreezing stability of cells vis-à-vis metabolite biosynthesis and accumulation definitely would support the prospect of using cryogenic techniques for the long-term preservation of such specialized cells.

REFERENCES

Barz, W., and Ellis, B. E. (1981). Plant cell cultures and their biotechnological potential. *Ber. Dtsch. Bot. Ges.* **94,** 1–26.

Butenko, R. G., Popov, A. S., Volkova, L. A., Chernyak, D. N., and Nosov, A. M. (1984). Recovery of cell cultures and their biosynthetic capacity after storage of *Dioscorea deltoidea* and *Panax ginseng* cells in liquid nitrogen. *Plant Sci. Lett.* **33,** 285–292.

Chen, T. H. H., Kartha, K. K., Constabel, F., and Gusta, L. V. (1984a). Freezing characteristics of cultured *Catharanthus roseus* (L.) G. Don cells treated with dimethylsulfoxide and sorbitol in relation to cryopreservation. *Plant Physiol.* **75,** 720–725.

Chen, T. H. H., Kartha, K. K., Leung, N. L., Kurz, G. W. G., Chatson, K. B., and Constabel, F. (1984b). Cryopreservation of alkaloid-producing cell cultures of periwinkle (*Catharanthus roseus*). *Plant Physiol.* **75,** 726–731.

Constabel, F. (1981). Advances in plant cell culture towards improved yields of secondary products (alkaloids). *Adv. Biotechnol. [Proc. Int. Ferment. Symp.], 6th, 1980,* pp. 109–115.

Dhoot, G. K., and Henshaw, G. G. (1977). Organization and alkaloid production in tissue cultures of *Hyoscyamus niger. Ann. Bot (London)* [N.S.] **41,** 943–949.

Diettrich, B., Popov, A. S., Pfeiffer, B., Neumann, D., Butenko, R. G., and Luckner, M. (1982). Cryopreservation of *Digitalis lanata* cell cultures. *Planta Med.* **46,** 82–87.

Diettrich, B., Haack, V., Popov, A. S., Butenko, R. G., and Luckner, M. (1985). Long-term storage in liquid nitrogen of an embryogenic cell strain of *Digitalis lanata. Biochem. Physiol. Pflanz.* **180,** 33–43.

Dougall, D. K., and Whitten, G. H. (1980). The ability of wild carrot cell cultures to retain their capacity for anthocyanin synthesis after storage at 140°C. *Planta Med., Suppl.,* pp. 129–135.

Kartha, K. K., ed. (1985). "Cryopreservation of Plant Cells and Organs." CRC Press, Boca Raton, Florida.

Kartha, K. K., Leung, N. L., Gaudet-LaPrairie, P., and Constabel, F. (1982). Cryopreservation of periwinkle, *Catharanthus roseus,* cells cultured *in vitro. Plant Cell Rep.* **1,** 135–138.

Mazur, P. (1969). Freezing injury in plants. *Annu. Rev. Plant Physiol.* **20,** 419–448.

Sala, F., Cella, R., and Rollo, F. (1979). Freeze-preservation of rice cells. *Physiol. Plant.* **45,** 170–176.

Seitz, U., Alfermann, A. W., and Reinhard, E. (1983). Stability of biotransformation capacity in *Digitalis lanata* cell cultures after cryogenic storage. *Plant Cell Rep.* **2,** 273–276.

Street, H. E. (1977). Cell (suspension) culture techniques. *In* "Plant Tissue and Cell Culture" (H. E. Street, ed.), pp. 61–102. Blackwell, Oxford.

Sugawara, Y., and Sakai, A. (1974). Survival of suspension cultured sycamore cells cooled to the temperature of liquid nitrogen. *Plant Physiol.* **54,** 722–724.

Watanabe, K., Mitsuda, H., and Yamada, Y. (1983). Retention of metabolic and differentiation potentials of green *Lavandula vera* callus after freeze-preservation. *Plant Cell Physiol.* **24,** 119–122.

Withers, L. A. (1978). The freeze-preservation of synchronously dividing cultured cells of *Acer pseudoplatanus. Cryobiology* **15,** 87–92.

Withers, L. A. (1985). Cryopreservation of cultured plant cells and protoplasts. *In* "Cryopreservation of Plant Cells and Organs" (K. K. Kartha, ed.), pp. 243–267. CRC Press, Boca Raton, Florida.

Withers, L. A., and Street, H. E. (1977). Freeze-preservation of plant cell cultures. *In*

"Plant Tissue Culture and Its Biotechnological Application" (W. Barz, E. Reinhard, and M. H. Zenk, eds.), pp. 226–244. Springer, Verlag, Berlin and New York.

Yamada, Y. (1984). Selection of cell lines for high yields of secondary metabolites. *In* "Cell Culture and Somatic Cell Genetics of Plants" (I. K. Vasil, ed.), Vol. 1, pp. 629–636. Academic Press, Orlando, Florida.

Zenk, M. H. (1978). The impact of plant cell culture on industry. *In* "Frontiers of Plant Tissue Culture 1978" (T. A. Thorpe, ed.), pp. 1–14. University of Calgary, Calgary, Canada.

CHAPTER **12**

Plant Regeneration

P. S. Rao

Plant Biotechnology Section
Bio-Organic Division
Bhabha Atomic Research Centre
Bombay 400 085, India

I. INTRODUCTION

The regeneration of plants from cultured cells and tissues is a key step in the application of tissue culture methodology for plant propagation and improvement. The development of efficient protocols for reproducible high frequency plant regeneration from cultured tissues would therefore assume great importance. This chapter attempts an analysis of plant regeneration as a whole, the modes of regeneration and techniques employed, and the factors which control and/or regulate plant regeneration, together with an account of the success achieved in regeneration of plants of medicinal value (see also Volume 3, this treatise, for several chapters on morphogenesis and plant regeneration).

II. GENERAL METHODOLOGY

Choice of Explants

Explant selection plays an important role in successful plant regeneration studies. Explants must generally be chosen from healthy, vigorous plants to obtain optimum results. In many instances the condition of the plant at the time of culture may influence the growth of explants in culture. Although all parts of the plant are capable of organogenesis under appropriate conditions, it is generally observed that immature tissues and organs are morphogenetically more plastic than the mature differentiated tissues. Shoot tips, axillary buds, seedling explants, immature embryos, and leaf and stem segments are particularly well suited as choice explants. In some plant groups such as legumes the response is genotype dependent.

Nutrient Medium

The composition of the nutrient medium is the next important parameter which must be optimized in order to obtain successful plant regeneration. The major constituents of the medium comprise inorganic and organic nutrients, carbon source, iron in the form of ferric citrate or chelate or EDTA (ferric sodium ethylenediaminetetracetate), vitamins, plant growth regulators, and in certain instances natural growth adjuvants such as deproteinized coconut milk, fruit pulp and juice, yeast extract (YE), and other plant extracts (see also Chapter 4, Volume 2, this treatise). These are usually sufficient for most plant tissues. Among the plant growth regulators, auxins [indole-3-acetic acid (IAA), naphthaleneacetic acid (NAA), 2,4-dichlorophenoxyacetic acid (2,4-D)] and cytokinins [kinetin (Kn), zeatin (Z), 6-benzylaminopurine (BAP), 6-γ,γ-dimethylallylaminopurine(2i-P)] are most often needed in culture, and their concentrations and ratio often determine the nature of growth and organogenesis.

Several media have been developed by various investigators, and some of the earliest formulations were those of White (1943) and Heller (1953). More recently, Murashige and Skoog's high salt medium (1962) or modifications thereof such as Eriksson (1965), Linsmaier and Skoog (1965), B_5 (Gamborg *et al.*, 1968), and Schenk and Hildebrandt (1972) have been used, the striking difference being essentially in quantity and form of nitrogen. A comparative account of the components of the various nutrient media is given by Bhojwani and Razdan (1983) and Ozias-Akins and Vasil (1985).

Culture Environment

Culture environment can greatly influence growth and organized development. The parameters include (1) pH of the medium, (2) humidity, (3) light, (4) temperature, and (5) the physical form of the medium. None of these factors has in the real sense been critically investigated although there exist sporadic reports. In general, the pH of the medium is set between 5.0 and 5.8; very little is known about the influence of the actual pH value of the medium on plant regeneration although it is known that pH values do change during culture. The effect of relative humidity (RH) is rarely investigated; however, there is one report that raising the RH in cultures of endive roots up to 98% influenced the development of buds (Bouniois and Margara, 1968). Light plays an important role in organized development (Thorpe, 1980). In tobacco callus, Siebert *et al.* (1975) observed that near-UV light stimulated shoot formation at low intensities whereas higher intensities were inhibitory. Not much is known about the temperature factor except that routinely cultures are maintained between 20 and 30°C.

III. MODES OF PLANT REGENERATION

Plant regeneration through tissue cultures occurs through one of two methods: (i) organogenesis and (ii) somatic embryogenesis. Many factors control these events, and a brief account is presented here.

A. Organogenesis

Organogenesis can be obtained either through direct differentiation of shoot buds from explants (direct morphogenesis) or through callus formation in explants and subsequent formation of shoots and roots. Organ formation *in vitro* was reported as early as 1939 when White (1939) observed shoot differentiation in the tobacco hybrid and Nobecourt (1939) that of root differentiation in callus cultures of carrot. Following these studies many attempts were made with tissues of diverse plant species which led to the elegant work of Skoog and Miller (1957), who demonstrated in tobacco pith callus that a balance between an auxin and a cytokinin determined the nature of organogenesis. A high ratio of cytokinin to auxin induced shoot differentiation whereas the reverse favored root formation; intermediate ratios induced callus formation. It was therefore clear that quantitative interactions brought about quantitative changes. This classical work laid the foundation for intensive studies on plant regeneration in a wide range of species. To date, plant regeneration has been reported in a large number of families of angiosperms (Vasil and Vasil, 1980; Evans *et al.*, 1981) belonging to diverse taxa including mono- and dicotyledonous species, and the list is ever expanding. Direct morphogenesis, that is, shoot bud differentiation directly from cultivated explants, or morphogenesis mediated through callus phase has been observed in many instances.

Factors Controlling Organogenesis

The organogenetic response of a particular explant is influenced by a variety of factors.

Explant Size. It is generally observed that explant size is critical. Small size explants have less regenerative ability compared to larger ones.

Physiological Age. The physiological age of the explant is also a critical factor. Raju and Mann (1970) have demonstrated that in *Echeveria,* young leaf explants initiated only roots, older leaves regenerated shoot buds, and leaves of medium age produced shoots and roots.

Season. Seasonal variations also influence regeneration. Explants *Solanum tuberosum* cultured during December–April were highly tuberogenic while those obtained during the rest of the season showed poor response (Fellenberg, 1963).

Oxygen Gradient. The dissolved oxygen concentration in culture has also been reported to have a promotive role in organogenesis of shoots in carrot cell cultures. Reduction in the oxygen gradient stimulated shoot formation whereas increased gradients promoted rooting (Kessel and Carr, 1972).

Light. In certain situations organogenesis is also influenced by the quality and the intensity of light. Haccius and Lakshmanan (1969) reported induction of embryos in callus cultures of *Nicotiana tabacum* exposed to high intensity light (10,000–15,000 lux). In tissue cultures of *Topinambour,* light from the blue region of the spectrum promoted shoot formation while red light favored rooting (Letouze and Beauchesne, 1969). Other factors which might influence organogenesis include sugars and environment.

B. Somatic Embryogenesis

The production of somatic embryos from isolated cell and tissue cultures and their regeneration into complete plants is by far the most convincing demonstration of the totipotency of plant cells. Somatic embryogenesis might occur directly as originating from single cells as in the case of carrot (Backs-Husemann and Reinert, 1970) or might occur indirectly, i.e., through an intervening callus phase reported in many plant species (for details, see Tisserat *et al.*, 1979; Kohlenbach, 1978; Sharp *et al.*, 1980; Vasil and Vasil, 1986). Embryogenesis has been reported in more than 100 species belonging to diverse families.

Factors Affecting Somatic Embryogenesis

Extensive investigations on somatic embryogenesis have indicated that two important parameters, namely, auxin and source of nitrogen, influence the process. Among the growth factors, 2,4-D in the range of 0.5–2.0 mg/liter has been found to be most useful in many systems. Other auxins such as NAA (Gleddie *et al.*, 1983), IAA (Bapat and Rao, 1979), or NAA and indolebutryric acid (IBA) (Jelaska, 1974) have also been used. Cytokinins normally do not stimulate embryogenesis; however, zeatin was especially promotive for embryogenesis in carrot (Fujimura and Komamine, 1980). Gibberellins generally inhibit somatic embryogenesis (Tisserat and Murashige, 1977; Fujimura and Komamine, 1975).

The role of nitrogen compounds in embryogenesis has been studied. Tazawa and Reinert (1969) observed that embryogenesis *in vitro* could be induced by both inorganic (KNO_3, $NH^+{}_4$, $NO_3{}^-$), and organic (amino acids, amides) compounds. Fujimura and Komamine (1975) studied the effects of various growth regulators on embryogenesis in carrot cell suspension cultures. Embryogenesis was inhibited by 2,4-D or IAA. Zeatin promoted embryogenesis; other cytokinins inhibited it. Kessel *et al.* (1974) reported that there is a critical level of dissolved oxygen in cell suspension cultures of carrot, and a relationship exists between dissolved oxygen, ATP, and embryogenesis; both O_2 and ATP enhanced embryogenesis.

Although numerous examples of somatic embryogenesis have been reported in various plant species, the formation of whole plants through embryogenesis process has been reported less frequently.

C. Plant Regeneration through Enhanced Axillary Branching

In a number of plant species, adventitious buds are produced *in vivo* from different organs such as root, leaf, and bulbs, and in such instances the rate of adventitious bud development can be considerably enhanced under culture conditions as in the case of *Begonia* (Reuther and Bhandari, 1981). Apical meristems under appropriate nutrient conditions may regenerate plants through axillary branching. Plants obtained through this method are genetically identical to mother plants, and pathogen-free plants may also be obtained. The size of the explant determines the rate of survival; larger shoot tips survive better. Axillary buds also regenerate shoots and eventually plants (Patel *et al.*, 1983).

IV. REGENERATION IN MEDICINAL PLANT TISSUE CULTURES

A wide variety of compounds of potential medicinal value are found within the plant kingdom, and approximately half the drugs in current use still originate from plants. Many plant species from which medicinal products can be derived have been the subject of micropropagation experiments (Table I).

Dioscorea

Dioscorea is a genus consisting of several widely distributed species some of which are cultivated for their edible tubers. Whereas tubers of about 15 species are known to contain steroidal sapogenin, diosgenin is the major precursor for several steroid hormones, cortisone and other corticosteroids, oral contraceptives, and several other drugs. The three main species which contain diosgenin in amounts commercially feasible to extract are *D. floribunda*, *D. composita*, and *D. deltoidea*. There is therefore considerable interest in large-scale multiplication of *Dioscorea* spp. as a cultivated crop to ensure continuous supplies. The normal method of propagation is through underground tubers but this is a rather slow process.

Lakshmi Sita *et al.* (1976) developed a method for rapid plant regeneration in *D. floribunda* by culturing nodal segments. In *D. deltoida*, Mascarenhas *et al.* (1976) used shoot tip, stem, leaf, tuber, and roots. Tuber explants developed multiple shoots in 4 weeks, and the shoots rooted. Plants obtained *in vitro* could be transferred to pots containing vermiculite and soil. Plantlet formation from hypocotyl-derived callus tissues of *D. deltoidea* has also been reported (Grewal and Atal, 1976). Culture of isolated shoot tips (including the first node excised from *in vitro* plantlets) resulted in profuse shoot formation.

Mantell *et al.* (1978) have reported plant regeneration and clonal multiplication in *D. alata* and *D. rotundata*, using nodal segments of plants grown under different photoperiods (16 and 12 hr day lengths). Nodal

TABLE I

Plant Regeneration in Medicinal Species

Plant	Explant source	Reference
Allium sativum	Stem, bulb, leaf	Abo-El-Nil (1977)
	Leaf	Havranek and Novak (1973)
	Bud tip	Bhojwani (1980)
	Shoot tip	Novak (1983)
	Meristem	Bovo and Mroginski (1985)
Aloe pretoriensis	Seed	Groenewald *et al.* (1976)
Ammi majus	Zygotic ovaries	Sehgal (1972)
	Hypocotyl	Grewal *et al.* (1976)
Angelica acutiloba	Embryogenic callus	Nakagawa *et al.* (1982)
Asparagus officinalis	Shoot, cladode	Reuther (1977a,b)
Atropa belladonna	Leaf sections	Eapen *et al.* (1978)
	Mesophyll protoplasts	Lorz and Potrykus (1979)
	Root	Thomas and Street (1970, 1972)
Bouvardia ternifolia	Seedling tissue, leaf tissue	Fernandez and Sanchez de Jimenez (1982)
Bupleurum falcatum	Leaf	Wang and Huang (1982)
Cannabis sativus	Leaf, stem, embryo, leaf petiole	Verzar-Petri *et al.* (1982)
Catharanthus roseus	Shoot tip	Takayama and Misawa (1982)
	Stem	Ramawat *et al.* (1978)
	Leaf, shoot	Dhruva *et al.* (1977)
	Various explants	Abou-Mandour *et al.* (1979)
Chrysanthemum cinerariaefolium	Shoot tip	Grewal and Sharma (1978), Wambugu and Rangan (1981)
C. morifolium	Leaf	Slusarkiewicz-Jarzina *et al.* (1982)
Cinchona ledgeriana	Seedlings	Hunter (1979)
C. ledgeriana, C. pubesens, C. succirubra	Shoot tip and cotyledons	Koblitz *et al.* (1983)
Coleus parviflorus	Leaf discs	Asokan *et al.* (1984)
Coptis japonica	Petiole	Syono and Furuya (1972), Ikuta *et al.* (1975)
Costus speciosus	Embryo	Pal and Sharma (1982)
Datura species	Protoplasts	Schieder (1980)
D. innoxia	Mesophyll protoplasts	Schieder (1975)
	Stem	Engvild (1973)
	Leaf discs	Iskander and Brossard-Chiriqui (1980), Sopory and Maheshwari (1976)
	Anther	Schieder (1978a)

(continued)

TABLE I

(Continued)

Plant	Explant source	Reference
D. innoxia + *D. discolor*	Protoplasts	Schieder (1978b)
D. innoxia + *D. stramonium*	Protoplasts	Schieder (1978b)
D. metel	Anther	Gupta and Babbar (1980)
Digitalis species	Various	Tewes *et al.* (1982)
D. lanata	Mesophyll protoplasts	Li (1981)
D. obscura	Root sections, hypocotyl	Perez-Bermudez *et al.* (1983)
D. purpurea	Seedlings	Hirotani and Furuya (1977)
Dioscorea species	Bulbil	Asokan *et al.* (1983)
		Mantell *et al.* (1978)
D. bulbifera	Stem and nodal tissue	Ammirato (1982)
		Forsyth and Van Staden (1982)
D. composita	Stem and nodal tissue	Datta *et al.* (1982)
D. deltoidea	Tuber	Mascarenhas *et al.* (1976)
	Hypocotyl	Grewal and Atal (1976)
D. floribunda	Nodal pieces	Lakshmi Sita *et al.* (1976)
Dubiosia myoporoides	Stem node	Kukreja and Mathur (1985)
	Foliar explants	Kukreja *et al.* (1986)
Eucalyptus citriodora	Shoot segment	Gupta *et al.* (1981)
	Hypocotyl and cotyledons	Lakshmi Sita (1982)
	Lignotubers	Aneja and Atal (1969)
Humulus lupulus	Leaf	Motegi (1976)
Hyoscyamus muticus	Shoot tip	Grewal *et al.* (1979)
	Protoplasts	Wernicke and Thomas (1980)
Iberis amara	Leaf, stem	Mudgal *et al.* (1981)
	Anther	Babbar *et al.* (1980)
Liquidamber styraciflua	Hypocotyl	Sommer and Brown (1980)
Mentha species	Axillary bud	Rech and Pires (1986)
Panax ginseng	Pith and root sections	Chang and Hsing (1978a,b, 1979, 1980a,b)
	Root sections, cotyledons	Choi *et al.* (1982)
Papaver orientale	Hypocotyl	Schuchmann and Wellmann (1983)
P. somniferum	Hypocotyl	Nessler (1982), Schuchmann and Wellmann (1983)
Parthenium argentatum	Roots, shoots	Staba and Nygaard (1983)
P. hysterophorus	Stem	Subramanian and Subba-Rao (1980)
Pergularia pallida	Stem, protoplasts	Bapat *et al.* (1986)
Physalis minima	Anther	George and Rao (1979)
	Leaf and stem	Bapat and Rao (1977)
Solanum dulcamara	Leaf	Zenkteler (1972)

(continued)

TABLE I

(Continued)

Plant	Explant source	Reference
S. khasianum	Leaf, protoplasts	Kowalczyk *et al.* (1983)
	Hypocotyl	Gunay and Rao (1982)
S. laciniatum	Leaf	Chandler and Dodds (1983)
S. melongena	Pith	Fassuliotis *et al.* (1981)
	Stem	Kamat and Rao (1978)
	Mesophyll protoplasts	Guri and Izhar (1984)
S. nigrum	Leaf	Bhatt *et al.* (1979)
S. sisymbriifolium	Stem pith	Fassuliotis (1975)
S. tuberosum	Tuber	Lam (1975)
	Leaf discs	Webb *et al.* (1983)
Tylophora indica	Stem	Rao *et al.* (1970)
	Protoplasts	Mhatre *et al.* (1984)
Withania somnifera	Anther	Vishnoi *et al.* (1979)

segments excised from 16-hr day plants consistently produced callus, root, and shoot growth, whereas the growth of segments taken from 12-hr day plants was variable and plantlet production was poor. The development of complete plantlets of *D. alata* and *D. rotundata* from nodal segments took 3–5 weeks.

In *D. deltoidea,* Chaturvedi *et al.* (1982) used shoot tip and single node cultures to obtain multibranched plants. These were cut up as single node explants, or the single nodes were rooted for planting in field. Direct shoot formation could also be induced from stem segments and leaves of *in vitro*-grown shoots. It was estimated that, using these methods, more than 160,000 true-to-type plants per year could be produced by *in vitro* culture of one elite mother plant.

Ammirato (1982) made an extensive study of growth and morphogenesis in tissue cultures of *D. floribunda, D. composita,* and *D. alata* and obtained *in vitro* formation of aerial tubers or bulbils in the axils of leaves in *D. alata* and *D. bulbifera.* Single node segments also formed tubers directly. Tubers formed *in vitro* were planted in soil and produced normal plants. Somatic embryos formed in callus cultures of *D. floribunda* derived from excised embryos and from nodal tissue cultures of *D. bulbifera.* In both species somatic embryos appeared when cultures were aged and on a medium lacking 2,4-D but containing glutamine and Z or abscisic acid (ABA). Transfer of somatic embryos to semisolid medium was required to obtain plantlet development. If embryos were placed en masse in groups of five or more, root and shoot development occurred on basal medium. Datta *et al.* (1982) obtained plants through regeneration of adventitious shoots in callus cultures of *D. composita*. Plant development has also been reported from culture of bulbil explants of *D. bulbifera* (white flesh genotype) and *D. alata* (white flesh and purple flesh genotype) (Asokan *et al.*, 1983). Genotypes differed in their response to growth regulator treatments. Shoot differentiation was preceded by callus development.

In *Dioscorea*, by conventional methods 8–10 plants can be produced from rhizomes of one living plant. The progeny thus obtained will take at least 3–4 years before it is fit for further multiplication through rhizome cuttings. The development of methods of plant regeneration employing tissue culture technique has obvious advantages for rapid and large-scale multiplication of the species.

Digitalis

Digitalis lanata and *D. purpurea* are important cardiac glycoside-synthesizing plant species; leaves are the main sites of cardenolide storage and formation (Reinhard and Alfermann, 1980). Leaf callus cultures of *D. purpurea* possess the potential for plant regeneration in which cardiac glycoside synthesis is restored (Hirotani and Furuya, 1977). Garve *et al.* (1980) investigated growth and morphogenesis in long-term cultures of *D. lanata* vis-à-vis cardenolide formation. Embryogenic cultures were initiated from various organs such as cotyledons, hypocotyl, roots of seedlings, as well as stem tissues. The regenerants exhibited similar range of cardenolide content as that of the intact plant. In leaf explants of *D. purpurea* shoot buds originated from the areas of leaf segments which were in contact with the surface of the medium (Rucker *et al.*, 1981).

Diettrich and Luckner (1980) reported plant regeneration in protoplast cultures of *D. purpurea* and attempted to select cell lines possessing a higher level of cardiac glycosides in the protoclones. Mesophyll protoplasts of *D. lanata* divided rapidly and developed into callus. Differentiation of shoots and plantlets occurred. Protoplasts being a single cell system, their regeneration into plants can be used for plant improvement via spontaneous or induced mutagenesis. In *D. obscura* (Perez-Bermudez *et al.*, 1983) also, the regeneration of roots, adventitious buds, and plants from hypocotyl, root, and cotyledon explants has been reported.

Ginseng

Ginseng (*Panax ginseng*) which yields ginsengosides has been extensively investigated. Somatic embryogenesis was reported by Butenko *et al.* (1968) in cultures derived from leaf, petiole, anthophore, and stem; however, complete plantlets did not regenerate. Chang and Hsing (1980a) obtained whole plant regeneration through embryogenesis in root-derived callus of ginseng. Callus tissues initiated from explants of mature root tissues produced numerous somatic embryos, and reculture of embryos resulted in the production of normal green plants, some of which flowered *in vitro* (Chang and Hsing, 1980b). Factors which control root and shoot formation in callus and leaflet cultures of ginseng have been investigated by Choi *et al.* (1982). A high percentage of embryoid and shoot formation was observed in root and cotyledon calli. Cotyledon calli, however, proved to be more suitable for morphogenesis than root calli.

Chrysanthemum

Plant regeneration in *Chrysanthemum cinerariaefolium* (a plant known to produce insecticidal pyrethrins) by shoot tip culture technique has been described by Grewal and Sharma (1978) and Wambuga and Rangan (1981).

Papaver

Ikuta *et al.* (1974) investigated 11 representative species of the Papaveraceae and reported induction of callus and redifferentiation of plantlets in 5 species: *Corydalis pallida, C. incisa, Papaver bracteatum, P. somniferum,* and *MacLeaya cordata.* Callus tissues derived from whole seedlings or stem, petiole, and hypocotyl differentiated shoot buds.

Tissues of opium poppy (*Papaver somniferum*) have been grown in culture as callus (Kamo *et al.*, 1980) and in liquid suspensions (Morris and Fowler, 1980). Most of these studies were directed toward examining of *in vitro* alkaloid metabolism. Regeneration of roots and shoots (Nessler and Mahlberg, 1979) and somatic embryogenesis and whole plant regenration has been documented (Nessler, 1982). This technique should be advantageous in recovering plants from mutant cell lines having altered alkaloid spectra. Somatic embryogenesis and plant regeneration has been reported in tissues of *Papaver orientale* (Schuchmann and Wellmann, 1983).

Parthenium

Some members of the Compositae contain sesquiterpene lactones, many of which are known to be allergenic and cytotoxic (Rodriquez *et al.*, 1976). Plants of *Parthenium argentatum* and *P. hysterophorus* are native to South and Central America and northern Mexico. *Parthenium argentatum* (Guayale) is a rubber-producing shrub while *P. hysterophorus* contains "parthenin," a major sesquiterpene lactone which is a potent contact allergin responsible for contact dermatitis. Staba and Nygaard (1983) established tissue cultures of nine different strains of *P. argentatum* using organ cultures. Excised root cultures developed small nodules which later formed shoots. Differentiation of shoots in *Parthenium* cultures has been reported by a number of researchers. Callus cultures derived from seedlings (Zavala *et al.*, 1980) or its cotyledons, stem, shoot tip, leaf blades, flowers, petioles, and roots (La Brecque, 1980; Subramaniam and SubbaRao, 1980; Wickham *et al.*, 1980; Dastoor *et al.*, 1981) are reported to readily develop roots and shoots. It has also been demonstrated that the leaf callus of *P. hysterophorus* is potentially capable of synthesizing allergenic principle(s) that are present in the normal plant (Subramaniam and SubbaRao, 1980).

Hop

Motegi (1976) has reported that leaf callus cultures of hop (*Humulus lupulus*) differentiated numerous shoot buds and plantlets.

Hemp

Callus cultures of hemp (*Cannabis sativa*) have been established by Verzar-Petri *et al.* (1982) from leaves. These differentiated roots and shoots. Loh *et al.* (1983) have investigated tissue cultures of *Cannabis* in relation to *in vitro* biotransformation of phenolics.

Catharanthus

Catharanthus species (periwinkle), have great medicinal value and produce more than 80 alkaloids, of which vincristine and vinblastine are being utilized as anticancer drugs. It has also been observed by many investigators that the synthesis of the alkaloids could be stimulated with the differentiation capacity of callus cultures. Dhruva *et al.* (1977) obtained shoot differentiation in leaf callus cultures of two varieties of *C. roseus*, one with pink flowers and another with white flowers, while Ramawat *et al.* (1978) reported similar results in stem callus cultures. Regeneration of plants from haploid and diploid callus of *Catharanthus roseus* has been described by Abou-Mandour *et al.* (1979). Phytochemical investigations have shown that the regenerated plants contain twice as much serpentine as stock plants. This clearly demonstrates the importance of cell culture for production of high yielding strains.

Constabel *et al.* (1982) have confirmed the earlier reports that periwinkle callus can be induced to form shoots and plants. In their studies a low frequency of multiple shoot formation was observed in young callus cultures, while in older tissues (2 years old) no shoot formation occurred. About 60–70% of shoots could be rooted and could be transplanted in soil. The alkaloid profiles of callus derived from original explants as well as the callus derived from regenerated shoots were identical. In addition, the regenerated shoots contained vindoline. These results are significant in the sense that the authors primarily determined a protocol for successful shoot regeneration in cell lines, rather than first determining the alkaloid-producing cell lines and then trying to find ways and means of regenerating plants from such cell lines.

Asclepiadaceae

The family Asclepiadaceae has many plants which are medicinally important. *Tylophora indica* has certain phenanthroindolizidine alkaloids which have curative principles for asthma. The morphogenetic potentialities of callus tissues of *T. indica* have been investigated (Rao *et al.*, 1970; Rao and Narayanaswamy, 1972). Tissue cultures established from leaf, stem, and root segments differentiated shoot buds 4 weeks after incubation. The differentiated shoot buds rooted, and complete plants were obtained. Benjamin *et al.* (1979) observed that whole callus cultures failed to synthesize the phenanthroindolizidine alkaloids, while in the regenerated plants established in the field the biosynthetic capacity was restored. In an extended study of the same species, Mhatre *et al.* (1984) successfully isolated protoplasts from the stem callus, and the protoplast-derived calli were regenerated into plants via the formation of somatic embryos/shoot buds (Fig. 1D and E).

Another member of the family Asclepiadaceae, *Pergularia pallida* also contains the phenanthroindolizidine ring compounds which are reported to have anticancer properties. Stem explants of *Pergularia* proliferated and developed into a callus in which shoot buds were formed. Protoplasts were also isolated and grown to calli, and plantlets were regenerated (Fig. 1A–C) via shoot bud formation (Bapat *et al.*, 1986).

Solanaceae

Many plants in the family Solanaceae are known to have medicinal properties, and the development of tissue culture techniques within the

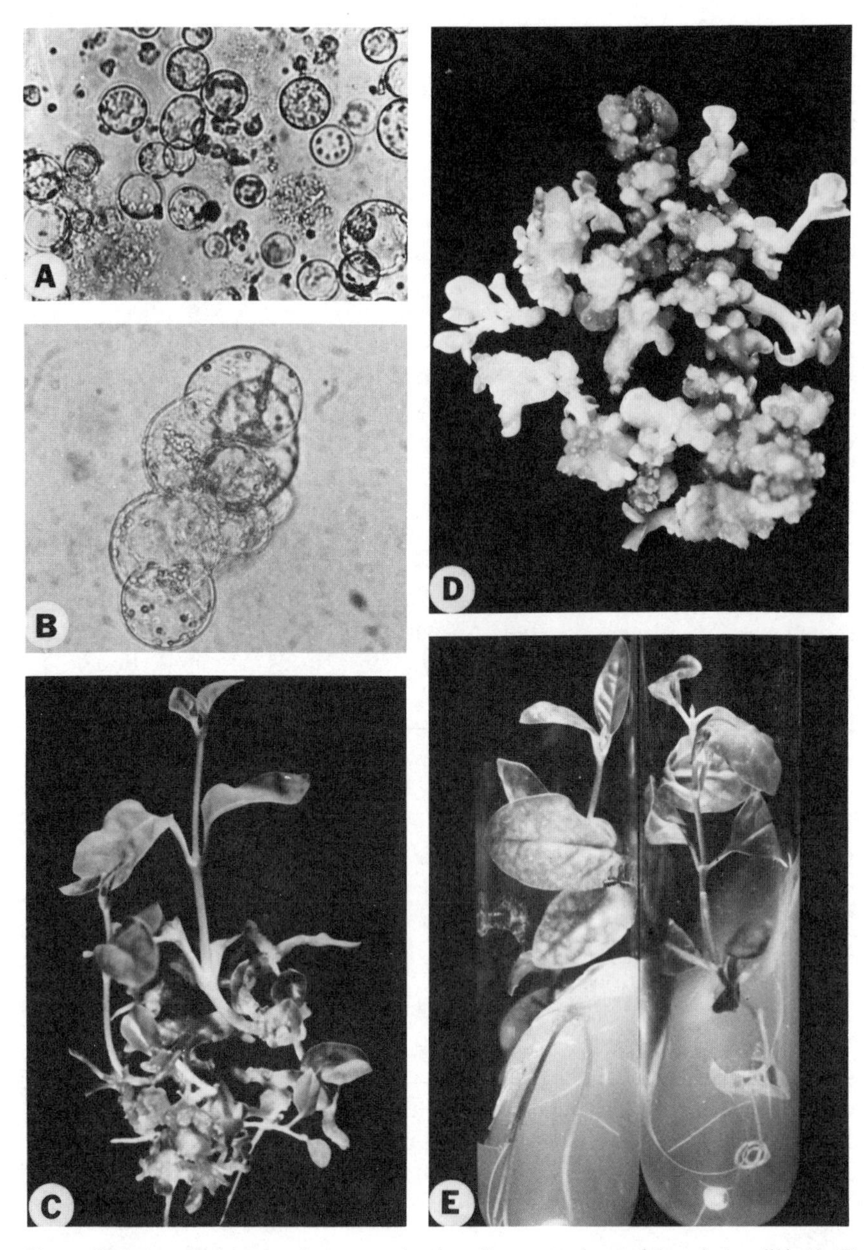

Fig. 1. Isolation, division, and regeneration in callus protoplasts of *Pergularia pallida* (A–C) and *Tylophora indica* (D and E). (A) Isolated protoplasts from stem callus. (B) Cell colony from isolated protoplasts. (C) Plants regenerated from protoplasts. (D) Regeneration of shoots in protoplast-derived callus. (E) Plants regenerated from protoplasts. (A–C. After Bapat *et al.*, 1986; D and E, after Mhatre *et al.*, 1984).

genus *Solanum* has indicated that certain species are very amenable for *in vitro* culture. Extensive studies on plant regeneration have been conducted using various explant sources. To cite few examples: leaf discs of *S. dulcamara* (Bhatt *et al.*, 1979), *S. laciniatum* (Davies and Dale, 1979; Chandler and Dodds, 1983), and *S. mammosum* (Kumar *et al.*, 1983); hypocotyl and protoplast of *S. melongena* (Kamat and Rao, 1978; Fassuliotis *et al.*, 1981). *S. nigrum* (Nehls, 1978), *S. sisymbriifolium* (Fassuliotis, 1975), and *S. tuberosum* (Shepard, 1982; Webb *et al.*, 1983).

Solanum khasianum is an important medicinal plant, the berries of which contain the steroidal glycoalkaloid solasodine (a precursor to diosgenin). Since this plant cannot be propagated by cuttings there is a great deal of interest in its multiplication through *in vitro* methods. Gunay and Rao (1982), have developed a technique for plant regeneration using hypocotyl explants of diploid and tetraploid species. Plantlets obtained were successfully established in soil. Kowalczyk *et al.* (1983) have also described plant regeneration from leaf discs and protoplast-derived callus of *S. khasianum*.

Tissue cultures of *Physalis minima*, a plant which yields physalin and 5,6-epoxyphysalin (C_{28}, 13,14-secosteroids), have been investigated rather extensively. Bapat and Rao (1977) reported regeneration of shoot buds and plantlets using auxins and cytokinins in factorial combinations. Sipahimalani *et al.* (1981) investigated the biosynthetic potential of the cultured tissues and regenerated plants (diploid and triploid). In differentiated callus derived from diploid plants, physalin content was 459 μg/g, of which physalin D was 430 μg, compared to callus obtained from triploid plants (912 μg/g) of which physalin D was 260 μg. The regenerated plants of diploid and triploid origin, however, showed a similar physalin profile.

Considerable investigations have been carried out in *Atropa belladonna* and *Datura* sp. from the point of view of plant regeneration, haploid induction, somatic hybridization through protplast fusion, and secondary product metabolism vis-à-vis the level of organization (Schieder, 1975, 1978a,b, 1980). Eapen *et al.* (1978) carried out morphogenetic and biosynthetic studies in tissue cultures of haploid and diploid plants of *Atropa belladonna*. Haploid callus retained its organogenetic potential whereas the diploid tissues showed a decline in the organogenetic potential during serial subculture and completely lost it by the eighth serial passage. The total alkaloid content in the calli and young shoot buds was low (8–20 μg/g) as compared to mature plants (1160 μg/g). The alkaloid spectrum of the shoot buds was comparable to that of unorganized calli. The regenerated flowering plants contained the principal belladonna alkaloids in quantities comparable to plants raised from seeds. RajBhandary *et al.* (1969) initiated root, callus, and cell suspension cultures of *Atropa belladonna* capable of plant regeneration. It was observed that the conditions leading to synthesis of principle alkaloids of belladonna are achieved only with the organization of roots.

In an extended study, factors influencing morphogenesis in excised roots and suspension cultures of *A. belladonna* have been investigated (Thomas and Street, 1972). If root cultures were maintained for periods longer than 28 days in culture, callus developed at the cut ends of roots spontaneously initiated shoot buds. Likewise, cell suspension cultures derived from root callus also showed the morphogenetic ability to form shoot buds and incipient plants or embryoids (Konar *et al.*, 1972).

Duboisia

Duboisia myoporoides (Solanaceae) is a potential raw material for the production of tropane alkaloids. Leaves of this plant contain 2–4% total alkaloids with nearly 60% hyoscine and 30% hyoscyamine. Poor seed germination and early loss of seed viability are the two major limiting factors for its commercial exploitation. Also, a wide range of variability is generally observed in seed-raised progenies. Considerable work has been carried out with different explants of *D. myoporoides* with a view to clonally propagate the valuable species. High frequency plant regeneration has been reported from stem nodal explants bearing an axillary bud. The plantlets obtained were successfully raised in large numbers to normal, adult plants under field conditions (Kukreja and Mathur, 1985). Extending this observation the same authors obtained *in vitro* complete plant regeneration via shoot bud formation from the foliar leaf explants. The entire process of producing plants to the greenhouse stage took about 4 months (Kukreja *et al.*, 1986).

Cinchona

Cinchona plants are of commercial interest because they contain alkaloids of the quinoline type. Quinine and quinidine are useful therapeutic agents. Quinine is antimalarial, and quinidine is used as a potent drug in cardiac arrythmiasis. *Cinchona* barks are also used as bitter principles in the beverage industry. Koblitz *et al.* (1983) have developed a procedure for *in vitro* clonal mass propagation of *C. ledgeriana* and *C. succirubra*. Apical meristems when cultured with cotyledons regenerated numerous shoots. For successful rooting of the shoots the investigators inverted the shoots on the rooting medium in such a way that the leaves lay on the surface of the medium and the base protruded into the air. Transplantation of rooted shoots into soil was not entirely successful owing to special requirements that *Cinchona* plants need for cultivation such as moist atmosphere with annual rain fall of no less than 210 cm. Hunter (1979) also established a method of regeneration of *C. ledgeriana* from seedlings.

Coptis

Rhizomes of *Coptis japonica* contain several isoquinoline alkaloids of which berberine and coptisine are pharmacologically important components used in abdominal and cardiac disorders. Roots of *Angelica acutiloba* are also medicinally important. Nakagawa *et al.* (1982) established callus as well as suspension cultures. In both species induction of somatic embryogenesis was possible in callus as well as cell suspension cultures. The somatic embryos developed into plants which proved to be remarkably uniform in both morphologic and chemical characters compared with sexually propagated species.

Organization and its role in the induction of some secondary products has been reported in many plant cell cultures. In tobacco tissue cultures, Tabata *et al.* (1970) reported plant regeneration and emphasized a correlation between bud formation and alkaloid synthesis. Similar results have been reported for tissue cultures producing volatile oils such as *Mentha* (Krikorian and Steward, 1969) and onion (Turnbull *et al.*, 1981). Dhoot and Henshaw (1977) established callus and cell suspension cultures from seedling parts such as roots, stems, and cotyledons of *Hyoscyamus niger* which formed roots and occasional shoots.

Scopalia

Scopalia japonica and *S. parviflora* are tropane alkaloid-producing plants. The rhizome extracts of both species are used as antispasmodics. Tabata *et al.* (1972) established callus cultures from stem and rhizome segments of *S. parviflora.* Many roots differentiated on the surface of callus tissues. In addition to root formation, some of the cultures formed shoots and occasionally plantlets which developed into normal plants when transplanted in soil. Wernicke and Kohlenbach (1975) used anther cultures for production of haploid plants of *S. carniotica, S. lurida,* and *S. physaloides.*

Allium

Organogenesis leading to the formation of plantlets has been demonstrated in garlic shoot and leaf callus cultures (Havranek and Novak, 1973). Abo-El-Nil (1977) reported organogenesis as well as embryogenesis in callus cultures initiated from stem tips, bulb leaf disks, and stem segments of *A. sativum*. Bovo and Mroginski (1985) have also reported the differentiation of plantlets by culturing 0.4–0.6-mm-long shoot meristems of *A. sativum*. Novak (1983) used shoot tip cultures of *A. sativum* and investigated the frequency of polyploid plants regenerated after colchicine application. Among the 140 regenerated plants almost 23% were solid tetraploids and 15% cytochimeras. Induction of polyploidy in shoot tip cultures has an important role in garlic breeding programmes.

Plants of *Ammi majus* contain coumarin ranging from 0.5 to 2.5%. Xanthotoxin, a furanocoumarin, is used in the treatment of leucoderma (vitilago) and psoriasis as well as in the manufacture of suntan lotions. Induction of polyembryony and development of shoots was reported in ovary cultures of *A. majus* (Sehgal, 1972). Subsequently Grewal *et al.* (1976) developed a method for mass multiplication by using hypocotyl segments. In this system the formation of somatic embryos was so prolific that any portion of a developing leafy shoot that came in contact with the medium again proliferated resulting in secondary embryogenesis.

Mentha

Mentha species were originally cultivated in eastern Asia, mainly in Japan and China, and their economic importance is due to the production of mint oil as a raw material for pharmaceutical and cosmetic uses as well as for flavoring foods, beverages, and tobacco. Few reports about the tissue culture of mint have been published, and these relate to conditions for cell suspension cultures and callus formation. Recently, conditions for rapid multiplication through the culture of axillary buds have been reported for several species: *M. arvensis, M. piperita, M. pulegium,* and *M. viridens*. Nodal segments from 1-year-old plants were grown and 15–20 shoots per explant with roots were obtained in 40 days (Rech and Pires, 1986).

ACKNOWLEDGMENT

I am grateful to Ms. Minal Mhatre for her unstinted help and cooperation during the preparation of the manuscript.

REFERENCES

Abo-El-Nil, M. M. (1977). Organogenesis and embryogenesis in callus cultrures of garlic (*Allium sativum* L.). *Plant Sci. Lett.* **9,** 259–264.

Abou-Mandour, A. A., Fisher, S., and Czygan. F. C. (1979). Regeneration von intakten pflanzen aus diploiden und haploiden kalluszellen von *Catharanthus roseus. Z. Pflanzenphysiol.* **91,** 83–88.

Ammirato, P. V. (1982). Growth and morphogenesis in cultures of the monocot yam. *Dioscorea. In* "Plant Tissue Culture 1982" (A. Fujiwara, ed.), pp. 169–170. Maruzen, Tokyo.

Aneja, S., and Atal, C. K. (1969). Plantlet formation in tissue culture from lignotubers of *Eucalyptus citriodora* Hook. *Curr. Sci.* **38,** 69–71.

Asokan, M. P., O'Hair, S. K.,and Litz, R. E. (1983). *In vitro* plant development from bulbil explants of two *Dioscorea* species. *HortScience* **18,** 702–703.

Asokan, M. P., O'Hair, S. K., and Litz, R. E. (1984). *In vitro* plant regeneration from leaf discs of Hausa potato (*Coleus parviflorus*). *HortScience* **19,** 75–76.

Babbar, S. B., Mittal, A., and Gupta, S. C. (1980). *In vitro* induction of androgenesis, callus formation and organogenesis in *Iberis amara* Linn. anthers. *Z. Pflanzenphysiol.* **100,** 409–414.

Backs-Husemann, D., and Reinert, J. (1970). Embryobildung durch isolierte Einzelzellen aus gewebekulturen von *Daucus carota. Protoplasma* **70,** 49–60.

Bapat, V. A., and Rao, P. S. (1977). Experimental control of growth and differentiation in organ cultures of *Physalis minima* Linn. *Z. Pflanzenphysiol.* **85,** 403–416.

Bapat, V. A., and Rao, P. S. (1979). Somatic embryogenesis and plantlet formation in tissue cultures of sandalwood (*Santalum album* L.). *Ann. Bot. (London)* [N.S.] **44,** 629–630.

Bapat, V. A., Mhatre, M., and Rao, P. S. (1986). Regeneration of plantlets from protoplast cultures of *Pergularia pallida. J. Plant Physiol.* **124,** 413–417.

Benjamin, B. D., Heble, M. R., and Chadha, M. S. (1979). Alkaloid synthesis in tissue cultures and regenerated plants of *Tylophora indica* Merr. (Asclepiadiaceae). *Z. Pflanzenphysiol.* **92,** 77–84.

Bhatt, P. N., Bhatt, D. P., and Sussex, I. M. (1979). Organ regeneration from leaf discs of *Solunum nigrum, S. dulcamara* and *S. khasianum. Z. Pflanzenphysiol.* **95,** 355–362.

Bhojwani, S. S. (1980). *In vitro* propagation of garlic by shoot proliferation. *Sci. Hortic. (Amsterdam)* **13,** 47–52.

Bhojwani, S. S., and Razdan, M. K., eds. (1983). "Plant Tissue Culture: Theory and Practice." Elsevier, Amsterdam.

Bouniois, A., and Margara, J. (1968). Recherches expérimentales sur la néoformation de bourgeons inflorescentiels ou végétatifs *in vitro* à partir d'explantats d'endive (*Cichorium intybus* L.). *Ann. Physiol. Veg.* **10,** 69–81.

Bovo, O. A., and Mroginski, L. A. (1985). Obtencion de plantas de ajo (*Allium sativum* L.) por cultivo *in vitro* de meristemas. *Phyton* **45,** 159–163.

Butenko, R. G., Grushvitskii, R. V., and Slepyan, L. I. (1968). Organogenesis and embryogenesis in tissue cultures of ginseng (*Panax ginseng*) and other *Panax* species. *Bot. Zh.* **53,** 906–911.

Chandler, S. F., and Dodds, J. H. (1983). Adventitious shoot initiation in serially subcultured callus cultures of *Solanum laciniatum. Z. Pflanzenphysiol.* **111,** 115–121.

Chang, W. C., and Hsing, Y. (1978a). Somatic embryogenesis of root callus of *Panax ginseng* C. A. Meyer, on a defined medium. *Natl. Sci. Counc. Mon. (Taipei)* **6,** 770–772.

Chang, W. C., and Hsing, Y. (1978b). Shoot formation in root callus of *Panax ginseng* C. A. Meyer *in vitro*. *Natl. Sci. Counc. Mon. (Taipei)* **6,** 1171–1173.

Chang, W. C., and Hsing, Y. (1979). Suspension cultures of root callus of *Panax ginseng* C. A. Meyer. *Natl. Sci. Counc. Mon. (Taipei)* **7,** 147–155.

Chang, W. C., and Hsing, Y. I. (1980a). Plant regeneration through somatic embryogenesis in root-derived callus of ginseng (*Panax ginseng* C. A. Meyer). *Theor. Appl. Genet.* **57,** 133–135.

Chang, W. C., and Hsing, Y. I. (1980b). *In vitro* flowering of embryoids derived from mature root callus of ginseng (*Panax ginseng*). *Nature (London)* **284,** 341–342.

Chaturvedi, H. C., Sharma, M., and Prasad, R. N. (1982). Morphogenesis, micropropagation and germplasm preservation of some economic medecinal plants. *In* "Tissue Culture of Economically Important Plants" (A. N. Rao, ed.), pp. 301–302. Proc. Intl. Symp., Singapore.

Choi. K. T., Kim, M. W., and Shin, H. S. (1982). Root and shoot formation from callus and leaflet cultures of ginseng (*Panax ginseng* C. A. Meyer). *In* "Plant Tissue Culture 1982" (A. Fujiwara, ed.), pp. 171–172. Maruzen, Tokyo.

Constabel, F., Gaudet-La-Prairie, P., Kurz, W. G. W., and Kutney, J. P. (1982). Alkaloid production in *Catharanthus roseus* cell cultures. XII. Biosynthetic capacity of callus from original explants and regenerated shoots. *Plant Cell Rep.* **1,** 139–142.

Dastoor, M. N., Schubert, W. W., and Peterson, G. R. (1981). Preliminary results of *in vitro* propagation of guayule. *J. Agric. Food Chem.* **29,** 686–688.

Datta, S. K., Datta K., and Datta, P. C. (1982). Propagation of yam *Dioscorea composita* through tissue culture. *In* "Tissue Culture of Economically Important Plants" (A. N. Rao, ed.), pp. 90–93. Proc. Intl. Symp., Singapore.

Davies, M. E., and Dale, M. M. (1979). Factors affecting *in vitro* shoot regeneration on leaf discs of *Solanum laciniatum* Ait. *Z. Pflanzenphysiol.* **92,** 51–60.

Dhoot, G. K., and Henshaw, G. G. (1977). Organisation and alkaloid production in tissue cultures of *Hyoscyamus niger*. *Ann. Bot. (London)* [N.S.] **41,** 943–947.

Dhruva, B., Ramakrishnan, T., and Vaidyanathan, C. S. (1977). Studies in *Catharanthus roseus* callus cultures, callus initiation and differentiation. *Curr. Sci.* **46,** 364–365.

Diettrich, B., and Luckner, M. (1980). Isolation and characterization of protoplast-derived clones from cell cultures of *Digitalis purpurea*. *Symp. Biol. Hung.* **22,** 340–348.

Eapen, S., Rangan, T. S., Chadha, M. S., and Heble, M. R. (1978). Morphogenetic and biosynthetic studies on tissue cultures of *Atropa belladonna* L. *Plant Sci. Lett.* **13,** 83–89.

Engvild, K. C. (1973). Shoot differentiation in callus cultures of *Datura innoxia*. *Physiol. Plant* **28,** 155–159.

Eriksson, T. (1965). Studies on the growth requirements and growth measurements of cell cultures of *Haplolappus gracilis*. *Physiol. Plant* **18,** 976–993.

Evans, D. A., Sharp, W. R., and Flick, C. E. (1981). Plant regeneration from cell cultures. *Hortic. Rev.* **3,** 214–314.

Fassuliotis, G. (1975). Regeneration of whole plants from isolated stem parenchyma cells of *Solanum sisymbriifolium*. *J. Am. Soc. Hortic. Sci.* **100,** 636–638.

Fassuliotis, G., Nelson, B. V., and Bhatt, D. P. (1981). Organogenesis in tissue culture of *Solanum melongena* cv. Florida. *Plant Sci. Lett.* **22,** 119–125.

Fellenberg, G. (1963). Uber die organbildung an *in vitro* kultivierten knollengewebe von *Solanum tuberosum*. *Z. Bot.* **51,** 113–141.

Fernandez, L., and Sanchez de Jimenez, E. (1982). *In vitro* culture of *Bouvardia ternifolia*. *Can. J. Bot.* **60,** 917–921.

Forsyth, C., and Van Staden, J. (1982). An improved method of *in vitro* propagation of *Dioscorea bulbifera*. *Plant Cell, Tissue Organ Cult.* **1,** 275–282.

Fujimura, T., and Komamine, A. (1975). Effects of various growth regulators on the embryogenesis in carrot cell suspension culture. *Plant Sci. Lett.* **5,** 359–364.
Fujimura, T., and Komamine, A. (1980). Mode of action of 2,4-D and zeatin on somatic embryogenesis in a carrot cell suspension culture. *Z. Pflanzenphysiol.* **99,** 1–8.
Gamborg, O. L., Miller, R. A., and Ojima, K. (1968). Nutrient requirements of suspension cultures of soybean root cells. *Exp. Cell Res.* **50,** 151–158.
Garve, R., Luckner, M., Vogel, E., Tewes, A., and Nover, L. (1980). Growth, morphogenesis and cardenolide formation in long term cultures of *Digitalis lanata. Planta Med.* **40,** 92–103.
George, L. and Rao, P. S. (1979). Experimental induction of triploid plants of *Physalis* through anther culture. *Protoplasma* **100,** 13–19.
Gleddie, S., Keller, W., and Setterfield, G. (1983). Somatic embryogenesis and plant regeneration from leaf explants and cell suspensions of *Solanum melongena* (eggplant). *Can. J. Bot.* **61,** 656–666.
Grewal, S., and Atal, C. K. (1976). Plantlet formation in callus cultures of *Dioscorea deltoidea* Wall. *Indian J. Exp. Biol.* **14,** 352–353.
Grewal, S., and Sharma, K. (1978). Pyrethrum plant (*Chrysanthemum cineriaefolium* Vis.) regeneration from shoot-tip culture. *Indian J. Exp. Biol.* **16,** 1119–1121.
Grewal, S., Sachdeva, U., and Atal, C. K. (1976). Regeneration of plants by embryogenesis from hypocotyl cultures of *Ammi majus* L. *Indian J. Exp. Biol.* **14,** 716–717.
Grewal, S., Koul, S., Ahuja, A., and Atal, C. K. (1979). Hormonal control of growth, organogenesis and alkaloid production in *in vitro* cultures of *Hyoscyamus muticus* Linn. *Indian J. Exp. Biol.* **17,** 558–561.
Groenewald, E. G., Koeleman, A., and Wessels, D. C. J. (1976). Callus formation and plant regeneration from seed tissue of *Aloe pretoriensis* Pole Evans. *Z. Pflanzenphysiol.* **75,** 270–272.
Gunay, A. L., and Rao, P. S. (1982). Plant regeneration from cultured hypocotyl explants of diploid and tetraploid *Solanum khasianum* Clarke. *Plant Cell Rep.* **1,** 202–204.
Gupta, P. K., Mascarenhas, A. F., and Jagannathan, V. (1981). Tissue culture of forest trees—Clonal propagation of mature trees of *Eucalyptus citroidora* Hook. by tissue culture. *Plant Sci. Lett.* **20,** 195–201.
Gupta, S. C., and Babbar, S. B. (1980). Enhancement of plantlet formation in anther cultures of *Datura metel* L. by prechilling of buds. *Z. Pflanzenphysiol.* **96,** 465–470.
Guri, A., and Izhar, S. (1984). Improved efficiency of plant regeneration from protoplasts of eggplant (*Solanum melongena* L.). *Plant Cell Rep.* **3,** 247–249.
Haccius, B., and Lakshmanan, K. K. (1969). Adventive embryonen–Embryoide adventiv Knospen. Ein Bertrag Zur Klairung der Bergriffe *Oesterr. Bot. Z.* **116,** 145–158.
Havranek. P., and Novak, F. J. (1973). The bud formation in the callus cultures of *Allium sativum* L. *Z. Pflanzenphysiol.* **68.** 308–318.
Heller, R. (1953). Recherches sur nutrition minérale des tissues végétaux cultives *in vitro. Ann. Sci. Nat., Bot. Biol. Veg.* [11] **14,** 1–223.
Hirotani, M., and Furuya, T. (1977). Restoration of cardenolide synthesis in redifferentiated shoots from callus cultures of *Digitalis purpurea. Phytochemistry* **16,** 610–611.
Hunter, C. S. (1979). *In vitro* culture of *Cinchona ledgeriana* L. *J. Hortic. Sci.* **54,** 111–114.
Ikuta, A., Syono, K., and Furuya, T. (1974). Alkaloids of callus tissues and redifferentiated plantlets in the Papaveraceae. *Phytochemistry* **13,** 2175–2179.
Ikuta, A., Syono, K., and Furuya, T. (1975). Alkaloids in plants regenerated from *Coptis* callus cultures. *Phytochemistry* **14,** 1209–1210.
Iskander, S. R., and Brossard-Chiriqui, D. (1980). Potentialites organogenes des disques foliaires du *Datura innoxia.* Mill., cultives *in vitro. Z. Pflanzenphysiol.* **98,** 245–254.

Jelaska, S. (1974). Embryogenesis and organogenesis in pumpkin explants. *Physiol. Plant* **31,** 257–261.

Kamat, M. G., and Rao, P. S. (1978). Vegetative multiplication of eggplants (*Solanum melongena*) using tissue culture techniques. *Plant Sci. Lett.* **13,** 57–65.

Kamo, K., Kimoto, W., Hsu, A., Bills, D., and Mahlberg, P. (1980). Alkaloids of *Papaver somniferum* L. in tissue cultures. *Bot. Soc. Am., Misc. Ser. Publ.* **158,** 57–58.

Kessel, R. H. J., and Carr, A. H. (1972). The effect of dissolved oxygen concentration on growth and differentiation of carrot tissue. *J. Exp. Bot.* **23,** 996–1007.

Kessel, R. H. J., Goodwin, C., and Philip, J. (1974). The relationship between dissolved oxygen concentration, ATP and embryogenesis in carrot (*Daucus carota*) tissue cultures. *Plant Sci. Lett.* **10,** 265–274.

Koblitz, H., Koblitz, D., Schmauder, H. P., and Gröger, D. (1983). Studies on tissue cultures of the genus *Cinchona* L. *In vitro* mass propagation through meristem-derived plants. *Plant Cell Rep.* **2,** 95–97.

Kohlenbach, H. W. (1978). Comparative somatic embryogenesis. *In* "Frontiers of Plant Tissue Culture 1978" (T. A. Thorpe, ed.). pp. 59–66. Univ. of Calgary Press, Calgary, Canada.

Konar, R. N., Thomas, E., and Street, H. E. (1972). The diversity of morphogenesis in suspension cultures of *Atropa belladonna* L. *Ann. Bot. (London)* [N.S.] **36,** 249–258.

Kowalczyk, T. P., Mackenzie, I. A., and Cocking, E. C. (1983). Plant regeneration from organ explants and protoplasts of the medicinal plant *Solanum khasianum* C. B. Clarke var. Chatterjeeanum Sengupta (syn. *Solanum viarum* Dunal). *Z. Pflanzenphysiol.* **111,** 55–69.

Krikorian, A. D., and Steward, F. C. (1969). Biosynthetic potentialities of tissue. *In* "Plant Physiology" (F. C. Steward, ed.), Vol. 5B, pp. 227–326. Academic Press, New York.

Kukreja, A. K., and Mathur, A. K. (1985). Tissue culuture studies in *Duboisia myoporoides* 1. Plant regeneration and clonal propagation by stem node cultures. *Planta Med.* **2,** 93–96.

Kukreja, A. K., Mathur, A. K., and Ahuja, P. S. (1986). Morphogentic potential of foliar explants in *Duboisia myoporoides* R. Br. (Solanaceae). *Plant Cell Rep.* **5,** 27–30.

Kumar, P. M., Ghose, S. K., Sen, S., and Sen, S. K. (1983). Regeneration of plants from protoplasts of *Solanum mammosum* L. *In* "Plant Cell Culture in Crop Improvement" (S. K. Sen and K. L. Giles, eds.), pp. 495–499. Plenum, New York.

La Brecque, M. (1980). Guayule bounces back. *Molybdenum Mosaic* **11,** 30–37.

Lakshmi Sita, G. (1982). Tissue culture of *Eucalyptus* species. *In* "Tissue Culture of Economically Important Plants" (A. N. Rao, ed.), pp. 180–184. Proc. Intl. Symp., Singapore.

Lakshmi Sita, G., Bammi, R. K., and Randhawa, G. S. (1976). Clonal propagation of *Dioscorea floribunda* by tissue culture. *J. Hortic. Sci.* **51,** 551–554.

Lam, S. L. (1975). Shoot formation in potato tuber discs in tissue culture. *Am. Potato J.* **52,** 103–106.

Letouze, R., and Beauchesne, G. (1969). Action d'éclairement monochromatiques sur la rhizogénèse de tissus de Topinambour. *C. R. Hebd. Seances Head. Sci.* **269,** 1528–1531.

Li, X. H. (1981). Plantlet regeneration from mesophyll protoplasts of *Digitalis Lanata* Ehrh. *Theor. Appl. Genet.* **60,** 345–347.

Linsmaier, E. M., and Skoog, F. (1965). Organic growth factor requirements of tobacco tissue culture. *Physiol. Plant* **18,** 100–127.

Loh, W. H. T., Hartsel, S. C., and Robertson, L. W. (1983). Tissue cultures of *Cannabis sativa* L. and *in vitro* biotransformation of phenolics. *Z. Pflanzenphysiol.* **111,** 395–400.

Lorz, H., and Potrykus, I. (1979). Regeneration of plants from mesophyll protoplasts of *Atropa belladonna*. *Experientia* **35,** 313–314.
Mantell, S. H., Haque, S. Q., and Whitehall, A. P. (1978). Clonal multiplication of *Dioscorea alata* and *D. rotunda* Poiryams by tissue culture. *J. Hortic. Sci.* **53,** 95–98.
Mascarenhas, A. F., Hendre, R. R., Nadgir, A. L., Ghugale, D. D., Godbole, D. A., Prabhu, R. A., and Jagannathan, V. (1976). Development of plantlets from cultured tissue of *Dioscorea deltoidea* Wall. *Indian J. Exp. Bot.* **14,** 604–606.
Mhatre, M., Bapat, V. A., and Rao, P. S. (1984). Plant regeneration in protoplast cultures of *Tylophora indica*. *J. Plant Physiol.* **115,** 231–235.
Morris, P., and Fowler, M. (1980). Growth and alkaloid content of cell suspension cultures of *Papaver somniferum*. *Planta Med.* **39,** 284–285.
Motegi, T. (1976). Induction of redifferentiated plant from hop leaf callus culture. *Proc. Crop Sci. Soc. Jpn.* **45,** 175–176.
Mudgal, A. K., Goel, S., Gupta, S. C., and Chopra, R. N. (1981). Regeneration of *Iberis amara* plants from *in vitro* cultured leaf and stem explants. *Z. Pflanzenphysiol.* **101,** 179–182.
Murashige, T., and Skoog, F. (1962). A revised medium for rapid growth and bioassays with tobacco tissue cultures. *Physiol. Plant.* **15.** 473–479.
Nakagawa, K., Miura, Y., Fukui, H., and Tabata, M. (1982). Clonal propagation of medicinal plants through the induction of somatic embryogenesis from the cultured cells. *In* "Plant Tissue Culture 1982" (A. Fujiwara, ed.), pp. 701–702. Maruzen, Tokyo.
Nehls, R. (1978). Isolation and regeneration of protoplasts from *Solanum nigrum* L. *Plant Sci. Lett.* **12,** 183–187.
Nessler, C. I. (1982). Somatic embryogenesis in the opium poppy. *Papaver somniferum. Physiol. Plant.* **55,** 453–458.
Nessler, C. I., and Mahlberg, P. (1979). Ultrastructures of laticifers in redifferentiated organs on callus from *Papaver somniferum* L. (Papaveraceae). *Can. J. Bot.* **57,** 675–685.
Nobecourt, P. (1939). Sur la perennite et l'augmentation de volume des cultures de tissus végétaux. *C. R. Seances Soc. Biol. Ses Fil.* **130,** 1270–1271.
Novak, F. J. (1983). Priduction of garlic (*Allium sativum* L.) tetraploids in shoot tip *in vitro* culture. *Z. Pflanzenzuecht.* **91,** 329–333.
Ozias-Akins, P., and Vasil, I. K. (1985). Nutrition of plant tissue cultures. *In* "Cell Culture and Somatic Cell Genetics of Plants" (I. K. Vasil, ed.). Vol. 2, pp. 129–147, Academic Press, Orlando, Florida.
Pal, A., and Sharma, A. K. (1982). Cytological and developmental aspects of callus in *Costus speciosus*. *In* "Tissue Culture of Economically Important Plants" (A. N. Rao, ed.), pp. 85–89. Proc. Intl. Symp., Singapore.
Patel, G. K., Bapat, V. A., and Rao, P. S. (1983). *In vitro* culture of organ explants of *Morus indica:* Plant regeneration and fruit formation in axillary bud culture. *Z. Pflanzenphysiol.* **111,** 465–468.
Perez-Bermudez, P., Cornejo, M. J., and Segura, J. (1983). *In vitro* propagation of *Digitalis obscura* L. *Plant Sci. Lett.* **30,** 77–82.
RajBhandary, S. B., Collin, H. A., Thomas, E., and Street, H. E. (1969). Root, callus and cell suspension cultures, from *Atropa belladonna* L. and *A. belladonna,* cultivar *lutea.* Doll. *Ann. Bot. (London)* **33;** 647–656.
Raju, M. V. S., and Mann, H. E. (1970). Regenerative studies on the detached leaves of *Echeveria elegans*. Anatomy and regeneration of leaves in sterile culture. *Can. J. Bot.* **48,** 1887–1891.
Ramawat, K. G., Bhansali, R. R., and Arya, H. C. (1978). Shoot formation in *Catharanthus roseus* (L.). G. Don, callus cultures. *Curr. Sci.* **47,** 93–94.

Rao, P. S., and Narayanaswamy, S. (1972). Morphogenetic investigations in callus cultures of *Tylophora indica*. *Physiol. Plant* **27,** 271–276.

Rao, P. S., Narayanaswamy, S., and Benjamin, B. D. (1970). Differentiation *ex ovulo* of embryos and plantlets in stem tissue cultures of *Tylophora indica*. *Physiol. Plant* **23,** 140–144.

Rech, E. L., and Pires, M. J. (1986). Tissue culture propagation of *Mentha* spp. by the use of axillary buds. *Plant Cell Rep.* **5,** 17–18.

Reinhard, E., and Alfermann, A. W. (1980). Biotransformation by plant tissue cultures. *Adv. Biochem. Eng.* **16,** 50–84.

Reuther, G. (1977a). Embryoide differenzierungsmuster im kallus der gattungen *Iris* and *Asparagus*. *Ber. Dtsch. Bot. Ges.* **90,** 417–437.

Reuther, G. (1977b). Adventitious organ formation and somatic embryogenesis in callus of *Asparagus* and *Iris* and its possible application. *Acta Hortic.* **78,** 217–224.

Reuther, G., and Bhandari, N. N. (1981). Organogenesis and histogenesis of adventitious organs induced on leaf blade segments of *Begonia–elatior* hybrids (*Begonia* × *hiemalis*) in tissue culture. *Gartenbauwissenschaft* **46,** 241–249.

Rodriguez, E., Towers, G. H. N., and Mitchell, J. C. (1976). Biological activities of sesquiterpene lactones. *Phytochemistry* **15,** 1573.

Rucker, W., Jentzsch, K., and Wichtl, M. (1981). Organ differentiation and glycoside formation in tissues of *Digitalis purpurea* L. cultured *in vitro*. The effect of various growth substances, basal media and light conditions. *Z. Pflanzenphysiol.* **102,** 207–220.

Schenk, R. U., and Hildebrandt, A. C. (1972). Medium and techniques for induction and growth of monocotyledonous and dicotyledonous plant cell cultures. *Can. J. Bot.* **50,** 199–204.

Schieder, O. (1975). Regeneration of haploid and diploid *Datura innoxia* Mill. Mesophyll protoplasts to plants. *Z. Pflanzenphysiol.* **76,** 462–466.

Schieder, O. (1978a). Haploids from *Datura innoxia* as a tool for the production of homozygous lines with high content of scopalamine and for the induction of mutants. *In* "Production of Natural Compounds by Cell Culture Methods" (A. W. Alfermann and E. Reinhard, eds.), pp. 330–336. Gessellschaft für Stahlen und Umwetforschung Mbtl, München.

Schieder, O. (1978b). Somatic hybrids of *Datura innoxia* Mill. + *D. discolor* Bernh and of *D. stramonium* L. var. *tatula* L. *Mol. Gen. Genet.* **162,** 113–119.

Schieder, O. (1980). Somatic hybrids between a herbaceous and two tree *Datura* species. *Z. Pflanzenphysiol.* **98,** 119–127.

Schuchmann, R., and Wellmann, E. (1983). Somatic embryogenesis in tissue cultures of *Papaver somniferum* and *P. orientale* and its relationship to alkaloid and lipid metebolism. *Plant Cell Rep.* **2,** 88–91.

Sehgal, C. B. (1972). *In vitro* induction of polyembryony in *Ammi majus* L. *Curr. Sci.* **41,** 263–264.

Sharp, W. R., Sondhal, M. R., Caldas, L. S., and Maraffa, L. S. (1980). The physiology of *in vitro* asexual embryogenesis. *Hortic. Rev.* **2,** 268–310.

Shepard, J. F. (1982). Cultiver dependent cultural refinements in potato protoplast regeneration. *Plant Sci. Lett.* **26,** 127–123.

Siebert, M., Wetherbee, P. J., and Job, D. D. (1975). The effects of light intensity and spectral quality on growth and shoot initiation in tabacco callus. *Plant Physiol.* **56,** 130–139.

Sipahimalani, A. T., Bapat, V. A., Rao, P. S., and Chadha, M. S. (1981). Biosynthetic potential of cultured tissues and regenerated plants of *Physalis minima* Linn. (Solanaceae). *J. Nat. Prod.* **14,** 114–118.

Skoog, F., and Miller, C. O. (1957). Chemical regulation of growth and organ formation in plant tissues cultivated *in vitro*. *Symp. Soc. Exp. Biol.* **11,** 118–131.
Slusarkiewicz-Jarzina, A., Zenkteler, M., and Podlewska, B. (1982). Regeneration of plants from leaves of *Chrysanthemum morifolium* Ram cv. *Bronze Bronholm,* in *in vitro* cultures. *Acta Soc. Bot. Pol.* **51,** 173–178.
Sommer, H. E., and Brown, C. L. (1980). Embryogenesis in tissue cultures of sweetgum. *For. Sci.* **26,** 257–260.
Sopory, S. K., and Maheshwari, S. C. (1976). Morphogenetic potentialites of haploid and diploid vegetative parts of *Datura innoxia. Z. Pflanzenphysiol.* **77,** 274–277.
Staba, E. J., and Nygaard, B. G. (1983). *In vitro* cultures of guayule. *Z. Pflanzenphysiol.* **109,** 371–378.
Subramanian, V., and SubbaRao, P. V. (1980). *In vitro* culture of the allergenic weed *Parthenium hysterophorus* L. *Plant Sci. Lett.* **17,** 269–277.
Syono, K., and Furuya, T. (1972). The differention of *Coptis* plants *in vitro* from callus cultures. *Experientia* **28,** 236.
Tabata, M., Yamamoto, H., and Hiraoka, N. (1970). Alkaloid production in the tissue cultures of some Solanaceous plants. *Colloq. Int. C.N.R.S.* **193,** 177–186.
Tabata, M., Yamamoto, H., Hiraoka, N., and Konoshima, M. (1972). Organisation and alkaloid production in tissue cultures of *Scopalia parviflora. Phytochemistry* **11;** 949–955.
Takayama, S., and Misawa, M. (1982). Mass propagation of ornamental plants through tissue cultures. *In* "Plant Tissue Culture 1982" (A. Fujiwara, ed.), pp. 681–682. Maruzen, Tokyo.
Tazawa, M., and Reinert, J. (1969). Extracellular and intracellular chemical environments in relation to embryogenesis *in vitro. Protoplasma* **68,** 157–173.
Tewes, A., Wappler, A., Peschke, E. M., Garve, R., and Nover, L. (1982). Morphogenesis and embryogenesis in long term cultures of *Digitalis. Z. Pflanzenphysiol.* **106,** 311–324.
Thomas, E., and Street, H. E. (1970). Organogenesis in cell suspension cultures of *Atropa belladonna* L. and *A. belladonna* cultiver *lutea* Doll. *Ann. Bot. (London)* [N.S.] **34,** 657–669.
Thomas, E., and Street, H. E. (1972). Factors influencing morphogenesis in excised roots and suspension cultures of *Atropa belladonna. Ann. Bot. (London) [N.S,]* **36,** 239–247.
Thorpe, T. A. (1980). Organogenesis *in vitro:* structural physiological and biochemical aspects. *Int. Rev. Cytol. Suppl.* **11A,** 71–111.
Tisserat, B., and Murashige, T. (1977). Effects of ethephon ethylene and 2,4-dichlorophenoxyacetic acid on asexual embryogenesis *in vitro. Plant Physiol.* **60,** 437–439.
Tisserat, B., Esan, E. B., and Murashige, T. (1979). Somatic embryogenesis in angiosperms. *Hortic. Rev.* **1,** 1–78.
Turnbull, A., Galpin, I. J., Smith, J. L., and Collin, H. A. (1981). Comparison of the onion plant (*Allium cepa*) and onion tissue culture. IV. Effect of shoot and root morphogenesis on flavour precursor synthesis in onion tissue culture. *New Phytol.* **87,** 257–268.
Vasil, I. K., and Vasil, V. (1980). Clonal propagation. *Int. Rev. Cytol., Supp.* **11A,** 145–173.
Vasil, I. K., and Vasil, V. (1986). Regeneration in cereal and other grass species. *In* "Cell Culture and Somatic Cell Genetics of Plants" (I. K. Vasil, ed.), Vol. 3, pp. 121–150. Academic Press, Orlando, Florida.
Verzar-Petri, G., Ladocsi, T., and Orosztan, P., (1982). Biosynthesis and localisation of cannabinoids in tissue cultures and sterile plantlets of *Cannabis sativa. In* "Plant Tissue Culture 1982" A. Fugiwara, ed. pp. 337–338. Maruzen, Tokyo.
Vishnoi, A., Babbar, S. B., and Gupta, S. C. (1979). Induction of androgenesis in anther cultures of *Withania somnifera. Z. Pflanzenphysiol.* **94,** 169–179.

Wambugu, F. M., and Rangan, T. S. (1981). *In vitro* clonal multiplication of pyrethrum (*Chrysanthemum cinerariaefolium* Vis.) by micropropagation. *Plant Sci. Lett.* **22,** 219–226.
Wang, P. J., and Huang, C. I. (1982). Production of saikosaponins by callus and redifferentiated organs of *Bupleurum falcatum* L. *In* "Plant Tissue Culture 1982" (A. Fujiwara, ed.), pp. 71–72. Maruzen, Tokyo.
Webb, K. J., Osifo, E. O., and Henshaw, G. G. (1983). Shoot regeneration from leaflet discs of six cultivars of potato (*Solanum tuberosum* subsp. *tuberosum*). *Plant Sci. Lett.* **30,** 1–8.
Wernicke, W., and Kohlenbach, H. W. (1975). Anther cultures in the genus *Scopalia. Z. Pflanzenphysiol.* **77,** 89–93.
Wernicke, W., and Thomas, E. (1980). Studies on morphogenesis from isolated plant protoplasts: Shoot formation from mesophyll protoplasts of *Hyoscyamus niger* and *Nicotiana tabacum Plant Sci. Lett.* **17,** 401–407.
White, P. R. (1939). Potentially unlimited growth of excised plant callus in an artificial medium. *Am. J. Bot.* **26,** 59–64.
White, P. R. (1943). "A Handbook of Plant Tissue Culture." Jacques Cattell Press, Lancaster, Pennsylvania.
Wickham, K., Rodreguez, E., and Arditti, J. (1980). Comparative phytochemistry of *Parthenium hysterophorus* L. (Compositae) tissue cultures. *Bot. Gaz. (Chicago)* **141,** 435–439.
Zavala, M. E., Biesboer, D. D., and Mahlberg, P. G. (1980). Callus induction and organogenesis on cultured tissue of guayule (*Parthenium argentatum* Gray). *Abstr. Int. Guayule Conf.*, 3rd.
Zenkteler, M. (1972). *In vitro* formation of plants from leaves of several species of the Solanaceae family. *Biochem. Physiol. Pflanz.* **163,** 509–512.

CHAPTER **13**

Two-Phase Culture

R. Beiderbeck
B. Knoop

Botanisches Institut der Universität Heidelberg
6900 Heidelberg, Federal Republic of Germany

I. INTRODUCTION

As early as 1963 Staba made a proposal to use plant cell cultures as a means to produce secondary metabolites of economic importance, e.g., drugs, pigments, flavors, fragrances. This proposal has been taken up numerous times but rarely with success. Up to 1982 about 30 different secondary substances had been successfully produced by cell cultures in amounts equal to or even higher than those found in the corresponding intact plant; however, up to the present only a single substance, shikonin, has been produced commercially (Staba, 1985). In many cases studied cultivated cells synthesized the secondary substances of interest only in trace amounts or not at all. The reasons for the disappointing results have been widely discussed, and some ways to overcome the limitations of production have been proposed (Teuscher, 1973; Berlin, 1983).

Secondary products may be either stored within the cells or excreted. A low yield of secondary substances released into the medium may be attributable to several factors. In those cases where low yield is due to feedback inhibition of membrane transport, biosynthesis, or gene activity, enzymatic or nonenzymatic degradation in the medium, or volatility of substances produced, it should be possible to increase the net production by the addition of an artificial site for accumulation and/or conservation of secondary substances in the culture medium. Using the "two-phase culture" (Beiderbeck, 1982) the concentrations of a secondary substance inside the protoplasts (C_p), in the culture medium (C_m), and in the accumulation/conservation phase (C_a) should approach an equilibrium depending on the affinity, the capacity, and the amount of the second phase material:

$$C_p \rightleftharpoons C_m \rightleftharpoons C_a$$

An ideal material for accumulating and conserving secondary substances has the following properties:

1. It is stable under conditions of autoclaving.
2. It is nontoxic and leaves the medium composition unchanged.
3. It dissolves or binds and stabilizes secondary substances released by the cells.
4. It offers a large surface per unit weight with high accumulation capacity.
5. It releases secondary substances during simple extraction procedures.
6. It binds desired substances in as specific a manner as possible.

It is easily recognized that the perfect second phase for the accumulation of a certain secondary substance has to be tailored to each substance. Until now only a few examples have become known which show second phases binding a vast number of cell products in a more or less nonspecific manner. Nevertheless, the fundamental concept can be demonstrated.

II. ACCUMULATION PHASES

A. Lipophilic Second Phase

Cell cultures with lipophilic second phases contain the nutrient medium (aqueous phase) and a small amount of lipophilic material in either liquid or solid form. Several studies have shown that plant cell cultures release lipophilic volatile substances into the gas phase. Some plant cells produce ethylene (LaRue and Gamborg, 1971; Huxter *et al.*, 1981), ethanol, and acetaldehyde (Thomas and Murashige, 1979). Hydrocarbons could be detected in the spent air of a suspension culture of *Matricaria chamomilla* crown gall cells passing through ethanol (Bisson *et al.*, 1983). These results led to the conclusion that the addition of a lipophilic phase to the culture medium can be helpful for secondary substance accumulation and detection.

1. Miglyol 812

Miglyol 812 (Dynamit Nobel) is a water-insoluble triglyceride of low viscosity composed of fatty acids with 8–10 carbons. In liquid media containing 10–12% of this material, transformed and nontransformed suspension cultures of *Matricaria chamomilla* (Beiderbeck, 1982; Bisson *et al.*, 1983), *Nicotiana tabacum*, and *Thuja occidentalis* (Berlin *et al.*, 1984) showed unimpaired growth in comparison with the conventional single-phase culture. Simultaneously, this Miglyol phase accumulated various secondary substances depending on the species of the cultured cells. For example, the *Matricaria* cell suspension yielded a mixture of UV-absorbing products which were synthesized and accumulated mostly during the log phase of cell growth. During the subsequent stationary phase these secondary products disappeared, probably metabolized by the starving cells (Beiderbeck, 1982) (Fig. 1). An analysis showed that during the first, second, or third week of the culture period and in the presence of Miglyol a great variety of different compounds (isolated by steam distillation) appeared in the cells, in the aqueous nutrient medium, and in the Miglyol. At any time the greatest number of different substances could be isolated from the Miglyol phase, but the pattern of substances

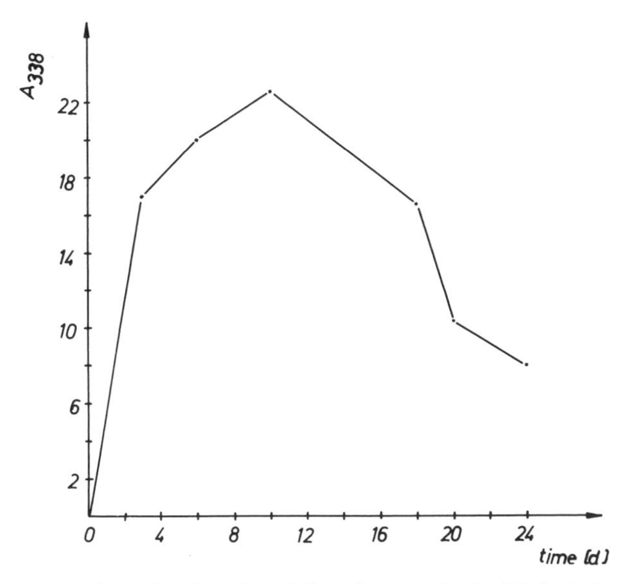

Fig. 1. The amount of UV-absorbing lipophilic substances in the Miglyol phase of a two-phase culture of *Matricaria chamomilla* cells during 24 days after inoculation (from Beiderbeck, 1982).

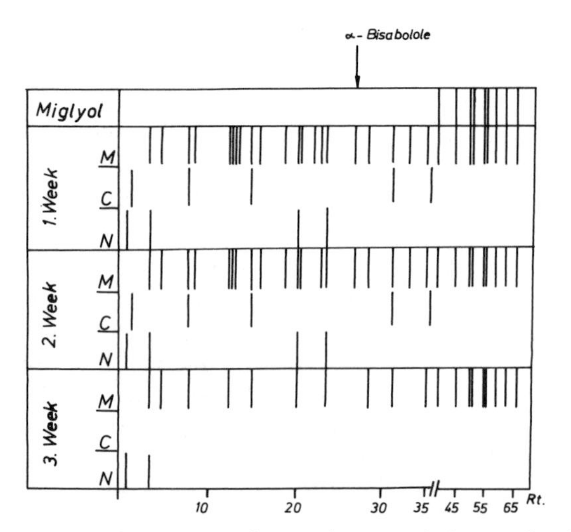

Fig. 2. Gas chromatography pattern of secondary metabolites isolated from Miglyol phases (M), cells (C), and aqueous nutrient media (N) of a suspension culture of *Matricaria chamomilla.* The cells were incubated in the presence of Miglyol during either the first, second, or third week of culture. At the top of the diagram is the pattern of Miglyol (after Bisson, 1983.)

varied in all three fractions from week to week. Some products were isolated during the entire 3-week period; others could be found during the first week only (Fig. 2). One such product was identified as the sesquiterpene alcohol α-bisabolole, a therapeutically important compound of the *Matricaria* flower, which has not been detected in conventional cell suspensions (Bisson *et al.*, 1983).

Similarly a suspension culture of *Thuja occidentalis* produced up to 3 mg/g dry weight/day of monoterpene in a two-phase culture with Miglyol compared to 0.8 mg as single-phase culture. More than 95% of these substances were recovered from the Miglyol phase and identified as α-pinene, β-pinene, myrcene, limonene, and terpinolene. Also labile iron–tropolonate complexes could be detected (Berlin *et al.*, 1984).

2. Liquid Paraffin

Liquid paraffin (10% v/v) as second phase with a *Matricaria* cell culture also accumulated UV-absorbing secondary substances from the culture medium. Interference with the cell growth was hardly detectable (Beiderbeck, 1982).

3. LiChroprep RP8

Instead of a liquid lipohilic phase, solid materials may also be introduced into the aqueous medium. One such material is LiChroprep RP8 (Merck, Darmstadt, West Germany), a silica gel (particle size 40–60 μm) with outer SiOH groups covalently bound to C_8 hydrocarbons. These hydrocarbons coat the silica gel particles as a monomolecular lipophilic layer. RP8 has been employed as a second phase in suspension cultures of *Pimpinella anisum* and *Valeriana wallichii*. Cells of *Pimpinella anisum* released small amounts of the phenylpropanoid anethole into the RP8 phase, whereas this compound could not be detected in single-phase cultures of the same cell line (Bisson, 1983). Similarly a cell line of *Valeriana wallichii* which did not produce detectable amounts of valepotriates in single-phase cultures produced a low level of these compounds after addition of RP8 (Becker and Herold, 1983). Another cell line normally containing 0.5% valepotriates (on a dry weight basis) inside the cells produced increased amounts in the presence of RP8 and imparted some of it into the lipophilic phase. Total enhancement was about 23%.

B. Solid Polar Phases

Beside the lipophilic compounds many products of plant cells are expected to be of polar character and hardly bound by lipophilic phases. Therefore, more polar adsorbents may serve as second phases.

1. XAD-4

XAD-4 (resin of Amberlite, Serva, Heidelberg, West Germany) is an adsorber resin on a polystyrene base and is reported to adsorb a great variety of diverse substances such as phenols, alcohols, and organic acids (Anonymous, 1984–1985). It is available as beads of 0.3–1.0 mm diameter and can be recovered from cell suspensions by repeated decanting. A solvent eluting a broad pattern of substances bound to XAD-4 is acetone.

The addition of varying amounts of XAD-4 (up to a concentration of 2% w/v) to different cell lines of *Nicotiana tabacum* does not impair growth of the cultures more or less. With increasing concentrations the cell lines behave differently: the growth of one line was reduced one-third with 4% XAD-4 in the medium, and the growth of another cell line

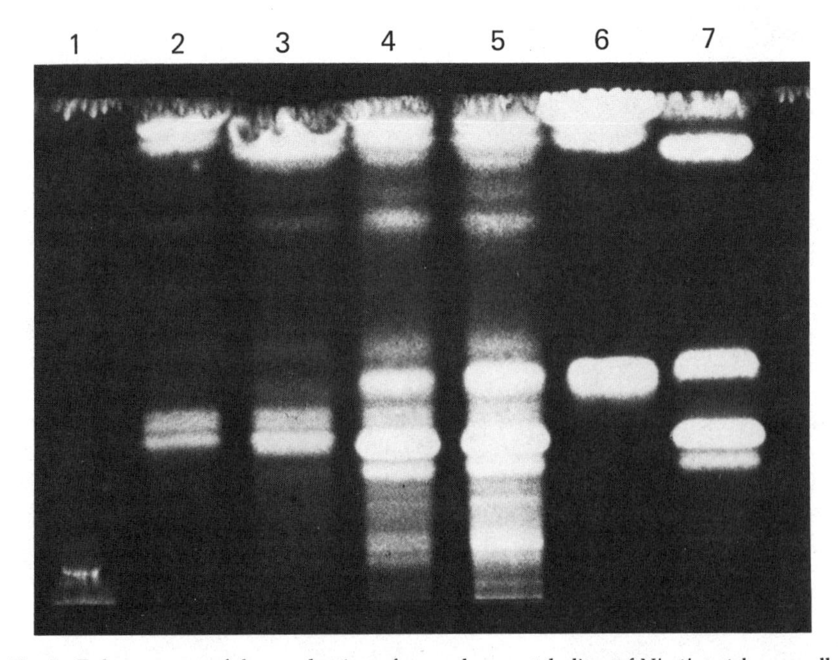

Fig. 3. Enhancement of the production of secondary metabolites of *Nicotiana tabacum* cells by the addition of increasing amounts of XAD-4. From left to right: 1, control without XAD-4; 2 to 5, increasing amounts of XAD-4 (0.2, 0.5, 1.5, and 3.0 g per 24-ml suspension); 6, mixture of reference substances consisting of chlorogenic acid (R_f 0.45), umbelliferone (0.95), and scopoletin (0.99); 7, acetone extract of cells. Fluorescence was recorded after spraying with Natural Compound Reagent (Roth) and irradiation at 365 nm (Maisch, 1985).

was even promoted by XAD-4 concentrations up to 12.5%. In any case the addition of XAD-4 enhanced the production of several secondary substances compared to the adsorbent-free control. One of these substances (Fig. 3, R_f 0.44) was identified by TLC and NMR as chlorogenic acid. XAD-4 (12.5%) increased its production more than 20-fold. As shown by TLC, the production of chlorogenic acid and several other substances varied: the production of some was enhanced, of others not at all (Maisch *et al.*, 1986).

2. Activated Charcoal

Activated charcoal (AC) has a high capacity to bind a wide spectrum of different substances. Although its binding mechanisms are not completely understood, AC is widely used in laboratories and technical processes (Mattson and Mark, 1971). Among the available variety of different AC qualities, beads of 2–3 mm diameter are most suitable as an addition to cell cultures since beads are easy to separate from cells.

In a single-phase culture of *Matricaria chamomilla* which releases coniferyl aldehyde into the medium, the addition of 0.8–4% (w/v) AC reduced the coniferyl aldehyde content of the aqueous phase depending on the amount of AC. On the other hand, compared to control cultures a 20- to 60-fold greater amount of coniferyl aldehyde could be recovered from the adsorbent (Knoop and Beiderbeck, 1983). The production of several other substances (unidentified) also showed substantial enhancement. Such an increased production of secondary metabolites may be traced partly to product accumulation in the AC and partly to a reduction of the rate of cell multiplication: after 1 week of culture with 0.8% AC as a second phase, the fresh weight of *Matricaria* cells was comparable to that of adsorbent-free controls whereas the output of coniferyl aldehyde was increased by a factor of 20. With 4% AC the final fresh weight was reduced to one-third that of controls, and the coniferyl aldehyde output per fresh weight was further increased (Knoop and Beiderbeck, 1983).

3. Other Adsorbents

Both polar adsorbents described show adsorbance of a wide spectrum of substances. This is an advantage when searching for total products but a disadvantage when only few products are desired. In the latter case the adsorbent should have an affinity restricted to a small group of specific substances. Moreover, the binding specifity of an adsorbent should affect the cells by accelerating biosynthesis of products to be finally bound. In order to learn more about adsorbent–product relationships, a series of other adsorbents was tested with regard to possible use in cell culture. Adsorbents available as beads were used directly, those available as fine powders had to be embedded into droplets of alginate to guarantee a cell-free harvest from the two-phase culture (Maisch *et al.*, 1986). In these experiments two additional adsorbents offered a good prospect for future use: XAD-7, an ester of acrylic acid, and Polyclar AT, a water-insoluble polyvinylpyrrolidone. Two other adsorbents, raw silk (as a protein) and Zeolith Taylor (a synthetic sodium aluminum silicate), severely affected culture growth and were difficult to extract. The magnesium trisilicate Florisil was unsuited for cell culture purposes since it increased the pH of the medium to an unacceptable range.

In contrast to the liquid second phases already described, the addition of solid adsorbents did reduce growth depending on the concentrations employed with only one exception: XAD-4, when applied to certain cell lines. This growth reduction may have several causes, two of which were clearly identified: (1) the adsorbent drives the pH of the medium to a range unacceptable to the cells (e.g., Florisil) and (2) the adsorbent withdraws essential components from the culture medium.

With AC the second case has been studied more thoroughly. AC has long been known to adsorb phytohormones and vitamins from nutrient media (Constantin *et al.*, 1977; Weatherhead *et al.*, 1979), causing deficiency symptoms in the cultivated cells. To avoid these effects the AC for a two-phase culture was pretreated with the phytohormones (naphthaleneacetic acid (NAA) and kinetin and/or vitamins in MS medium (Murashige and Skoog, 1962). After this preloading the AC beads were washed and transferred to hormone-free fresh MS medium. During autoclaving and subsequent shaking a new equilibrium of the preloaded compounds formed between solid and liquid phase. By varying the amount of NAA, kinetin, and vitamins during the preloading procedure, a series of experiments led to calibration curves which allowed adjustment of the NAA, kinetin, and vitamin content in the final culture medium to a predetermined value. The use of such preloaded AC in cell cultures of *Nicotiana tabacum* led to a curing of several deficiency symptoms caused by pure AC: the protein content per fresh weight, the average and maximum cell sizes, as well as the cytological image of the culture could be fully normalized (Table I). With 4% AC the fresh weight increase was reduced to 7.7% of control. After preloading this AC with auxin and cytokinin it could be restored to more than 50% (Beiderbeck and Knoop, 1984). The preloaded AC thus serves as a hormone and/or vitamin source for the cells.

Furthermore, it could be demonstrated that AC preloaded with phytohormones and vitamins was still able to adsorb considerable amounts of secondary metabolites from spent culture medium. When cells of *Nicotiana tabacum* were grown in the presence of AC preloaded with different amounts of phytohormones the pattern of secondary substances eluted from the AC samples showed great differences when examined by TLC. Depending on the NAA content of the AC (kinetin constant) there appeared a number of new fluorescent substances. Other products could be found after varying the kinetin content of the AC (NAA constant) (B. Knoop and R. Beiderbeck, unpublished). Obviously, culture media with AC carrying different concentrations of phytohormones support different pathways of secondary substances (Zenk *et al.*, 1984). Consequently preloaded AC with different hormone and vitamin concentrations may be used to control production of secondary metabolites in mass cultures.

TABLE I

Restoration of Normal Growth in Cultures of *Nicotiana tabacum* after Preloading Activated Charcoal with NAA[a]

Medium and AC treatment	Protein content of cells (mg/g fresh weight)	Cell length, maximum/most frequent (μm)
MS	1.27 (0.22)[b]	800/150
MS + AC	0.99 (0.35)	1150/220
MS − H	0.85 (0.36)	1170/370
MS − H + AC pretreated with		
0 mg/liter NAA	0.82 (0.15)	1050/270
50 mg/liter NAA	1.14 (0.19)	710/170
100 mg/liter NAA	1.28 (0.11)	500/130
200 mg/liter NAA	1.47 (0.18)	390/110
300 mg/liter NAA	2.06 (0.56)	390/110

[a] AC, Activated charcoal, 0.8 g per 24-ml culture; MS, complete medium with hormones (NAA + kinetin); H, hormones.

[b] Standard deviation in brackets. For details see Beiderbeck and Knoop (1984).

III. DISCUSSION

The two-phase culture may offer several benefits to tissue culturists. In many cases plant cell suspensions may yield secondary substances which normally escape detection because of very low production levels, rapid degradation, or volatility. Two-phase culture will help to detect such "hidden" capacity for synthesis. Moreover, the yield of substances which are produced in low amounts during conventional single-phase culture may be enhanced by the introduction of artificial accumulation and/or conservation sites.

Up to the present only a very limited number of second phases has been tried experimentally. No general rules about their influence on plant cells and about the prospect of their use can be formulated. It is evident that any second phase accumulates a characteristic pattern of substances (Maisch, 1985). In the future this pattern may be planned in advance by an intelligent choice or construction of the second phase, e.g., adsorbents commonly used in affinity chromatography. Also, very few solvents have been used until now to recover secondary substances from second phases. Preliminary experiments have shown that fractional extraction is possible. As only a limited array of detection methods has been employed, many substances which may have escaped detection should be found by more detailed search.

Finally, the two-phase culture should be extended to other organisms: bacteria, algae, fungi, and animal cells are possible candidates for successful exploitation. Preliminary experiments with the ascomycete *Penicillium purpurogenum* in the presence of XAD-4 indicated a substantial increase of dye production (Maisch *et al.*, 1986).

As far as plant cell suspension cultures are concerned the two-phase culture may be combined with most other methods known to enhance secondary metabolite production. Many cell lines and culture conditions which did not show secondary substance production before may be given a second look using two-phase culture. A further extension of this method could consist in the application of second phases in continuous culture systems.

All experiments reported here describe the improvement of the production of secondary metabolites of suspension cultures without any preceding selection. Another way to increase productivity is the selection of high-producing cell lines from low-producing cultures. Rapid and simple selection procedures for colorless and nonfluorescent compounds which also preserve the cell material are scarce. Here the principle of two-phase culture (adsorbent culture) can be used with advantage for substances which are released from the cells into the medium.

Filters covered with a thin layer of an adsorbent serve as selective fields for high-producing cell lines (Knoop and Beiderbeck, 1985). Small cell aggregates are plated onto such filters with a cell carrier foil in between (Fig. 4). Released cell products pass through this foil and are locally bound to the adsorbent. After the intact cell aggregates are transferred to fresh medium together with the carrier foil for further cultivation, the filters can be dried and developed in such a way that a stream of an eluent transports the bound substances to the adsorbent-free filter back. Here they can be detected by routine methods, i.e., UV fluores-

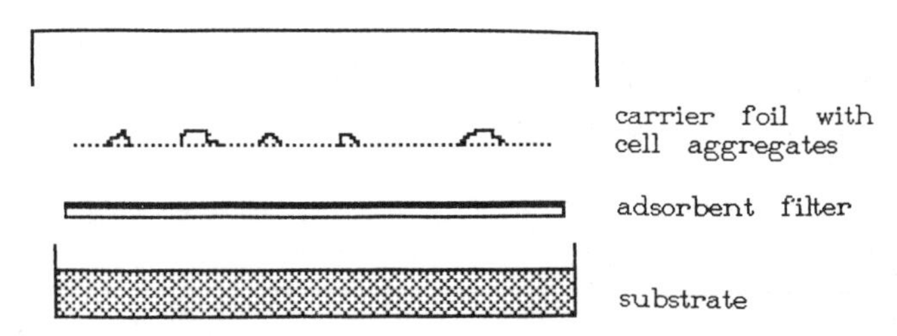

Fig. 4. Schematic arrangement of an adsorbent filter plate used to isolate high-producing cell strains.

cence or staining with an adequate reagent. The suitability of this new method was demonstrated by the selection of a tobacco cell strain releasing an uncommon pattern of substances compared to the original suspension (Thumann *et al.*, 1987).

REFERENCES

Anonymous. (1984–1985). "Serva-Feinbiochemica für die Forschung." Catalogue of Serva, Heidelberg, Federal Republic of Germany.

Becker, H., and Herold, S. (1983). RP8 als Hilfsphase zur Akkumulation von Valepotriaten aus Zellsuspensions-Kulturen von *Valeriana wallichii*. *Planta Med.* **49,** 191–192.

Beiderbeck, R. (1982). Zweiphasenkultur—Ein Weg zur Isolierung lipophiler Substanzen aus pflanzlichen Suspensionskulturen. *Z. Pflanzenphysiol.* **108,** 27–30.

Beiderbeck, R., and Knoop, B. (1984). Ein Adsorbens als Speicher für Phytohormone in einer pflanzlichen Suspensionskultur. *Z. Naturforsch., C: Biosci.* **39C,** 45–49.

Berlin, J. (1983). Naturstoffe aus pflanzlichen Zellkulturen. *Chiuz.* **17,** 77–84.

Berlin, J., Witte, L., Schubert, W., and Wray, V. (1984). Determination and quantification of monoterpenoids secreted into the medium of cell cultures of *Thuja occidentalis*. *Phytochemistry* **23,** 1277–1279.

Bisson, W. (1983). Untersuchungen zur Bildung, Zusammensetzung und Akkumulation von ätherischem Öl in Zellkulturen von *Matricaria chamomilla* L., unter besonderer Berücksichtigung systemfremder Akkumulationsorte (Zweiphasensystem). Dissertation, Universität Heidelberg.

Bisson, W., Beiderbeck, R., and Reichling, J. (1983). Die Produktion ätherischer Öle durch Zellsuspensionen der Kamille in einem Zweiphasensystem. *Planta Med.* **47,** 164–168.

Constantin, M. J., Henke, R. R., and Mansur, M. A. (1977). Effect of activated charcoal on callus growth and shoot organogenesis in tobacco. *In Vitro* **13,** 293–296.

Huxter, T. J., Thorpe, T. A., and Reid, D. M. (1981). Shoot initiation in light- and dark-grown tobacco callus: The role of ethylene. *Physiol. Plant* **53,** 319–326.

Knoop, B., and Beiderbeck, R. (1983). Adsorbenskultur—Ein Weg zur Steigerung der Sekundärstoffproduktion in pflanzlichen Suspensionskulturen. *Z. Naturforsch., C: Biosci.* **38C,** 484–486.

Knoop, B., and Beiderbeck, R. (1985). Adsorbent filter—A tool for the selection of plant suspension culture cells producing secondary substances. *Z. Naturforsch., C: Biosci.* **40C,** 297–300.

LaRue, T. A. G., and Gamborg, O. L. (1971). Ethylene production by plant cell cultures. *Plant Physiol.* **48,** 394–398.

Maisch, R. (1985). Steigerung der Sekundärstoffproduktion in pflanzlichen Zellkulturen durch reversible Adsorption. Diplomarbeit, Universität Heidelberg.

Maisch, R., Knoop, B., and Beiderbeck, R. (1986). Adsorbent culture of plant cell suspensions with different adsorbents. *Z. Naturforsch. C. Biosci.* **41C,** 1040–1044.

Mattson, J. S., and Mark, H. B. (1971). "Activated Carbon." Dekker, New York.

Murashige, T., and Skoog, F. (1962). A revised medium for rapid growth and bioassays with tobacco tissue cultures. *Physiol. Plant.* **15,** 473–497.

Staba, E. J. (1963). The biosynthetic potential of plant tissue cultures. *Dev. Microbiol.* **4,** 193–198.

Staba, E. J. (1985). Milestones in plant tissue culture systems for the production of secondary products. *J. Nat. Prod.* **48,** 203–209.

Teuscher, E. (1973). Probleme der Produktion sekundärer Pflanzenstoffe mit Hilfe von Zellkulturen. *Pharmazie* **28,** 6–18.

Thomas, D. S., and Murashige, T. (1979). Volatile emissions of plant tissue cultures: Identification of the major components. *In Vitro* **15,** 654–658.

Thumann, J., Knoop, B., and Beiderbeck, R. (1987). Selection of secondary substance producing variants by an improved adsorbent filter method. *Biol. Plant. (Prague)* (in press).

Weatherhead, M. A., Burdon, J., and Henshaw, G. G. (1979). Effects of activated charcoal as an additive plant tissue culture media part. *Z. Pflanzenphysiol.* **94,** 399–405.

Zenk, M. H., Schulte, U., and El-Shagi, H. (1984). Regulation of anthrachinone formation by phenoxyacetic acids in *Morinda* cell cultures. *Naturwissenschaften* **71,** 266.

CHAPTER **14**

Continuous Culture of Plant Cells

J. Stefan Rokem

Department of Applied Microbiology
Institute of Microbiology
The Hebrew University—Hadassah Medical School
Jerusalem 91 010, Israel

I. INTRODUCTION

The use of the continuous culture technique enables the control of growth in defined equilibrium conditions and was initially suggested by Monod (1950) and Novick and Szilard (1950). This technique has been applied to the study of the physiology, biochemistry, and genetics of prokaryotic (bacteria and blue–green algae) and eukaryotic (fungi, animal, and plant) cells. The ability to grow plant cells under continuous culture conditions in suspensions has enabled the study of the cytology, physiology, and biochemistry of plants, as well as production of plant biomass and plant metabolites under more controlled conditions. The aim of this chapter is to review and evaluate the various continuous culture techniques employed as well as to asses the future potential of this technique for the study of various aspects of phytochemistry. The literature on this subject has been reviewed by several authors (King and Street, 1977; Street, 1976) and more recently by Wilson (1980), who covered work based mainly on the chemostat principle up to 1978, and by Kurz and Constabel (1981), who covered the literature up to 1977. For the basic theory and general applications of the continuous culture technique, see the work by Herbert and co-workers (Herbert *et al.*, 1956).

II. CHARACTERISTICS OF PLANT CELLS

The structure and physiology of plant cells imposes restrictions on the study of these cells using the continuous culture technique:

1. The size of plant cells varies from 20 to 150 μm in diameter, and they are 30–100 times larger than bacterial cells and 8–10 times bigger than most fungal cells. The larger size of plant cells and their tendency to form aggregates results in quick settling, when agitation is not adequate. The proportion of aggregates to free cells varies for different cultures. Approximately 60% of the cells of *Morinda citrifolia* were either single or clusters of two cells, when grown as a suspension (Zenk *et al.*, 1975). This is considered a high percentage of free cells. Various attempts to break up aggregates of *Rosa* sp. (Paul's Scarlet rose) and *Glycine max* cells by chemical, enzymatic, and physical means did not give sustained disaggregation without affecting the apparent biochemical state of the culture (Kubek and Shuler, 1978). For biomass and metabolite production, aggregation may not be harmful, but for metabolic and physiological studies balanced growth of uniform cell material is desirable.

2. A characteristic of importance for continuous culture growth is the sensitivity of plant cells to shear stress. The plant cell wall has high tensile strength but low shear resistance.

3. The metabolic rate of plant cells is lower, when compared to microbial cells, resulting in relatively long doubling times (20–100 hr). All these characteristics influence and limit the configurations of the growth vessels that can be used for plant cell growth.

4. Continuous culture experiments extend for long periods of time, and contamination can be a serious problem. To avoid contamination rigorous aseptic techniques are required. In the study of *Nicotiana tabacum* cells, to avoid contamination, 60 μg penicillin G was added to the chemostat shortly after inoculation. The penicillin was diluted out before steady state was reached, and the culture did not become contaminated (Sahai and Shuler, 1984). However, it is possible to propagate plant cells without addition of antibiotics. In a study of diosgenin production by *Dioscorea deltoidea*, cells were grown in a chemostat without antibiotic addition, and the chemostat was uncontaminated for 1 year (B. Tal, personal communication).

III. VARIOUS TECHNIQUES FOR CONTINUOUS CULTURE OF PLANT CELLS

Two main methods for continuous culture have been described, the chemostat and turbidostat techniques. These techniques were used for growth of plant cells (Street, 1976). Plant species grown with these techniques are listed in Table I.

TABLE I

Plant Species Grown in Continuous Culture

Plant species	Continuous culture technique[a]	Reference
Acer pseudoplatanus	C, T	Wilson *et al.* (1971), King and Street (1977), King (1977)
Galium mollugo	C	Wilson and Marron (1978), Wilson and Balagué (1985)
Phaseolus vulgaris	C	Bertola and Klis (1978)
Catharanthus roseus	C	Balagué and Wilson (1982)
Nicotiana tabacum	C	Kato *et al.* (1976, 1980), Sahai and Shuler (1982, 1984)
Dioscorea deltoidea	C	Tal and Goldberg (1982)
Mentha piperita	C	Tal *et al.* (1983)
Daucus carota	SC	Dougall and Weyrauch (1980)
Ocimum basilicum	T	Dalton (1983)
Asparagus officinalis	T	Peel (1982)
Spinacia oleracea	T	Dalton (1980)
Chenopodium rubrum	C	Husemann (1983)
Glycine max	C	Kurz (1973)

[a] C, Chemostat; T, turbidostat; SC, semicontinuous culture.

A. The Chemostat Technique (Principle—fixed rate of new medium input)

Most continuous culture studies with plant cells have been undertaken using the chemostat technique (Wilson, 1980). With this technique the cells are grown with one nutrient as the limiting growth factor, whose concentration determines the amount of biomass that is obtained. Fresh medium is added at a constant rate, and the cell concentration will, if the criteria for ideal mixing and homogeneous growth are fulfilled, reach a steady state where the cell concentration is independent of time. The chemostat enables balanced growth where cell composition remains constant for a set of growth conditions (growth rate, medium composition, temperature, etc.).

Results of studies of *Acer pseudoplatanus* cells grown in a chemostat culture showed that a balanced state of growth is possible for a plant cell suspension. Cell dry weight, DNA and RNA amounts, respiration rate (King *et al.*, 1973; King and Street, 1977), nitrogen assimilation rate and activity of enzymes involved in nitrogen assimilation (Young, 1973), carbohydrate metabolism (Fowler and Clifton, 1975), and level of free amino acids (Street *et al.*, 1975) were all constant during steady-state growth. It was also found that a plant cell suspension of *Acer pseudo-platanus* taken through a series of steady states (by changing the dilution rate) will return to the same physiological and morphological characteristics as obtained earlier for a specific dilution rate (Street, 1976).

From the model of Monod (1950) three kinetic factors were defined; (1) μ_{max}, the maximum specific growth rate (hr^{-1}), (2) K_s, substrate saturation constant (mole concentration enabling half the maximum growth rate, and (3) Y, yield coefficient (grams cells formed/gram substrate utilized). There are no comprehensive, systematic, or comparative studies of these factors even though different plant species have been grown at steady states, using the chemostat technique (Table I).

Growth yields were determined at fixed growth rates. The yield coefficient for glucose can be calculated for *Acer pseudoplantanus* cells as approximately 0.37 g cell dry weight/g glucose (King, 1977). For *Nicotiana tabacum* a yield of between 0.36 and 0.5 g cell dry weight/g sucrose was obtained (Kato *et al.*, 1976). The yield of *Catharanthus roseus* on sucrose was reported as 0.46 g cell dry weight/g sucrose utilized (Balagué and Wilson, 1982), and a yield value of 0.45 g cell dry weight/g glucose could be calculated for *Nicotiana tabacum* cells in a more recent investigation (Sahai and Shuler, 1984). The biomass yield of *Dioscorea deltoidea* was 0.4 g cell dry weight/g sucrose utilized in both batch and continuous culture (Tal and Goldberg, 1982). From the above results it can be concluded that the growth efficiency of plant cells is similar to that of microbial cells, where yield values range from 0.4 to 0.5 g of cells/g carbohydrate carbon source.

The K_s value has been determined only for *Acer pseudoplatanus* cells growing on various limiting nutrients. The K_s value was 0.13 m*M* for NO_3^- (King *et al.*, 1973), 32 μM for phosphate (Wilson, 1980), and 0.5 m*M* for glucose (King, 1977). These K_s values are larger than those reported for microbial cells, but there are not enough data to draw any conclusions from this comparison. The maximum growth rate of plant cells is usually determined from batch growth kinetics, using similar conditions as in the continuous culture growth system.

Steady-state growth rates used in chemostats range from 20% to close to 100% of μ_{max}. Cells of *Nicotiana tabacum* were grown in conventional fermentors where the growth rate was increased stepwise (0.05 day^{-1}) in order to avoid the sudden fall of cell mass at the initiation stages of continuous culture operation (Kato *et al.*, 1980). It was finally fixed at 0.42 day^{-1} (a generation time of 39.5 hr). The increase of the growth rate from 0.08 day^{-1} to 0.24 day^{-1} in one step resulted in the washout of *Dioscorea deltoidea* cells from the growth vessel (Fig. 1) (Tal and Goldberg, 1982). The chemostat is ideal for physiological and biochemical studies

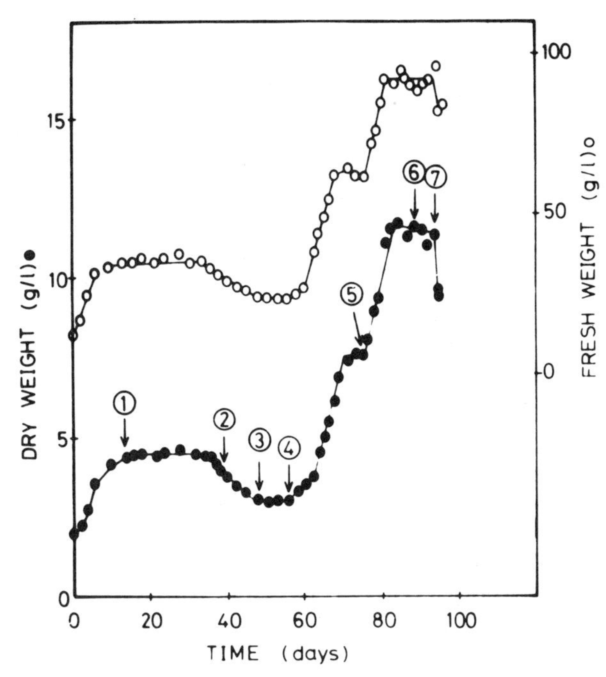

Fig. 1. Effect of nitrogen sources on growth of *Dioscorea deltoidea* cells in a chemostat. Cells were incubated in MS medium with 30 g sucrose/liter and 500 mg KNO_3/liter but without NH_4NO_3. The aeration rate was 0.18 volumes of air per volume of fermentor volume per minute (VVM). Arrows indicate the following: 1, Dilution rate was fixed at 0.08 day^{-1}; 2, KNO_3 (1.5 g/liter) was added to the culture broth in the growth vessel; 3, NH_4NO_3 (625 mg/liter) was added to the culture broth in the growth vessel; 4, a solution containing KNO_3 (1.9 g/liter) and NH_4NO_3 (1.65 g/liter) was added to the culture broth in the growth vessel and to the medium reservoir; 5, the aeration rate was changed to 0.3 VVM; 6, the aeration rate was changed to 0.377 VVM; 7, the dilution rate was increased to 0.24 day^{-1}. (From Tal and Goldberg, 1982, with permission of George Thieme Verlag.)

of plant cells grown at steady state, but only *Acer pseudoplatanus* cells have been thoroughly investigated (see the reviews of Street, 1976; Wilson, 1980).

B. The Turbidostat Technique (Principle—fresh medium input intermittently to maintain a fixed cell density in the suspension)

In the turbidostat the continuous determination of cell density is performed by optical means. Cell density is the determining factor for fresh medium addition. When the cell density rises above a preset value, fresh medium is added and the medium addition is balanced by automatic harvest of used medium and cells by a constant level device. The turbidostat allows work at low cell densities. Using this technique the cells are grown very close to their maximal growth rate. Therefore the turbidostat is a valuable tool for studying regulatory mechanisms of metabolism. The turbidostat has not been used as frequently as the chemostat in studies of plant cells (Table I).

In a study of *Acer pseudoplatanus* cells a growth rate corresponding to a mean generation time of 70 hr was reported (Wilson *et al.*, 1971). In another study, photoautotrophic growth of *Asparagus officinalis* was obtained using the turbidostat technique, and the regulation of higher plant photosynthesis at the cellular level was studied (Peel, 1982). Increase of the photosynthetically active radiation resulted in: (1) a threefold increase in the average specific growth rate, (2) no change in the chlorophyll content, and (3) an increase in the carotenoid content. The maximal growth rate obtained for photoautotrophic growth (0.36 day^{-1}, corresponding to a generation time 46 hr) compares favorably with growth rates in heterotrophic cultures (Peel, 1982). Further studies using this technique could give useful information regarding the development, operation, and regulation of higher plant photosynthesis at the cellular level under controlled conditions. Bligny (1977) studied cells of *Acer pseudoplatanus* grown by the turbidostat technique. The optical density measurements were made by an interesting method, with two optical fibers outside the growth vessel. This circumvented the fouling problems that may occur when medium and cells flow through an optical device.

C. Others: Continuous Phased Systems (Synchronous growth)

The chemostat technique has also been used in efforts to obtain synchronous growth. In a phased growth system the goal is to obtain cells that are in the same stage of cell development, and ideally the cells divide at the same time. This growth technique enables the study of enzymes or metabolites that are present at a certain stage in the development of the cell.

Synchronous growth was achieved by flushes of nitrogen or ethylene gas at regular intervals to soybean suspension cultures growing as chemostat cultures (Constabel *et al.*, 1974, 1977). The application of nitrogen flushes did not adversely affect the cultures while the mitotic index decreased dramatically and the cells divided at a rate of one division per 24 hr. Increasing the intervals of N_2 treatment to 30 hr increased the division cycle to 30 hr. The synchrony of the cells was reported to be even better when the cells were treated with 3% ethylene for 3 hr followed by 3% CO_2 for 3 hr and 30 hr of areation before the next treatment with ethylene (Constabel *et al.*, 1974, 1977). This technique has not yet been fully utilized in the efforts to understand the various aspects of the mitotic cycle.

IV. FERMENTOR DESIGN

Due to limitations imposed by the nature of growth of plant cells in suspension, conventional vessels for microbial propagation are, in many cases, not applicable for mass propagation or continuous culture studies of plant cells. The attempts to cultivate plant cells on a large scale have resulted in the design of equipment that is also suitable for the continuous culture of plant cells (Wagner and Vogelmann, 1977; Wilson, 1978; Fowler, 1982; Ulbrich *et al.*, 1985). Certain cells like *Nicotiana tabacum* (Kato *et al.*, 1976) and *Mentha piperita* (Tal *et al.*, 1983) were able to grow in conventional fermentors. This was possible probably because of the higher resistance of these cells toward shear stress. Many other cells require, in order to grow, vessel designs where the shear stress is lower and settling out does not occur while at the same time the oxygen demand of the cells is satisfied. *Dioscorea deltoidea* cells could not grow in a conventional fermentor during the logarithmic growth phase, whereas after the logarithmic growth phase the cells were not destroyed by the impeller action in a conventional fermentor (Tal *et al.*, 1983).

Plant cells require less oxygen per volume and unit time due to their intrinsically lower growth rate as compared to microorganisms. A Q_{O_2} of 7.8 μl $_{O_2}$ mg dry weight^{-1} hr^{-1} was found for *Acer pseudoplatanus* cells (Givan and Collin, 1967) as compared to over 100 μl O_2 mg dry weight^{-1} hr^{-1} for microbial cells in the early exponential growth phase. Air bubble-generated turbulence with the help of low speed mixing (Wilson *et al.*, 1971) or air bubbles alone for mixing (Kurz, 1973; Wilson, 1976) was shown to be adequate to supply the plant cells with the oxygen required.

One vessel configuration for continuous culture was described by Wilson and co-workers (Wilson *et al.*, 1971) (Fig. 2). By the addition of an

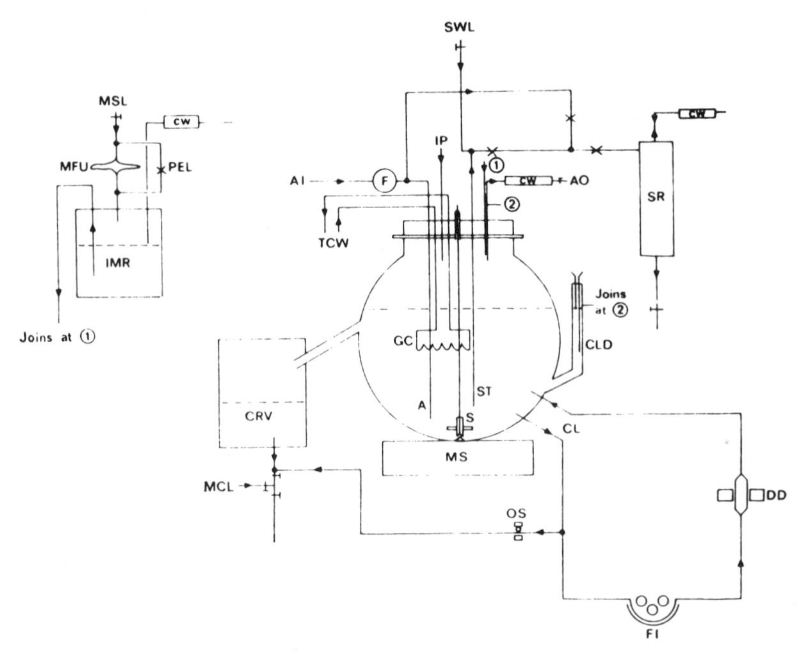

Fig. 2. Flow diagram for open continuous culture system (chemostat). Arrows indicate directions of flow. X indicates clip on silicone rubber tubing line. A, Aerator; AI, air inlet; AO, air outlet; CL, circulation line; CLD, constant level device; CRV, culture receiving vessel; CW, nonabsorbent cotton wool filter; DD, density detector; F, miniature airline filter; FI, flow inducer; GC, glass coil; IMR, intermediate medium reservoir; IP, inoculation port; MCL, mercuric chloride solution line; MFU, medium filter unit; MS, magnetic stirrer motor; MSL, medium filter unit; MS, magnetic stirrer motor; MSL, medium supply line; OS, outlet solenoid valve; PEL, pressure equalizing line; S, stirrer; SR, sample receiver; ST, sample tube; SWL, sterile water line; TCW, temperature controlling water supply. (From Wilson *et al.*, 1971, with permission of Oxford University Press.)

electronic module for monitoring cell density and a medium input solenoid valve, this system will also perform as a turbidostat.

Kurz (1973) described a system where a pulsed air stream produces large air bubbles entering at the bottom of a column (Fig. 3). The pulses of compressed air provide both agitaion and aeration and help to break up cell aggregates. The setup is fitted with a dual pump for the addition and withdrawal of medium from the vessel to obtain a constant flow. With this design cultures containing aggregates of no more than four cells of *Glycine max* and *Triticum monococcum* were obtained (Kurz, 1973).

Another type of bubble column fermentor was described by Wilson (1976) where the mixing also was by aeration (Fig. 4). The vessel is an inverted cone, allowing for aeration and mixing of the whole culture

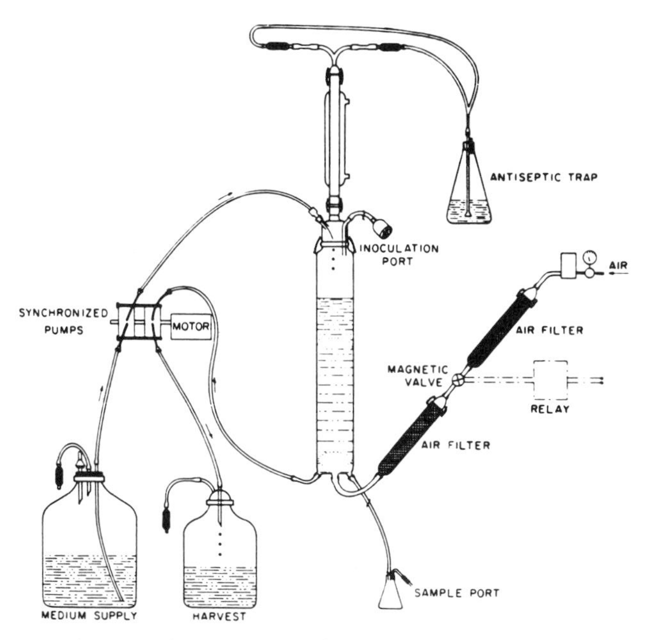

Fig. 3. Schematic drawing of a continuous culture fermentor for plant cells as proposed by Kurz. (From Kurz, 1973, with permission of Academic Press.)

fluid. The culture volume is kept constant by a siphon overflow. This type of fermentor is relatively easy to assemble and has been used by various groups (Tal and Goldberg, 1983; Tal *et al.*, 1983; Sahai and Shuler, 1982, 1984; Wilson, 1976; Wilson and Marron, 1978; Wilson and Balagué, 1985). Air-lift vessels have been introduced for the growth of plant cells (Townsley *et al.*, 1982; Ulbrich *et al.*, 1985); however, there are so far no reports of continuous culture experiments using air-lift fermentors.

V. THEORY AND PRACTICE OF CONTINUOUS CULTURE PRINCIPLES

Examples from the literature show that in most cases the growth of plant cells using the continuous culture technique follows the kinetics as described by Monod (1950). The extensive research performed with *Acer pseudoplatanus* in chemostats (Street, 1976) showed conclusively that these cells grow according to the proposed theory of the continuous

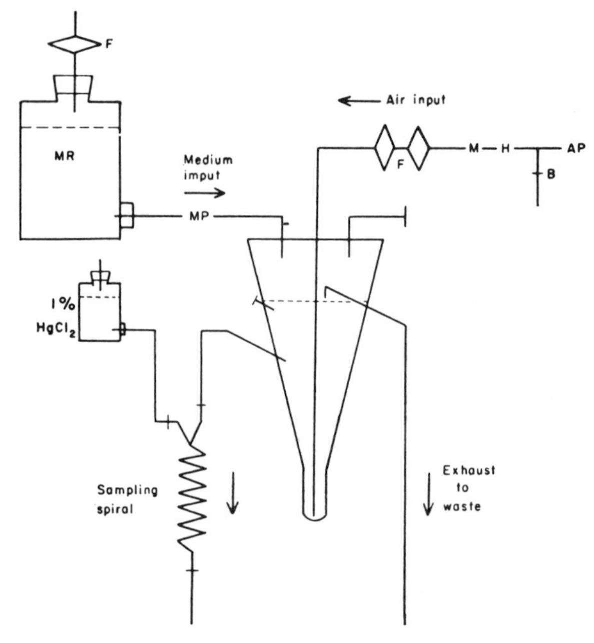

Fig. 4. Schematic drawing of a continuous culture fermentor for plant cells as proposed by Wilson. AP, Air pump; B, air bleed; F, air filter; H, humidifer; M, air flow meter; MP, medium pump; MR, medium reservoir. (From Wilson, 1976, with permission of Academic Press.)

culture. In more recent investigations other plant species were also shown to adhere to the continuous culture theory. Growth of *Dioscorea deltoidea* cells in a "Wilson" type fermentor followed Monod's kinetics when changes in the growth medium were made (Fig. 5). The growth responses of *Catharanthus roseus* and *Galium mollugo* to a step up and step down in input sucrose concentration followed a time course in accordance with Monod's kinetics (Balagué and Wilson, 1982; Wilson and Balagué, 1985).

To resolve the problems caused by the settling tendencies of plant cells, a "mixing factor" was introduced (Sahai and Shuler, 1982). The difference between the exit cell concentration (C_e) and the average cell concentration (C_{av}) were estimated for *Nicotiana tabacum* and *Glycine max* cells grown in chemostats of the "Wilson" type. The vessel design, placement of air inlet tube, outlet port, nozzle size, cell concentration, and aggregation were all factors that influenced the "mixing factor." A correlation between specific growth rate μ, C_e/C_{av}, and dilution rate D was derived for the case of incomplete mixing, so that the kinetics followed that described by Monod.

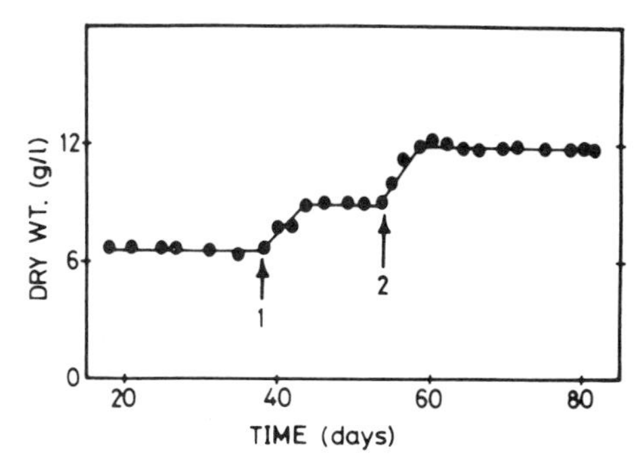

Fig. 5. Growth of *Dioscorea deltoidea* cells in a continuous culture. Cells were inoculated into a "Wilson" type fermentor (Fig. 4) containing 450 ml of MS medium (Murashige and Skoog, 1962) with 15 g/liter sucrose, half of the concentration of Mg, P, and N, and 0.1 ppm of 2,4-dichlorophenoxyacetic acid. The dilution rate was 0.24 day^{-1} with an aeration rate of 0.33 VVM. Arrow 1, S_o sucrose increased to 30 g/liter; arrow 2, S_o $MgSO_4$ and S_o KH_2PO_4 increased to 0.75× of the normal in MS medium. (From Tal *et al.*, 1983, with permission of Springer-Verlag.)

Wall growth, shear stress, and lysis are other obstacles to ideal continuous culture operation. Wall growth (biofilm formation) (Bryers and Characklis, 1982) influences the growth kinetics in the chemostat. The cells on the wall are continuously reinoculating the growth medium resulting in growth kinetics which do not adhere to the theory. When wall growth is present, plant cells in the continuous culture are able to grow at growth rates which are higher than the μ_{max} obtained in batch cultures for similar conditions. Wilson and Marron (1978) reported a doubling time of 25 hr for growth of *Galium mollugo* cells in a chemostat whereas the minimal doubling time in batch was 35 hr. On a small scale (lab fermentors) this effect is especially evident since the area-to-volume ratio is high. For reliable kinetic data wall growth should be eliminated, although in practice this might be impossible. A decrease in the Ca^{2+} concentration was shown to reduce wall growth of *Agrostemma githago* (Takayama *et al.*, 1977) and *Nicotiana tabacum* cells (Sahai and Shuler, 1984) to a large extent. Aggregation results in cell heterogeneity and introduces undesirable biochemical differentiation. Cell lines should be selected where aggregation is minimized. Prolonged subculture of suspensions as well as medium modifications may lead to more homogeneous cell suspensions.

In a study by Dougall and Weyrauch (1980), carrot cells were grown at semicontinuous conditions in shake flasks. The medium was added and withdrawn on a daily basis to avoid technical problems such as instability of pumping rate, maintenance of culture volume, and wall growth. With phosphate limitation the biomass steady states at various growth rates did not follow classical Monod kinetics, but could be described by a model proposed by Nyholm (1976). In this model, limitations by "conservative" substrates, not utilized immediately on entering the culture (like phosphate), will lead to a growth rate-dependent biomass yield. For glucose limitation (a nonconservative substrate) the yield did not change with dilution rate (Dougall and Weyrauch, 1980). Growth rate-dependent biomass yield was also obtained for phosphate-limited growth of *Ocimum basilicum* (Dalton, 1983).

VI. BIOMASS PRODUCTION

A continuous fermentation is ideal for biomass production for the following reasons:

1. It is possible to grow the cells under highly defined and constant growth conditions.
2. There is an efficient utilization of the medium, and it is possible to develop a well-balanced medium that is fully utilized.
3. The productivity (gram cell dry weight per liter fermentor day) of a continuous process is greater than a batch process and can save fermentor volume and capital investment. The reduction in turnaround time reduces operation costs.
4. The growth period can be extended much longer than in batch culture.
5. There is a greater potential to control and monitor steady-state growth than batch growth.
6. The cells obtained from steady-state growth are of constant composition.
7. Theoretically, if growth conditions are well balanced, the only products should be cells, CO_2, H_2O, and heat.

Several studies have been performed to look into the possibility of using plant cell suspensions as an alternative way to produce plant biomass (Byrne and Koch, 1962; Tulecke *et al.*, 1965; Tulecke, 1966). The use of plant cell suspensions as food was suggested (Byrne and Koch, 1962); however, due to high cost the feasibility for food production by plant cell tissue was suggested for only very specific conditions (e.g., space travel).

The use of *Nicotiana tabacum* suspension cultures as raw material for cigarettes was studied by Kato and co-workers (Kato *et al.*, 1976, 1980).

To obtain cells of suitable quality for cigarettes the final nitrogen content has to be low as compared to that normally obtained in *Nicotiana tabacum* suspension grown cells. A one-stage continuous culture of *Nicotiana tabacum* cells using a medium with a low nitrogen content resulted in large changes in the structure of the cells and a low biomass yield. A two-stage continuous culture system was developed using two conventional 60-liter fermentors equipped with two flat blade turbine impellers. In the first fermentor the cells were grown in a medium containing as nitrogen sources 1.71 g/liter KNO_3 and 1.485 g/liter NH_4NO_3, which is normal for plant tissue culture media. In the second stage the nitrogen concentration was reduced to one-fifth that in the medium of the first stage. Tobacco cells were obtained from the second fermentor at a rate of 6.6 g/liter day and with a nitrogen content fit for use in cigarettes (Kato *et al.*, 1980).

VII. METABOLITE PRODUCTION

Much of the recent interest in plant cell suspension cultures is due to their potential use for production of high-value metabolites found only in plants. These metabolites are presently extracted from whole plant material. In the whole plant these compounds are often present at low concentrations in special parts of the plant and together with compounds of similar structure. For a recent review on metabolites produced by plant cells, see Misawa (1985). Most of the studies for metabolite production have been done using batch culture systems (Rokem and Goldberg, 1985). In some cases metabolites are formed after there is no further increase in biomass, whereas in other cases synthesis of the metabolite occurs simultaneously with cell division and the maximum accumulation is achieved in the stationary phase (Rokem *et al.*, 1985).

In order to obtain a better understanding of the control of metabolite production, plant cells have been studied using the continuous culture technique, specifically observing metabolite production. However, the chemical and physical conditions for growth and product formation are usually different (Zenk *et al.* 1977). To obtain a large amount of metabolite a large amount of cells should be present. Therefore the one-stage continuous culture can be used to improve the growth conditions to obtain optimal amounts of biomass. *Dioscorea deltoidea* cells were grown on MS medium (Murashige and Skoog, 1962), and the effects of the concentrations and ratio of nitrogen sources were investigated, using the chemostat, for their effect on growth (Fig. 1). With this method it was possible to optimize growth with regard to the nitrogen sources. The method is valid for the development of an optimal medium for growth as regards all the nutrients, whereas the production of many metabolites occur after growth has ceased.

The cells growing at steady-state conditions can be used to inoculate either batch cultures (with other nutritional and growth conditions) or a successive continuous culture (Tal and Goldberg, 1982). In the plant cells studied the amount of metabolite formed in a one-stage continuous culture is usually much lower than that obtained in batch culture. For example, *Dioscorea deltoidea* produced 0.1%/dry weight of cells of diosgenin in the chemostat, whereas 1.8% diosgenin was obtained by growing the cells in a batch culture (Tal and Goldberg, 1982). Anthraquinone formation by *Galium mollugo* was 7–30 times lower than that obtained in batch cultures (Wilson and Marron, 1978). No alkaloids were formed by *Catharantus roseus* cells when the cells were grown in a chemostat with sucrose limitation, whereas when grown under phosphate limitation alkaloids were synthesised (Balagué and Wilson, 1982).

For studies on metabolite production, a two-stage continuous culture can give valuable information on the parameters which are of importance for metabolite formation. Growth and production phases can be separated, and the first stage can be optimized for cell growth and the second for product formation. The amount of ethanol-soluble phenolics formed by *Nicotiana tabacum* was studied using a two-stage continuous culture working on the chemostat principle (Sahai and Shuler, 1984). Cells in the second stage were morphologically distinct from those obtained in the first stage and produced much higher levels of phenolics per cell dry weight than either in the first stage or in a single-stage unit (Fig. 6). In the first stage 10.7 μg/ml of total phenolics was obtained as compared to 19.6 μg/ml in the second stage (in batch cultures 17.6 μg/ml was obtained). In a single-stage unit working at a dilution rate similar to the global dilution rate of the two-stage system, 13.3 μg/ml of total soluble phenolics was obtained. The productivity was also higher for the two-stage process (0.076 μg/ml hr) as compared to the one-stage process (0.053 μg/ml hr). These results indicate that the ability of tobacco cells to synthesize phenolics is related to the physiological state of the cells and not directly to their growth rate.

Investigating diosgenin formation by *Dioscorea deltoidea* using a two-stage continuous culture system, Tal and co-workers determined the crucial role of phosphate concentration during growth on product formation (Tal *et al.*, 1983). The phosphate concentration during the growth phase had a large influence on the later production of diosgenin, where-

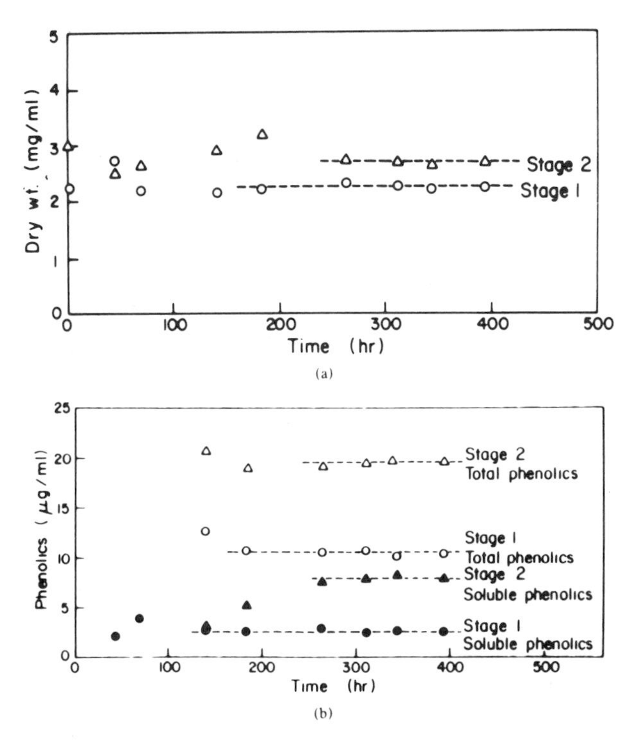

Fig. 6. (a) Growth (measured as dry weight) and (b) phenolics production of *Nicotiana tabacum* in a two-stage continuous culture maintained at 26°C. Dilution rate of first stage was 0.0083 hr^{-1} and of second stage, 0.0075 hr^{-1}. No additional nutrients were added to the second stage. (From Sahai and Shuler, *Biotech. Bioeng.*, copyright © 1984, reprinted by permission of John Wiley & Sons, Inc.).

as if the phosphate was added in the production phase there was no increase in diosgenin concentration. The absence of sugar in the spent medium leaving the first fermentor led to the search for an intermediate already present during growth, and an immediate steroidal precursor to diosgenin was isolated and characterized (Tal *et al.*, 1984). With this system, logarithmically growing cells could be harvested continuously from the first stage, and independently the diosgenin precursor could be converted in one enzymatic step to the final product.

The above results clearly show that a two-stage system has great potential for study of enhancement of production by (1) precursor feeding to the second stage (where it is not toxic to growth), (2) slow feeding of essential nutrients directly to the second stage, and (3) the ability to find the conditions (which often are different from growth) beneficial for product formation (see also Chapter 13, this volume). It is noteworthy that the only process for plant metabolites commercialized (production of shikonin by *Lithospermum erythrorhizon*, Curtin, 1983) utilizes a two-stage (batch) process. The use of the continuous culture technique for actual production of plant metabolites could also be advantageous. The reactor volumes could be decreased, the growth and product stages could be optimized separately, and control of the process would be easier. However, there are also potential problems involved in the continuous long-term growth of plant cells. The main problem would probably be the production stability of the cell line in use. Some plant cells have been shown to require continual selection to keep production ability, whereas other plants were more stable (Zenk, 1978). Even though production of metabolites with continuous culture might not be realized, the continuous culture method is an invaluable research tool for determination of the conditions for use in production by batch culture techniques.

VIII. FUTURE PROSPECTS

The continuous culture technique for growth and study of plant cells has been used by a comparatively small number of research groups. The potential of this technique has not yet been fully explored. The assumption is that plant cells behave similarly to microbial cells, but comprehensive data from different plant species are still needed. Studies of the influence of various substrates, and especially that of phytohormones, as limiting nutrients could give us a better understanding of their role in the life of a plant cell. Quantitative characteristics of growth are still lacking, for example, data on the maintenance energy and the ATP yield of plant cells.

The continuous culture technique should be used not only for basic studies but also for finding solutions of more applied problems. Wall growth and aggregate formation might be utilized to find conditions where even a one-stage continuous step may lead to stable production of metabolites. Some metabolites are excreted into the medium under certain growth conditions (alkaloids from *Catharanthus roseus*, Pétiard, 1980; terpenoids from *Thuja occidentalis*, Witte *et al.*, 1983). It should be possible to devise methods for continuous extraction of metabolites while keeping the cells growing under continuous conditions, as proposed for *Thuja occidentalis* terpenoids (Witte *et al.*, 1983).

ACKNOWLEDGMENT

My thanks to Dr. M. Platt, Dr. M. Matilsky, and Dr. I. Goldberg for their comments on the manuscript.

REFERENCES

Balagué, C., and Wilson, G. (1982). Growth and alkaloid biosynthesis by cell suspensions of *Catharanthus roseus* in a chemostat under sucrose and phosphate limiting conditions. *Physiol. Veg.* **20,** 515–522.

Bertola, M. A., and Klis, F. M. (1978). *In* "4th Intl. Congr. of Plant Tissue Culture," Abstr., p. 73. Univ. of Calgary, Calgary, Canada.

Bligny, R. (1977). Growth of suspension cultured *Acer pseudoplatanus* cells in automatic culture units of large volume. *Plant Physiol.* **59,** 502–505.

Bryers, J. D., and Characklis, W. G. (1982). Processes governing primary biofilm formation. *Biotechnol. Bioeng.* **24,** 2451–2476.

Byrne, A. F., and Koch, R. B. (1962). Food production from submerged culture of plant tissue cells. *Science* **135,** 215–216.

Constabel, F., Kurz, W. G. W., Chatson, K. B., and Gamborg, O. L. (1974). Induction of partial synchrony in soybean suspension cultures. *Exp. Cell Res.* **85,** 105–112.

Constabel, F., Kurz, W. G. W., Chatson, K. B., and Kirkpatrick, J. W. (1977). Partial synchrony in soybean cell suspension cultures induced by ethylene. *Exp. Cell Res.* **105,** 263–268.

Curtin, M. E. (1983). Harvesting profitable products from plant tissue culture. *Bio/Technology* **1,** 649–657.

Dalton, C. C. (1980). Photo-autotrophy of spinach cells in continuous cultures: Photosynthetic development and sustained photo-autotrophic growth. *J. Exp. Bot.* **31,** 791–784.

Dalton, C. C. (1983). Photosynthetic development of *Ocimum basilicum* cells on transition from phosphate to fructose limitation. *Physiol. Plant.* **59,** 623–626.

Dougall, D. K., and Weyrauch, K. W. (1980). Growth and anthocyanin production by carrot suspension cultures grown under chemostat conditions with phosphate as the limiting nutrient. *Biotechnol. Bioeng.* **22,** 337–352.

Fowler, M. W. (1982). The large scale cultivation of plant cells. *Prog. Ind. Microbiol.* **16,** 207–229.

Fowler, M. W., and Clifton, A. (1975). Rhytmic oscillations in carbohydrate metabolism during growth of Sycamore (*Acer pseudoplatanus* L.) cells in continuous (chemostat) culture. *Biochem. Soc. Trans.* **3,** 395–398.

Givan, C. V., and Collin, H. A. (1967). Studies on the growth in culture of plant cells. II. Changes in respiration rate and nitrogen content associated with the growth of *Acer pseudoplatanus* L. cells in suspension cultures. *J. Exp. Bot.* **18,** 321–331.

Herbert, D., Ellsworth, R., and Telling, R. C. (1956). The continuous culture of bacteria: A theorethical and experimental study. *J. Gen. Microbiol.* **14,** 601–622.

Husemann, W. (1983). Continuous culture and growth of photo-autotrophic cell suspensions from *Chenopodium rubrum*. *Plant Cell Rep.* **2,** 59–62.

Kato, A., Kawazoe, S., Iijima, M., and Shimizu, Y. (1976). Continuous culture of tobacco cells. *J. Ferment. Technol.* **54,** 82–87.

Kato, A., Asakura, A., Tsuji, K., Ikeda, F., and Iijima, M. (1980). Biomass production of nitogen-reduced tobacco cells in two-stage continuous culture. *J. Ferment. Technol.* **58,** 375–382.

King, P. J. (1977). Studies on the growth in culture of plant cells. XXII. Growth limitation by nitrate and glucose in chemostat cultures of *Acer pseudoplatanus*. *J. Exp. Bot.* **28,** 142–155.

King, P. J., and Street, H. E. (1977). Growth patterns in cell cultures. *In* "Plant Tissue and Cell Culture" (H. E. Street, ed.), pp. 269–337. Blackwell, Oxford.

King, P. J., Mansfield, K. J., and Street, H. E. (1973). Control of growth and cell division in plant cell suspension cultures. *Can. J. Bot.* **51,** 1807–1823.

Kubek, D. J., and Shuler, M. L. (1978). On the generality of methods to obtain single-cell plant suspension cultures. *Can. J. Bot.* **56,** 2521–2527.

Kurz, W. G. W. (1973). A chemostat for single cell cultures of higher plants. *In* "Tissue Culture: Methods and Applications" (P. F. Kruse, Jr. and M. K. Patterson, Jr., eds.), pp. 359–363. Academic Press, New York.

Kurz, W. G. W., and Constabel, F. (1981). Continuous culture of plant cells. *In* "Continuous Cultures of Cells (P. H. Calcott, ed.), Vol. 2, pp. 141–157. CRC Press, Boca Raton, Florida.

Misawa, M. (1985). Production of useful plant metabolites. *Adv. Biochem. Eng.* **31,** 59–88.

Monod, J. (1950). La technique de culture continuée. Théorie et application. *Ann. Inst. Pasteur, Paris* **79,** 390–410.

Murashige, T., and Skoog, F. (1962). A revised medium for rapid growth and bioassays with tobacco tissue cultures. *Physiol. Plant.* **15,** 473–497.

Novick, A., and Szilard, L. (1950). Description of the chemostat. *Science* **112,** 715–716.

Nyholm, N. (1976). A mathematical model for microbial growth under limitation by conservative substrates. *Biotechnol. Bioeng.* **18,** 1043–1056.

Peel, E. (1982). Photo-autotrophic growth of suspension cultures of *Asparagus officinalis* L. cells in turbidostats. *Plant Sci. Lett.* **24,** 147–155.

Pétiard, V. (1980). Mise en évidence d'alcaloïdes dans le milieu nutritif de cultures de tissues de *Catharanthus roseus* G. Don. *Physiol. Veg.* **18,** 331–337.

Rokem, J. S., and Goldberg, I. (1985). Secondary metabolites from plant cell suspension cultures: Methods for yield improvement. *Adv. Biotechnol. Processes* **4,** 241–274.

Rokem, J. S., Tal, B., and Goldberg, I. (1985). Methods for increasing diosgenin production by *Dioscorea* cells in suspension cultures. *J. Nat. Prod.* **48,** 210–222.

Sahai, O. P., and Shuler, M. L. (1982). On the nonideality of chemostat operation using plant cell suspension cultures. *Can. J. Bot.* **60,** 692–700.

Sahai, O. P., and Shuler, M. L. (1984). Multistage continuous culture to examine secondary metabolite formation in plant cells: Phenolics from *Nicotiana tabacum*. *Biotechnol. Bioeng.* **26,** 27–36.

Street, H. E. (1976). Applications of cell suspension cultures. In "Applied and Fundamental Aspects of Plant Cell, Tissue, and Organ Culture" (J. Reinert and Y. P. S. Bajaj, eds.), pp. 649–667. Springer-Verlag, Berlin and New York.

Street, H. E. (1977). Cell suspension culture techniques. *In* "Plant Tissue and Cell Culture" (H. E. Street, ed.), pp. 55–99. Blackwell, Oxford.

Street, H. E., Gould, A. R., and King, P. J. (1975). Nitrogen assimilation and protein synthesis in plant cell cultures. *Proc. 50th Annu. Meet. Soc. Exp. Biol.*

Takayama, S., Misawa, M., Ko, K., and Misato, T. (1977). Effect of cultural conditions on the growth of *Agrostemma githago* cells in suspension culture and the concomitant production of anti-plant virus substance. *Physiol. Plant.* **41,** 313–320.

Tal, B., and Goldberg, I. (1982). Growth and diosgenin production by *Dioscorea deltoidea* cells in batch and continuous culture. *Planta Med.* **44,** 107–110.

Tal, B., Rokem, J. S., and Goldberg, I. (1983). Factors affecting growth and product formation in plant cells grown in continuous culture. *Plant Cell Rep.* **2,** 219–222.

Tal, B., Tamir, I., Rokem, J. S., and Goldberg, I. (1984). Isolation and characterization of an intermediate steroid metabolite in diosgenin biosynthesis in suspension cultures of *Dioscorea deltoidea* cells. *Biochem. J.* **219,** 619–624.

Townsley, P. M., Webster, F., Kutney, J. P., Salisbury, P., Hewitt, G., Kawamura, N., Choi, L., and Jacoli. G. G. (1983). The recycling air lift fermentor for plant cells. *Biotechnol. Lett.* **5,** 13–18.

Tulecke, W. (1966). Continuous cultures of higher plant cells in liquid media: The advantages and potential use of a phytostat. *Ann. N.Y. Acad. Sci.* **139,** 162–175.

Tulecke, W., Taggart, R., and Colavito, L. (1965). Continuous culture of higher plant cells in liquid media. *Contrib. Boyce Thompson Inst.* **23,** 33–46.

Ulbrich, B., Wlesner, W., and Arens, H. (1985). Large scale production of rosmarinic acid from plant cell cultures of *Coleus blumei* Benth. *In* "Primary and Secondary Metabolism of Plant Cell Cultures" (K. H. Neumann, W. Barz, and E. Reinhard, eds.), pp. 293–303. Springer-Verlag, Berlin and New York.

Wagner, F., and Vogelmann, H. (1977). Cultivation of plant tissue cultures in bioreactors and formation of secondary metabolites. *In* "Plant Tissue Culture and its Biotechnological Application" (W. Barz, E. Reinhard, M. H. Zenk, eds.), pp. 245–252. Springer-Verlag, Berlin and New York.

Wilson, G. (1976). A simple and inexpensive design of chemostat enabling steady state growth of *Acer pseudoplatanus* L. cells under phosphate limited conditions. *Ann. Bot. (London)* [N.Y.] **40,** 919–932.

Wilson, G. (1978). Growth and product formation in large scale and continuous culture systems. *In* "Frontiers of Plant Tissue Culture 1978" (T. A. Thorpe, ed.), pp. 169–177. Univ. of Calgary, Calgary, Canada.

Wilson, G. (1980). Continuous culture of plant cells using the chemostat principle. *Adv. Biochem. Eng.* **16,** 1–25.

Wilson, G., and Balagué, C. (1985). Biosynthesis of anthraquinone by cells of *Galium mollugo* L. grown in a chemostat with limiting sucrose or phosphate. *J. Exp. Bot.* **36,** 485–493.

Wilson, G., and Marron, P. (1978). Growth and anthraquinone biosynthesis by *Galium mollugo* L. cells in batch and chemostat culture. *J. Exp. Bot.* **29,** 837–851.

Wilson, S. B., King, P. J., and Street, H. E. (1971). Studies on the growth in culture of plant cells. XII. A versatile system for the large scale batch or continuous culture of plant cell suspensions. *J. Exp. Bot.* **22,** 177–207.

Witte, L., Berlin, J., Wray, V., Schubert, W., Kohl, W., Hofle, G., and Hammer, J. (1983). Monoterpenes and diterpenes from cell cultures of *Thuja occidentalis. Planta Med.* **49,** 216–221.

Young, M. (1973). Studies on the growth in culture of plant cells. XVI. Nitrogen assimilation during nitrogen-limited growth of *Acer pseudoplatanus* L. cells in chemostat culture. *J. Exp. Bot.* **24,** 1172–1185.

Zenk, M. H. (1978). The impact of plant cell culture on industry. *In* "Frontiers of Plant Tissue Culture 1978" (T. A. Thorpe, ed.), pp. 1–14. Univ. of Calgary, Calgary, Canada.

Zenk, M. H., El-Shagi, H., and Schulte, U. (1975). Anthraquinone production by cell suspension cultures of *Morinda citrifolia*. *Planta Med., Suppl.*, pp. 79–101.

Zenk, M. H., El-Shagi, H., Arens, H., Stöckigt, J., Weiler, E. W., and Deus, B. (1977). Formation of the indole alkaloids serpentine and ajmalicine in cell suspension cultures of *Catharanthus roseus*. *In* "Plant Tissue Culture and Its Biotechnological Application" (W. Barz, E. Reinhard. and M. H. Zenk, eds.), pp. 27–43. Springer-Verlag, Berlin and New York.

CHAPTER **15**

Use of Immunoassays in the Detection of Plant Cell Products

Heather A. Kemp
Michael R. A. Morgan

AFRC Institute of Food Research
Norwich Laboratory
Norwich NR4 7UA, England

I. INTRODUCTION

Immunoassays are based on the interaction of an antibody with its antigen. The interaction is an extremely specific one, with closely related materials being unable to bind to a particular antibody, and is usually of high affinity. Immunoassay methods are, therefore, able to be specific and sensitive. Sample preparation techniques are often minimal, since antibodies can pick out the appropriate antigens in complex biological matrices. In addition, since immunoassays are usually technically simple and performed in a batch-wise manner, sample throughput is relatively high and rapid. Consequently these types of assay are often cheaper than more conventional methods of analysis. The unique combination of properties makes immunoassay ideally suited to analysis of plant products, especially for cell culture where sample numbers are high.

An immunoassay can be applied to any compound able to bind to an antibody. Antibodies can be produced against compounds covering a wide range of sizes from molecular weights as low as 100 to whole cells. However, in practice some types of material, such as lipid or polysaccharide, are less good at stimulating antibody production as others, such as protein. Compounds of molecular weights above 1,000–5,000 are able to stimulate antibody production directly, i.e., they are immunogenic. Smaller substances need to be bound chemically to larger ones (almost always a protein) before appropriate antibody production can be stimulated. Such small molecules are known as haptens, and it is with this group of low molecular weight, nonimmunogenic compounds that this chapter will be concerned.

II. METHODOLOGY OF RADIOIMMUNOASSAY

The first type of immunoassay developed was a radioimmunoassay (RIA) which was first used in 1959 by Yalow and Berson to measure the polypeptide hormone insulin in plasma. The potential of the technique was soon realized in medical research and quickly developed for the measurement of peptide and protein hormones. Later it was extended to include the measurement of haptens (Oliver *et al.*, 1968) which enabled steroid and drug immunoanalysis. The first use of an immunoassay in plant science came in 1969 (Fuchs and Fuchs, 1969) with the measurement of indoleacetic acid and gibberellic acid.

The principle of radioimmunoassay is very simple and is based on the competition of a constant amount of labeled antigen with unlabeled antigen (the unknown) for a limited number of specific antibody binding sites (Fig. 1). The more unknown that is present, the less labeled antigen is able to be antibody bound. Separation of the free and antibody-bound phases enables the calculation of the amount of unlabeled antigen present in the sample with reference to a standard curve. The strength of binding of the antibody to the antigen is very important. It increases the sensitivity and decreases reaction times.

Low molecular weight compounds (i.e., <1000–5000) such as phytochemicals are not immunogenic themselves and so have to be coupled to a protein of high molecular weight before immunization. In this manner antibodies will be produced to the low molecular weight hapten as well as to the protein. The way in which the hapten is covalently linked

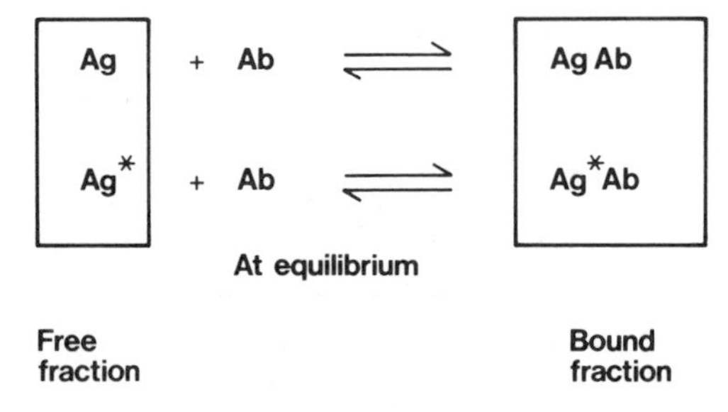

Fig. 1. The principle of immunoassay. Ag, Antigen; Ab, antibody; Ag*, labeled antigen.

to the protein determines the specificity of the antibody. The reaction usually involves linking a carboxylic acid group to an amino group on the protein either by the carbodiimide reaction (Goodfriend *et al.*, 1964) or by the mixed anhydride procedure (Erlanger *et al.*, 1957). If there is no carboxylic acid group available on the hapten then one can be produced by modification of a suitable functional group, usually a hydroxyl or a keto group. The antibody will recognize most strongly the part of the hapten which is distal to the site of conjugation to the protein, and will show less specificity for compounds that differ from the immunizing hapten only at or near to the conjugation site. This is a crucially important factor to consider when producing protein–hapten conjugates. Other ways of coupling haptens are available such as the periodate method which can be used for carbohydrate residues (Butler and Chen, 1967) and the heterobifunctional reagent *m*-maleimidobenzoic acid *N*-hydroxysuccinimide (Sohda *et al.*, 1985). This reagent will couple amino groups to free thiol residues. For a review of conjugation methods for haptens, see Erlanger (1980).

Production of suitable antisera can sometimes be a lengthy procedure, especially for haptens. However the procedure is fairly simple involving the immunization of animals, usually rabbits, with the hapten– protein conjugate mixed in Freund's adjuvant which prolongs and enhances the immune response. The route of administration can be subcutaneous, intramuscular, intradermal and even intraperitoneal, intravenous, or into the lymph nodes. Procedures for immunization using as little as 100–500 μg of conjugate per animal—as well as larger amounts—have been described. After a booster injection, not less than 4 weeks after the primary injection, the animal is bled and the serum checked for the required antibody content, avidity, and specificity. It has frequently been noted that if the animal is left for a much longer period, i.e., months or even years, between the primary injection and the booster then the antiserum obtained after the booster is of a very high titer and affinity (Morris, 1985). If a good antiserum is not obtained immediately, however, then the animals can be boosted further as this usually produces an increase in antibody levels.

The antibodies produced are polyclonal—the antiserum consisting of a large number of different antibodies against the hapten, each the product of different cell lines, and each with different properties relative to the hapten. Recently it has been possible to produce monoclonal antibodies, identical antibodies as the product of one cell line or clone (Kohler and Milstein, 1975). Two of the advantages of monoclonal antibodies are that they are available in unlimited amounts and that their method of production facilitates obtaining an antibody of desired specificity and/or avidity. Monoclonal antibodies also confer advantages on assay methodology development. They are, however, more expensive and time consuming to produce than polyclonal sera. For hapten assays, the impact of monoclonal antibody technology has been less dramatic than on protein immunoassays.

A number of different radioisotopes have been used to label antigens in radioimmunoassay (Skelley *et al.*, 1973). The iodine radionuclides ^{133}I and ^{125}I were the first to be used, and they have the advantage that they can be counted directly and quickly in a γ counter. Also the immunoassays using these labels are more sensitive than the alternatives using ^{3}H or ^{14}C, and the labeled compounds can be synthesized in most laboratories whereas most ^{3}H- and ^{14}C-labeled antigens cannot. ^{3}H and ^{14}C radioisotopes were introduced for compounds which could not be iodinated, but as they are both β emitters they must be mixed with scintillation cocktails before their radioactivity can be determined; counting is slow and costly and can be susceptible to quenching effects.

After incubation of the antibody with labeled and unlabeled antigen it is necessary to separate the antibody-bound and free phases before the radioactivity in one of these phases is determined. It is important for obvious reasons that the separation system used does not disturb the equilibrium of the reaction. This can sometimes occur with adsorptive techniques, especially if the antibody avidity is low or if the procedure is not performed as rapidly as possible. There are several types of methods commonly used:

Chemical precipitation techniques, such as ammonium sulfate or polyethylene glycol precipitation of bound phase which must be removed by centrifugation.

Adsorptive methods, the most common being adsorption of the free phase with dextran-coated charcoal and centrifugation.

Immunological methods, such as the use of a species-specific second antibody (an antibody raised in one animal able to bind to the antibodies of another) or protein A (a material able to bind to the antibodies of several animal species). The precipitated antibody-bound phases are removed by centrifugation.

Solid phase techniques which utilize either immobilized antigen or antibody to absorb the complementary phase. The techniques can be particulate in nature (e.g., Sepharose beads, magnetic particles, or cellulose particles) or continuous (e.g., large surfaces such as the surfaces of disks, tubes, or microtitration plates). Some of these procedures have the obvious advantage of avoiding the need for time-consuming centrifugation steps.

The actual performance of a radioimmunoassay involves reacting sample or standard (antigen) and a fixed amount of labeled antigen with a limited amount of antibody in a tube for a set period of time (Fig. 2).

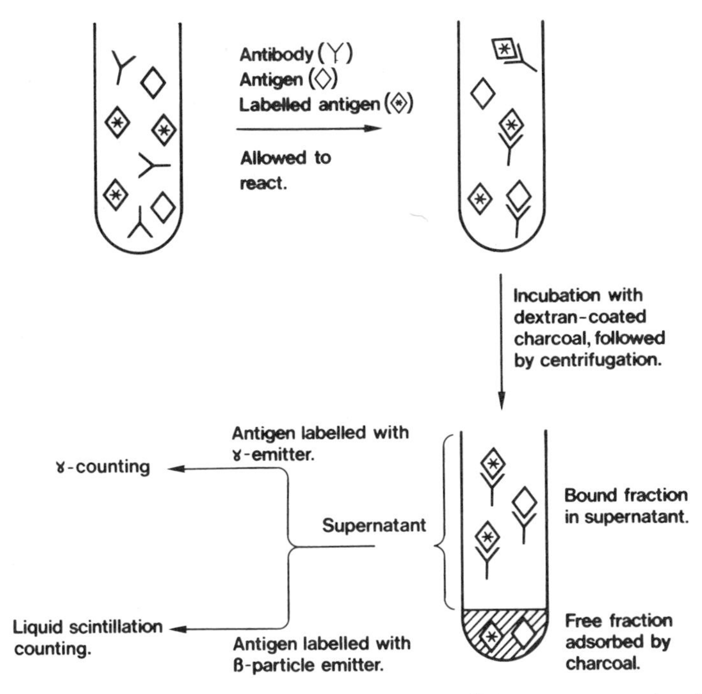

Fig. 2. Diagrammatic representation of the stages in a radioimmunoassay using dextran-coated charcoal to separate bound and free phases.

After equilibrium is reached, the phases are separated. If dextran-coated charcoal is used, it is added in suspension to the tube, which is mixed and allowed to stand for a short period of time before centrifugation. A longer time period is required prior to centrifugation when ammonium sulfate is used to separate the phases. The radioactivity present in either supernatant or pellet can be determined. If a β-emitter is used, the phase selected (normally the supernatant) is mixed with scintillation fluid and counted. If a γ source is present, the spun-down pellet is normally counted directly by placing the tube in a γ counter.

III. USE OF RADIOIMMUNOASSAY IN THE MEASUREMENT OF PHYTOCHEMICALS

Most of the immunoassays used in phytochemical detection have been radioimmunoassays employing tritium labels (Weiler, 1983, 1984). Assays have been developed for all of the growth hormones which, since they only occur in the microgram per kilogram range, are very difficult to quantify by conventional methods. Weiler (1979) developed a radioimmunoassay for free and conjugated abscisic acid using a tritiated label and $NH_4)_2SO_4$ precipation. The assay required a simple methanol extraction, with no further clean up, and had a limit of detection of 66 pg. It was used to monitor the distribution of abscisic acid in more than 100 species of plants. Gibberellic acid and indoleacetic acid have been measured using an immunoassay which did not employ a radioactive tracer, but where antigen inhibited the inactivation of modified bacteriophage by antiserum (Fuchs *et al.*, 1971).

A range of sensitive and specific immunoassays have been developed for a large number of structurally different secondary plant products. Compounds which have been measured in plant tissues or juice by RIA using a tritium label include vindoline (Westekemper *et al.*, 1980), solasidine (Weiler *et al.*, 1980), limonin (Weiler and Mansell, 1980), sennosides (Atzorn *et al.*, 1981), naringin (Jourdan *et al.*, 1983), digoxin and related compounds (Weiler and Zenk, 1976), and quinine and quinidine (Morgan *et al.*, 1985a). Digoxin was also measured with an ^{125}I label which gave increased sensitivity over the tritium label. Arens *et al.* (1978) used tritiated tracers in RIAs to determine the distribution of the ajmalicine and serpentine in plants. It was also used to select plants with high alkaloid levels for breeding and culture purposes. Potato glycoalkaloids have been assessed by RIA (Matthew *et al.*, 1983) in clinical studies but not in plant material.

A more rapid, semiquantitative type of radioimmunoassay was developed by Weiler and Zenk (1979) for mass screening of digoxin and nicotine in plant extracts. The assay was performed in multiwell test plates using an ^{125}I tracer. The bound and free phases were separated using charcoal and the plate centrifuged. Radioactivity in the charcoal was then determined by placing the plate on a film packet and exposing it overnight. Radioactivity was transformed into fluorescent light which was detected with a sheet of X-ray film. The intensity of the spot on the film was proportional to the radioactivity present. The authors called this an autoradiographic immunoassay (ARIA) and claimed that it could be used to process more than 10^4 samples per day.

RIAs can prove extremely useful for monitoring secondary plant products in cell culture where large numbers of samples must be screened quickly. There seems to have been little use of the technique for this purpose, however, although an alternative use of RIA in cell culture was reported by Tanahashi *et al.* (1984). They used an RIA to study the conversion of loganin to secologanin by monitoring loganin in plant cells and culture medium. The assay, which involved ^3H-labeled loganin and an ammonium sulfate precipitation, had a low cross-reaction with secologanin.

IV. METHODOLOGY OF ENZYME-LINKED IMMUNOSORBENT ASSAYS (ELISAs)

Although radioimmunoassay is a useful technique, it has major limitations when applied to plant cell products. Increasingly the use of radioactivity presents more problems associated with its safe handling and disposal. Radioactive tracers in this field are usually not commercially available and have to be synthesized. Such syntheses are easier in chemical terms for ^{125}I derivatives, but their lack of stability and comparatively higher potency presents further problems. Detection of ^{3}H and ^{14}C is time consuming and costly. An attractive alternative to radioisotopes is the use of nonisotopic, enzyme labels, first introduced by Van Weeman and Schuurs (1971) and by Engvall and Perlmann (1971). Enzyme immunoassays present none of the safety problems associated with radioactivity and utilize stable reagents able to be stored for long periods. One format of enzyme immunoassay, the ELISA, can be performed in the plastic wells of microtitration plates (Voller *et al.*, 1974), and the advantages of this procedure have made it widely utilized. Since one of the assay components is immobilized to the walls of the plate, separation of bound and free phases is rapid and simple. The assays are readily amenable to automation, if required, but can also be carried out with a minimum of equipment. In enzyme assays where end point detection is colorimetric, then this can be assessed by eye, or with computer-linked, automated plate readers.

The enzyme to be used in an ELISA must fulfill certain requirements. It must be reasonably inexpensive, available in a highly purified form, and suitable for linkage to proteins and haptens. To produce a large enough signal it must have a high turnover number and also make use of a substrate which produces a stable, soluble, easily measured product, preferably highly colored.

The two most common approaches used for enzyme immunoassay involve either (i) enzyme labeling of hapten and immobilization of antibody or (ii) immobilization of hapten and enzyme labeling of antibody. As will be explained later, the latter format has certain advantages.

Usually the enzyme label of haptens is made by producing a conjugate using any of the methods mentioned in Section II. If the enzyme is used to

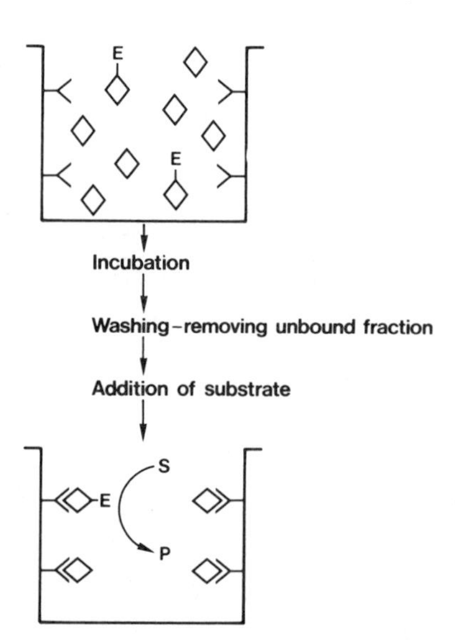

Fig. 3. The direct ELISA performed on a microtitration plate. Y, Immobilized antibody; ◇, antigen; ◇, enzyme-labeled antigen; S, substrate; P, product.

label the antibody, the coupling methods are often cross-linking techniques (O'Sullivan, 1984). By far the most popular method of cross-linking is the glutaraldehyde method (Avrameas *et al.*, 1978), but the bifunctional reagent *m*-maleimidobenzoic acid *N*-hydroxysuccinimide is useful for enzymes that possess sulfhydryl residues. Among the enzymes mostly used as labels are the following: horseradish peroxidase, alkaline phosphatase, β-galactosidase, and glucose-6-phosphate dehydrogenase.

In the ELISA procedure requiring enzyme labeling of hapten, usually called the direct ELISA, the antibody is adsorbed onto the plastic wells of the plate. This immobilization is a simple, noncovalent adsorption. The sample and a fixed amount of the enzyme–hapten conjugate are added to the well and compete for the antibody binding sites (Fig. 3). The amount of label bound to the well after washing is inversely proportional to the amount of antigen in the sample. However, this type of assay suffers from the disadvantages that it requires large amounts of antibody and that the enzyme is added in the presence of sample extract which may affect its activity. Furthermore, an enzyme–hapten conjugate has to be prepared for each hapten assayed, a difficult synthesis to optimize for maximum assay sensitivity while retaining maximum enzyme activity.

The second type of ELISA, the indirect method, overcomes these problems. In this case a readily synthesized protein–hapten conjugate is

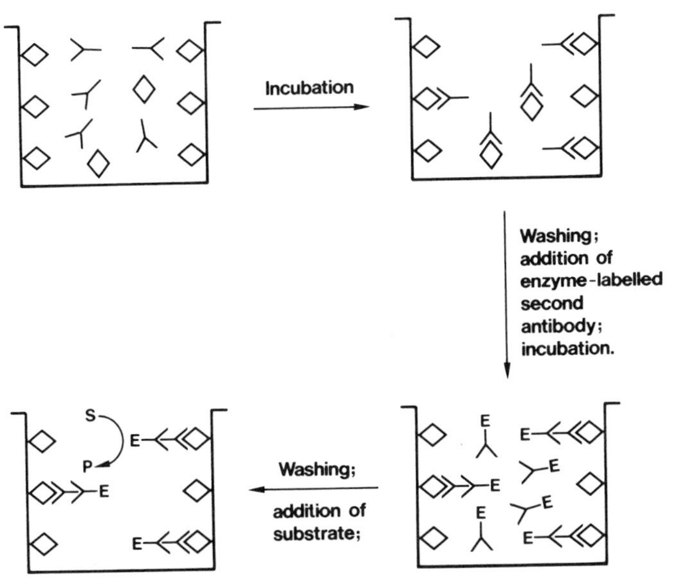

Fig. 4. The indirect ELISA performed on a microtitration plate, Y, Primary antibody; ◇, antigen; ◇, immobilized antigen; ≻–E, enzyme-labeled second antibody; S, substrate; P, product.

adsorbed to the wells of the plate. The antibody and the sample are added to the well, and the antibody becomes distributed between the solid-phase antigen and the antigen in solution according to the amount of sample present (Fig. 4). After washing, the amount of antibody bound to the well is quantified by adding a second antibody which recognizes IgG of the species used to raise the first antibody. In the indirect ELISA it is the second antibody that is conjugated to an enzyme. Many such preparations are commercially available. Again the amount of enzyme present is proportional to the amount of antigen in the sample.

V. USE OF ELISA IN THE MEASUREMENT OF PHYTOCHEMICALS

Several of the ELISAs of the direct type using antibody-coated plates and labeled haptens have been reported for plant growth hormones. Weiler *et al.* (1981) reported an ELISA for 3-indoleacetic acid in intact and decapitated coleoptiles with a detection limit of 3–4 pg. Abscisic acid and its glucosyl ester have been determined in plant suspension cultures and tissue culture using alkaline phosphatase–abscisic acid (Weiler, 1982), and two ELISAs have been developed for gibberellins (Atzorn and Weiler, 1983). In the gibberellin (GA) assays two antisera were raised, one against GA, and one against GA_4 and GA_7. The ELISAs were found to be 100–200 times more sensitive than radioimmunoassays using the same antisera, but less precise.

In our laboratory at the Institute of Food Research (Norwich), ELISAs have been developed for two groups of phytochemicals. These are the determination of total glycoalkaloids in potatoes and of secondary plant metabolites produced in cell culture. This latter group includes quassin from *Quassia amara* (Simaroubaceae) and quinine and quinidine from *Cinchona* (Rubiaceae).

For the glycoalkaloid assay, antiserum was raised to α-solanine–bovine serum albumin (Morgan *et al.*, 1983) which was prepared by the periodate cleavage method through the sugar residues (Fig. 5). As the only difference between α-solanine and α-chaconine is in the sugar residues, the antiserum recognized both equally well. Since in commercially available tubers 95% or more of the glycoalkaloid fraction consists of these two compounds, the assay was suitable for total glycoalkaloid determination without the need for sample hydrolysis. Elevated levels of these naturally occurring compounds have been associated with an unacceptable bitter taste and possibly with cases of poisoning in man and animals (Morgan and Coxon, 1987). The assay developed was of the

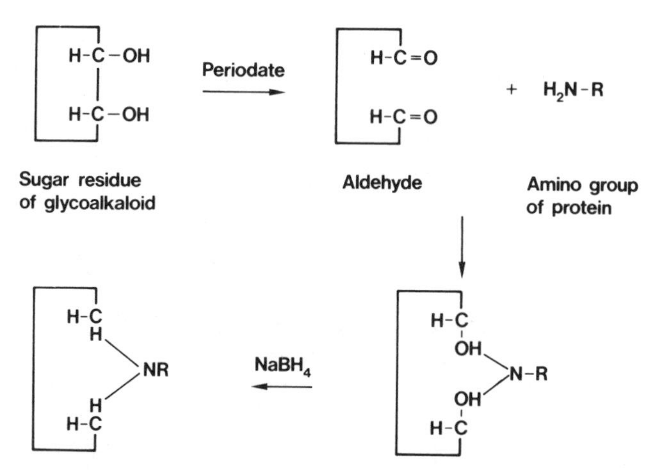

Fig. 5. Conjugation of α-solanine and α-chaconine (glycoalkaloids) to protein by the periodate cleavage method.

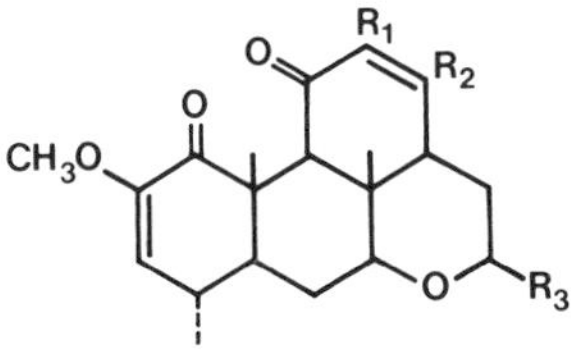

	R_1	R_2	R_3
Quassin	OCH_3	CH_3	=O
Neoquassin	OCH_3	CH_3	HOH
12-hydroxyquassin	OH	CH_3	=O
18-hydroxyquassin	OCH_3	CH_2OH	=O

Fig. 6. Structure of quassin and related compounds.

indirect type, and a solanine–keyhole limpet hemocyanin conjugate was used to coat the microtitration plates. Results using the ELISA compared very well with two conventional chemical assays (Morgan *et al.*, 1985b). The ELISA was much simpler to carry out, with an easy sample preparation and could assay large numbers of samples much more rapidly.

The same assay design was used for detection of quassin. In this case a nonspecific antiserum was required which would recognize the metabolites neoquassin, 18-hydroxyquassin, and 12-hydroxyquassin as well as quassin. They are natural bittering agents with insecticidal and therapeutic properties as well as being potential food additives. The differences between these metabolites are restricted to the C and D rings (Fig. 6), and so conjugates were prepared by opening the lactone ring and introducing a carboxylic acid group which was then coupled to protein (Robins *et al.*, 1985). It was hoped that the resulting antiserum would only recognize changes in the A and B rings and would therefore recognize the four metabolites. This was also assisted by the fact that the preparation of quassin used was impure and contained the three other metabolites of interest in different amounts. By using this preparation the two hydroxyl-substituted forms were conjugated with protein also. The antiserum produced did cross-react well with the metabolites that differed in the C and D rings, though as a result the antibody avidity for the compounds of interest was decreased and so the detection limit of the assay was relatively low (5 ng). A more specific antiserum was also produced using the hemisuccinate of 18-hydroxyquassin coupled to protein by the mixed anhydride procedure (Robins *et al.*, 1985), and this gave a detection limit of 5 pg of quassin when used in an indirect ELISA.

Considerable interest has been shown in the production of *Cinchona* alkaloids such as quinidine and quinine by plant cell culture as they are both economically important, quinine as an antimalarial drug and food bittering agent and quinidine as an antiarrhythmic agent and food bittering agent. Quinine and quinidine differ from each other only in their three-dimensional configuration; quinine has (8*S*,9*R*) stereochemistry and quinidine has (8*R*,9*S*). Conjugates were made of both compounds following a formation of 9-hemisuccinylquinine and 9-hemisuccinylquinidine (Morgan *et al.*. 1985a). The former was coupled to protein directly while the latter was conjugated using a bridge made by inserting 6-amino-*N*-hexanoic acid between hapten and protein. The antisera produced using these conjugates were very specific in that they did not recognize alkaloids of the oposite configuration; quinine did not cross-react with the anti-quinidine antisera and vice versa. ELISA cross-reactions with other compounds were also low.

Radioactive tracers of quinine and quinidine were produced using [^{3}H[acetic anhydride, and RIAs were developed. A comparison of the properties of the RIAs and ELISAs for quinine and quinidine set up using the same antisera preparations was extremely interesting (Morgan *et al.*, 1985a). For quinine, antiserum dilution was 1 in 10,000 for RIA but 1 in 100,000 for ELISA. For quinidine the dilutions were 1 in 5,000 and 1 in 50,000 for RIA and ELISA, respectively. The standard curves showed limits of detection of 50 pg and 1 ng for quinine and quinidine RIAs, and 10 pg and 100 pg, respectively, for the ELISAs. The 10-fold increase in sensitivity for ELISA over RIA is illustrated for quinidine in Fig. 7. Cross-reactions in the ELISAs were much lower than observed in the RIAs. For example, 10,11-dehydroquinine showed a cross-reaction of 35% with the RIA but only 2.7% in the ELISA. These figures were most likely to be due to the increased dilution of antibodies able to be used in the ELISA system, reducing the influence of less avid, less specific antibodies. The presence of inherent methodological causes may also contribute to these observations.

VI. CONCLUSIONS

Immunoassays are eminently suited to the rapid, specific, and sensitive analysis of large numbers of samples, such as those generated by cultures of plant cells. The setting up of an immunoassay requires considerable expertise in order to produce the key reagent of the technique, the antibody. It is not, therefore, well suited to all situations. However,

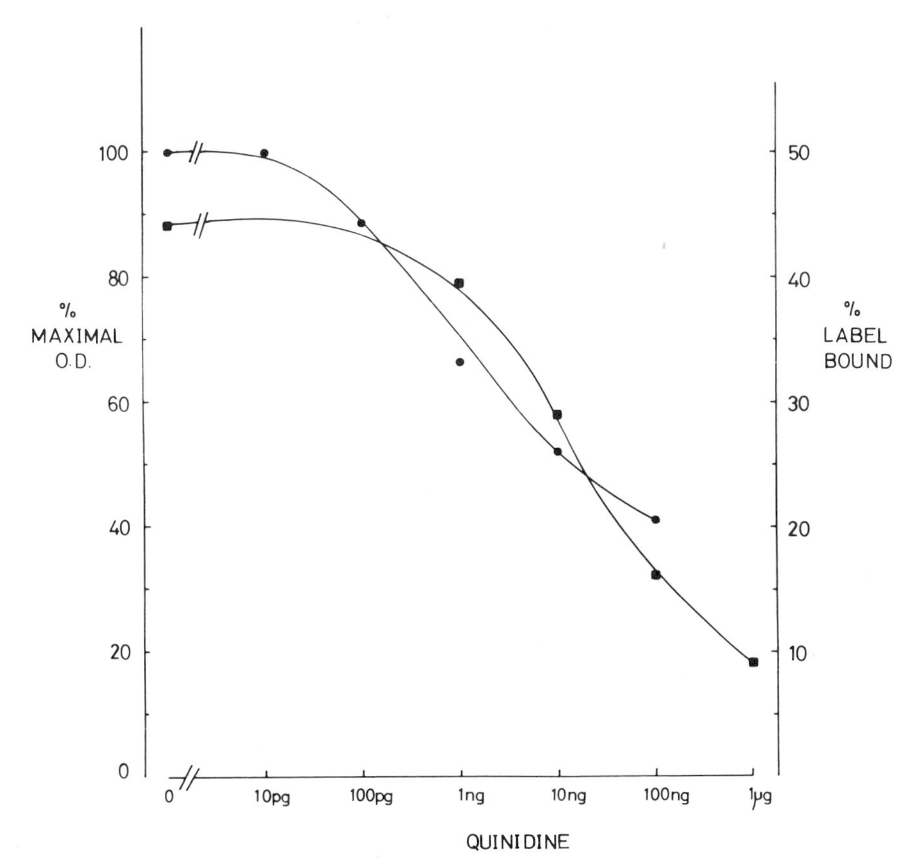

Fig. 7. Standard curves for quinidine obtained by ELISA (●) and by RIA (■). Curves show changes in optical density (for the ELISA) or changes in binding of radiolabel (for the RIA) against mass of quinidine per microtitration plate well (ELISA) or tube (RIA).

the methods for antibody production, particularly against haptenic compounds, allow considerable manipulation of antibody characteristics. Therefore it is sometimes possible to produce antibodies able to analyze groups of all compounds, as well as antibodies recognizing one structure only. Radioimmunoassay, the first of the many immunoassay formats to be described, has been applied to the determination of a number of plant cell products. However, it seems that the technique of enzyme immunoassay, and in particular the ELISA version performed on microtitration plates, has a number of advantages over RIA without the disadvantages of working with radiolabels. Many enzyme immunoassays for different plant cell products have been described, and this format seems likely to be the dominant one for the foreseeable future.

ACKNOWLEDGMENT

The authors wish to acknowledge Dr. R. Robins (AFRC Institute of Food Research, Norwich Laboratory) for helpful discussion.

REFERENCES

Arens, H., Stöckigt, J., Weiler, E. W., and Zenk, M. H. (1978). Radioimmunoassays for the determination of the indole alkaloids ajmalicine and serpentine in plants. *Planta Med.* **34,** 37–46.

Atzorn, R., and Weiler, E. W. (1983). The immunoassay of gibberellins. II. Quantitation of GA_3, GA_4 and GA_7 by ultra-sensitive solid-phase enzyme immunoassays. *Planta* **159,** 7–11.

Atzorn, R., Weiler, E. W., and Zenk, M. H. (1981). Formation and distribution of sennosides in *Cassia angustifolia,* as determined by a sensitive and specific radioimmunoassay. *Planta Med.* **41,** 1–14.

Avrameas, S., Ternynck, T., and Guesdon, J.-L. (1978). Coupling of enzymes to antibodies and antigens. *Scand. J. Immunol.* **8,** 7–23.

Butler, V. P., and Chen, J. P. (1967). Digoxin-specific antibodies. *Proc. Natl. Acad. Sci. U.S.A.* **57,** 71–78.

Engvall, E., and Perlmann, P. (1971). Enzyme-linked immunoadsorbent assay (ELISA). Quantitative assay of immunoglobulin G. *Immunochemistry* **8,** 871–874.

Erlanger, B. F. (1980). The preparation of antigenic hapten–carrier conjugates: A survey. *In* "Methods in Enzymology" (H. Van Vunakis and J. J. Langone, eds.), Vol. 70, Part A, pp. 85–103. Academic Press, New York.

Erlanger, B. F., Borek, F., Beiser, S. M., and Lieberman, S. (1957). Steroid-protein conjugates. I. Preparation and characterization of conjugates of bovine serum albumin with testosterone and with cortisone. *J. Biol. Chem.* **228,** 713–727.

Fuchs, S., and Fuchs, Y. (1969). Immunological assay for plant hormones using specific antibodies to indoleacetic acid and gibberellic acid. *Biochim. Biophys. Acta* **192,** 528–530.

Fuchs, S., Haimovich, J., and Fuchs, Y. (1971). Immunological studies of plant hormones. *Eur. J. Biochem.* **18,** 384–390.

Goodfriend, T. L., Levine, L., and Fasman, G. D. (1964). Antibodies to bradykinin and angiotensin: A use of carbodimides in immunology. *Science* **144,** 1344–1346.

Jourdan, P. S., Weiler, E. W., and Mansell, R. L. (1983). Radioimmunoassay for naringin and related flavanone 7-neohesperidosides using a tritiated tracer. *J. Agric. Food Chem.* **31,** 1249–1255.

Kohler, G., and Milstein, C. (1975). Continuous cultures of fixed cells producing antibodies of predefined specificity. *Nature (London)* **256,** 495–497.

Matthew, J. A., Morgan, M. R. A., McNerney, R., Chan, H. W.-S., and Coxon, D. T. (1983). Determination of solanidine in human plasma by radioimmunoassay. *Food Chem. Toxicol.* **21,** 637–640.

Morgan, M. R. A., and Coxon, D. T. (1987). Re-examining tolerances: Glycoalkaloids in potatoes. *In* "Natural Toxicants" (D. H. Watson, ed.), pp. 221–230. Ellis Horwood, Ltd., Chichester, England.

Morgan, M. R. A., McNerney. R., Matthew, J. A., Coxon, D. T., and Chan. H. W.-S. (1983). An enzyme-linked immunosorbent assay for total glycoalkaloids in potato tubers. *J. Sci. Food Agric.* **34,** 593–598.

Morgan, M. R. A., Bramham, S., Webb, A. J., Robins, R. J., and Rhodes, M. J. C. (1985a). Specific immunoassays for quinine and quinidine: Comparison of radioimmunoassay and enzyme-linked immunosorbent assay procedures. *Planta Med.* **51,** 237–241.

Morgan, M. R. A., Coxon, D. T., Bramham, S., Chan, H. W.-S., van Gelder, W. M. J., and Allison, M. J. (1985b). Determination of the glycoalkaloid content of potato tubers by three methods including enzyme-linked immunosorbent assay. *J. Sci. Food Agric.* **36,** 282–288.

Morris, B. A. (1985). Principles of immunoassay. *In* "Immunoassays in Food Analysis" (B. A. Morris and M. N. Clifford, eds.), pp. 21–52. Elsevier, Amsterdam.

Oliver, G. C., Parker, B. M., Brasfield. D. L., and Parker, C. W. (1968). The measurement of digitoxin in human serum by radioimmunoassay. *J. Clin. Invest.* **47,** 1035–1042.

O'Sullivan, M. J. (1984). Enzyme immunoassay. *In* "Practical Immunoassay. The State of the Art" (W. R. Butt, ed.), pp. 37–70. Dekker, New York.

Robins, R. J., Morgan, M. R. A., Rhodes, M. J. C., and Furze, J. M. (1985). Cross reactions in immunoassays for small molecules: Use of specific and non-specific antisera. *In* "Immunoassays in Food Analysis" (B. A. Morris and M. N. Clifford, eds.), pp. 197–211. Elsevier, Amsterdam.

Skelley, D. S., Brown, L. P., and Besch, P. K. (1973). Radioimmunoassay. *Clin Chem. (Winston-Salem, N.C.)* **19,** 146–186.

Sohda, M., Fujiwara, K., Saikusa, H., Kitagawa, T., Nakamura, N., Hara, K., and Tone, H. (1985). Sensitive enzyme immunoassay for the quantification of aclacinomycin A using β-D-galactosidase as a label. *Cancer Chemother. Pharmacol.* **14,** 53–58.

Tanahashi, T., Nagakura, N., Inouye, H., and Zenk, M. H. (1984). Radioimmunoassay for the determination of loganin and the biotransformation of loganin to secologanin by plant cell cultures. *Phytochemistry* **23,** 1917–1922.

Van Weeman, B. K., and Schuurs, A. (1971). Immunoassay using antigen–enzyme conjugates. *FEBS Lett.* **15,** 232–236.

Voller, A., Bidwell, D. E., Huldt, G., and Engvall, E. (1974). A microplate method of enzyme-linked immunosorbent assay and its application to malaria. *Bull. W. H. O.* **51,** 209–211.

Weiler, E. W. (1979). Radioimmunoassay for the determination of free and conjugated abscisic acid. *Planta* **144,** 255–263.

Weiler, E. W. (1982). An enzyme-immunoassay for *cis*-(+)-abscisic acid. *Physiol. Plant.* **54,** 510–514.

Weiler, E. W. (1983). Immunoassay of plant constituents. *Biochem. Soc. Trans.* **11,** 485–495.

Weiler, E. W. (1984). Immunoassay of plant growth regulators. *Annu. Rev. Plant Physiol.* **35,** 85–95.

Weiler, E. W., and Mansell, R. L. (1980). Radioimmunoassay of limonin using a tritiated tracer. *J. Agric. Food Chem.* **28,** 543–545.

Weiler, E. W., and Zenk, M. H. (1976). Radioimmunoassay for the determination of digoxin and related compounds in *Digitalis lanata. Phytochemistry* **15,** 1537–1545.

Weiler, E. W., and Zenk, M. H. (1979). Autoradiographic Immunoassay (ARIA): A rapid technique for the semiquantitative mass screening of haptens. *Anal. Biochem.* **92,** 147–155.

Weiler, E. W., Kruger, H., and Zenk, M. H. (1980). Radioimmunoassay for the determination of the steroidal alkaloid solasodine and related compounds in living plants and herbarium species. *Planta Med.* **39,** 112–124.

Weiler, E. W., Jourdan, P. S., and Conrad, W. (1981). Levels of 3-indoleacetic acid in intact and decapitated coleophiles as determined by a specific and sensitive solid-phase enzyme immunoassay. *Planta* **153,** 561–571.

Westekemper, P., Wiecyorek, V., Gueritte, F., Langlois, N., Langlois, Y., Potier, P., and Zenk, M. H. (1980). Radioimmunoassay for the determination of the indole alkaloid vindoline in *Catharanthus. Planta Med.* **39,** 24–37.

Yalow, R. S., and Berson, S. A. (1959). Assay of plasma insulin in human subjects by immunological methods. *Nature (London)* **184,** 1648.

Index

G

H

I

W

X

Z